Mammalian Cell Membranes

VOLUME THREE

Mammalian Cell Membranes VOLUME THREE

Surface Membranes of Specific Cell Types

Edited by

G. A. Jamieson Ph.D., D.Sc.
Research Director
American Red Cross Blood Research Laboratory
Bethesda, Maryland, USA

and

Adjunct Professor of Biochemistry
Georgetown University Schools of Medicine and Dentistry
Washington, DC, USA

and

D. M. Robinson Ph.D.
Professor of Biology, Georgetown University

and

Member, Vincent T. Lombardi Cancer Research Center
Georgetown University Schools of Medicine and Dentistry
Washington, DC, USA

BUTTERWORTHS
LONDON · BOSTON
Sydney · Wellington · Durban · Toronto

THE BUTTERWORTH GROUP

UK
Butterworth & Co (Publishers) Ltd
London: 88 Kingsway, WC2B 6AB

AUSTRALIA
Butterworths Pty Ltd
Sydney: 586 Pacific Highway, Chatswood, NSW 2067
Also at Melbourne, Brisbane, Adelaide and Perth

SOUTH AFRICA
Butterworth & Co (South Africa) (Pty) Ltd
Durban: 152–154 Gale Street

NEW ZEALAND
Butterworths of New Zealand Ltd
Wellington: 26–28 Waring Taylor Street, 1

CANADA
Butterworth & Co (Canada) Ltd
Toronto: 2265 Midland Avenue,
Scarborough, Ontario, M1P 4S1

USA
Butterworths (Publishers) Inc
Boston: 19 Cummings Park,
Woburn, Mass. 01801

First published 1977

© Butterworth & Co (Publishers) Ltd 1977

ISBN 0 408 70773 9

Library of Congress Cataloging in Publication Data (Revised)
Main entry under title:

Mammalian cell membranes.

   Includes bibliographical references and index.
   CONTENTS: v. 1. General concepts.   v. 2. The
diversity of membranes.   v. 3. Surface membranes of
specific cell types.   v. 4. Membranes and cellular functions.
   1. Mammals—Cytology.   2. Cell membranes.
I. Jamieson, Graham A., 1929–   II. Robinson,
David Mason, 1932–   [DNLM: 1. Cell membrane.
2. Mammals. QH601 M265]
QL739.15.M35   599′.08′75   75-33317
ISBN 0-408-70773-9

Filmset and printed Offset Litho in Great Britain by
Cox & Wyman Ltd, London, Fakenham and Reading

Contents

Contributors

C. R. AUSTIN
Physiological Laboratory, University of Cambridge, Cambridge, CB3 2EG, England

LUIGI M. DE LUCA
Differentiation Control Section, Lung Cancer Branch, National Cancer Institute, National Institutes of Health, Bethesda, Maryland 20014, USA

CATHERINE HICKEY-WILLIAMS
American National Red Cross Blood Research Laboratory, 9312 Old Georgetown Road, Bethesda, Maryland 20014, USA

G. A. JAMIESON
American National Red Cross Blood Research Laboratory, 9312 Old Georgetown Road, Bethesda, Maryland 20014, USA

MARTI JETT
American National Red Cross Blood Research Laboratory, 9312 Old Georgetown Road, Bethesda, Maryland 20014, USA

ROBERT J. MCLEAN
Department of Biological Sciences, State University College, Brockport, New York 14420, USA

WINIFRED G. NAYLER
Cardiothoracic Institute, 2 Beaumont Street, London, W1N 2DX, England

C. A. PASTERNAK
Department of Biochemistry, St George's Hospital Medical School, University of London, Blackshaw Road, Tooting, London, SW17 0QT, England

JERE P. SEGREST
Departments of Pathology and Biochemistry, University of Alabama in Birmingham, The Medical Center, Birmingham, Alabama 35294, USA

DAVID F. SMITH
Biochemisches Institut der Universität Freiburg im Breisgau, D78 Freiburg IBR, West Germany

LIST OF CONTRIBUTORS

G. R. STRICHARTZ
Department of Physiology and Biophysics, State University of New York at
Stony Brook, Stony Brook, New York 11794, USA

EARL F. WALBORG, JR.
Department of Biochemistry, The University of Texas System Cancer Center,
M.D. Anderson Hospital and Tumor Institute, Houston, Texas 77025, USA

Preface

This series on 'MAMMALIAN CELL MEMBRANES' represents an attempt to bring together broadly based reviews of specific areas so as to provide as comprehensive a treatment of the subject as possible. We sought to avoid producing another collection of raw experimental data on membranes, rather have we encouraged authors to attempt interpretation, where possible, and to express freely their views on controversial topics. Again, we have suggested that authors should not pay too much attention to attempts to avoid all overlap with fellow contributors in the hope that different points of view will provide greater illumination of controversial topics. In these ways, we hope that the series will prove readable for specialists and generalists alike.

The first volume, entitled *General Concepts*, served to introduce the subject and covered the essential aspects of physical and chemical studies which have contributed to our present knowledge of membrane structure and function. The second volume, *The Diversity of Membranes*, was concerned with specific types of intra- and extracellular membranes. This third volume, *Surface Membranes of Specific Cell Types*, as its title indicates, reviews the knowledge that we have of the surface membranes of the various cell types which have been studied in any detail to this time. *Membranes and Cellular Functions* will be covered in Volume 4, which will deal with ultrastructural, biochemical and physiological aspects. Since the cell surface represents the point of interaction with the cellular environment, Volume 5, entitled *Responses of Plasma Membranes*, addresses itself to the way in which external influences are mediated by the plasma membrane.

As editors, our approach to our responsibilities has been rather permissive. With regard to nomenclature and useful abbreviations, we have used 'cell surfaces' and 'plasma membranes' where appropriate rather than 'cell membranes' since this last is nonspecific. Both British and American usage and spelling have been utilized depending upon personal preference of the authors and editors with, again, no attempt at rigid adherence to a particular style.

While the title of the series is 'MAMMALIAN CELL MEMBRANES', we have encouraged authors to introduce concepts and techniques from non-mammalian systems which may be useful in their application to eukaryotic cells. The aim of this series is to provide a background of information and, hopefully, a stimulation of interest to those investigators working in, or about to enter, this burgeoning field.

Finally, the editors would like to acknowledge the dedication and resourcefulness of their secretary and editorial assistant, Mrs Alice R. Scipio, in the coordination and preparation of these volumes.

G. A. JAMIESON
D. M. ROBINSON

1

The erythrocyte: topomolecular anatomy of MN-glycoprotein

Jere P. Segrest

Departments of Pathology and Biochemistry, University of Alabama in Birmingham, The Medical Center, Birmingham, Alabama

1.1   INTRODUCTION

Biological membranes represent thin sheets of aqueous discontinuity that provide a complex barrier to a variety of ionic and polar molecules. Lipids and proteins, in approximately equal proportions, form the bulk of the dry weight of most isolated membranes, such as red cell ghosts (Korn, 1969). Two key problems concerning membranes that are, as yet, only partially resolved are the precise organization of the lipids and proteins of biological molecules into aqueous barriers and the relationship of this organization to membrane function.

1.1.1   Lipid organization in biological membranes

Artificial phospholipid bilayers have many of the properties of membranes. It is reasonably clear now, on the basis of X-ray diffraction studies of lipid model membrane systems (Levine, Bailey and Wilkins, 1968) and of membranes themselves, both *in vivo* and *in vitro* (Engelman, 1971), that phospholipid bilayers form a significant portion of the structure of most biological membranes.

The formation of the phospholipid bilayer can be best understood by consideration of the closely related phenomenon of micelle formation. Molecules such as sodium dodecyl sulfate which have both a polar or charged end and a nonpolar or hydrophobic end (i.e. are amphipathic) in general tend to form spherical- or ellipsoidal-shaped aggregates when suspended in aqueous media (Tanford, 1974). These aggregates or micelles are composed of the amphipathic molecules oriented with their hydrophobic ends directed

1

toward the micelle center and away from the surrounding water, and their polar ends directed outward and exposed to the water.

The driving forces in micelle formation are the hydrophobic interactions, which are of considerable biological importance, being a major factor in protein folding and the annealing of double-stranded DNA. Although our understanding of this force is still basically qualitative, it appears that hydrophobic interactions are the result of the propensity of water to seek to minimize interfacial surfaces with anything other than itself or other polar molecules (Tanford, 1973); water has a higher negative free energy when interacting with itself or polar molecules than with nonpolar molecules. For example, the aversion of water to an interaction with air produces surface tension, i.e. a minimization of the water–air interfacial area.

Phospholipid molecules are amphipathic molecules. However, owing to their geometry (their hydrophobic portions are equal to or greater in cross-sectional area than the polar ends), they tend to form multilamellar structures in aqueous solutions rather than simple micelles*. Each lamella or sheet is formed of a bilayer of phospholipid molecules, with the nonpolar portion oriented toward the middle away from the water to form an aqueous discontinuity.

Electron spin resonance (ESR) (Kornberg and McConnell, 1971a), nuclear magnetic resonance (NMR) (Chapman et al., 1968) and differential thermal calorimetry (Steim et al., 1969), among other techniques, have provided additional information about the physical state of membrane lipids.

The finding that the hydrocarbon interior of phospholipid bilayers, including those associated with many biological membranes, is often fluid under appropriate conditions of temperature (McConnell, Wright and McFarland, 1972) is having profound effects upon our understanding of many membrane-associated phenomena, such as movement of receptor molecules transverse to the cell surface (Frye and Edidin, 1970). Below a certain critical temperature, called the *liquid crystalline transition temperature* (Phillips, Ladbrooke and Chapman, 1970), these same bilayer interiors become semicrystalline (i.e. have a wax-like consistency).

Factors which affect the transition temperature include the degree of saturation and chain length of the phospholipid hydrocarbon chains. Other lipids, especially cholesterol, also have a profound effect on the fluidity of bilayers. From these facts alone, it is clear that the lipid composition of membranes can profoundly influence their bulk properties [including, to a small degree, ionic permeability (Papahadjopoulos, 1973)]. Because of the extremely hydrophobic nature of the long hydrocarbon chains of phospholipid molecules, bilayers are quite stable structures.

Bretscher (1972) has proposed that the erythrocyte membrane has an asymmetrical distribution of phospholipids across the plane of its bilayer. In his model, phosphatidylcholine and sphingomyelin form the outer leaflet of the erythrocyte membrane bilayer and phosphatidylethanolamine and phosphatidylserine form the inner leaflet.

---

* The geometry of a sphere explains this phenomenon. The inward-directed portion of an amphipathic molecule will pack much more easily if it has a small cross-sectional area compared with its polar end. A bilayer, on the other hand, should be most stable if both ends are approximately of equal area. If lecithin is hydrolyzed to lysolecithin, with one-half the nonpolar cross-sectional area of lecithin, micelles rather than bilayers are formed in water.

It is reasonable to infer that the colligative properties of the phospholipid molecules will largely determine the physical nature of the membrane as a result of hydrophobic interactions, the strength of which is illustrated by the low, almost negligible rate of phospholipid flip-flop (i.e. movement or exchange of individual phospholipid molecules across a bilayer) observed in bilayers by ESR studies (Kornberg and McConnell, 1971b).

1.1.2   Protein organization in biological membranes

It is apparent that knowledge of the gross physical state of lipids in biological membranes cannot explain many of such important properties as mediation of cell–cell interactions, the presence of specific receptors and antigenic binding sites on cell surfaces, message transmission across membranes, and active transport. Proteins, which represent close to 50 percent of the dry weight of most membranes, must have an important role in these phenomena, either directly (for example, as carrier molecules) or indirectly (for example, by local organization of lipids). Therefore, the structure and organization of membrane proteins relative to lipids may be the key to the understanding of membrane function.

One approach to understanding this organization is to utilize the principle of dominance of hydrophobic interactions. Nonpolar regions of membrane proteins will have a strong tendency to be excluded from the aqueous phase and to bury themselves in the membrane interior (Singer, 1971; Tanford, 1974). An even more useful principle is to assume that polar regions of proteins, particularly if charged, will be excluded from the hydrophobic region. Resistance to passage of charged amino acid residues into this region of low dielectric constant will be significant even if the polar regions are neutralized with counter-ions (Singer, 1971; Tanford, 1974). The low degree of phospholipid flip-flop has its basis in this principle (Kornberg and McConnell, 1971b). Use of these simple rules implies that knowledge of the amino acid sequence of a membrane protein can provide important guidelines as to the way in which it may be associated with the membrane lipids.

Although lipids may dominate the physical nature of membranes, it is equally clear that proteins can modulate the organization, and, therefore, the properties of the lipids of membranes and artificial bilayers. This modulation can be diffuse, in the way that cholesterol has a generalized effect on phospholipid bilayers (Chapman, 1968), or it can be multifocal, in the way that intramembranous particles seen by freeze-etch electron microscopy are discrete (Branton, 1969).

An example of protein having a diffuse effect on phospholipid bilayer structure is the interaction of the plasma apolipoproteins (delipidated protein components of the lipoproteins) with phospholipid vesicles in which these vesicles undergo a morphological alteration upon addition of apoprotein (Hoff, Morrisett and Gotto, 1973).

In a reciprocal fashion, the lipids of a lipoprotein complex such as a membrane can affect the structure and properties of the protein moieties. When plasma apolipoproteins (Lux et al., 1972) or the hydrophobic peptide of the human erythrocyte membrane glycoprotein (to be discussed later) interact with phospholipid vesicles, conformational changes are produced in these proteins. Similar conformational changes are probably involved in the activation of enzymes such as cytochrome oxidase by association with phospholipids (Vanderkooi *et al.*, 1972).

The state of organization of the proteins of membranes has been studied by a variety of physical techniques including X-ray diffraction (Blaurock, 1972), circular dichroism (CD) (Lenard and Singer, 1966), fluorescent probe spectroscopy (Metcalfe, Metcalfe and Engelman, 1971), proton magnetic resonance (PMR) (Glaser *et al.*, 1970) and ESR (Tourtellotte, Branton and Keith, 1970). In general these techniques have been unsuccessful in elucidating precise details of protein organization in biological membranes because of the statistical nature of the information gained, though they have been far more successful in studies of membrane lipid organization.

Because of the heterogeneity and asymmetrical distribution of membrane proteins, techniques providing average statistical values are basically unsuitable for studying most intact biological membranes. What is required to study the topomolecular anatomy of membrane proteins *in situ* is the reconstitution of membranes from well characterized, homogeneous constituents. This is the approach that has been utilized in the studies to be described in this chapter.

1.1.3   Erythrocyte membrane

Membrane models generally assume a common topomolecular pattern for all, or most, biological membranes. The accepted model for many years was the unit membrane (Robertson, 1964), which was based in a large part upon electron microscopic and X-ray diffraction studies of the myelin nerve sheath. Currently the fluid-mosaic model of Singer and Nicolson (1972) serves as the basic membrane paradigm. However, there is reason to suspect that the topomolecular patterns of biological membranes vary, as for example in the purple membrane (Blaurock and Stoeckenius, 1971).

It is important when investigating general principles of membrane organization to select reasonably representative membranes for study. The erythrocyte membrane, in addition to being convenient for study, has many of the properties of the fluid-mosaic model. X-ray diffraction studies suggest that a phospholipid bilayer is a major component of the erythrocyte membrane (Wilkins, Blaurock and Engelman, 1971). Integral membrane proteins in the erythrocyte can move laterally in the plane of the membrane under certain conditions (Tillack, Scott and Marchesi, 1972).

Another advantage of using the erythrocyte membrane is that the classification and characterization of the major polypeptide chains associated with it have begun. The classification of the major polypeptide chains of the human erythrocyte ghost prepared by Fairbanks, Steck and Wallach (1971) will be used here. The polypeptides are numbered from I to VI in the direction of decreasing molecular weight as determined by polyacrylamide gel electrophoresis in sodium dodecyl sulfate.

Four of these proteins are designated *peripheral proteins* on the basis of their solubilization by ionic manipulations, i.e. polypeptides I, II (mol. wt in excess of 200 000), polypeptide V (mol. wt 41 000) and polypeptide VI (mol. wt 36 000). Polypeptides I and II appear to be identical to the protein termed *spectrin* (Marchesi and Steers, 1968) or *tektin* (Masia and Ruby, 1968). Polypeptide VI has subsequently been shown to be the enzyme glyceraldehyde-3-phosphate dehydrogenase (Carraway and Shin, 1972).

Three of the human erythrocyte membrane polypeptides are designated *integral proteins* by Fairbanks, Steck and Wallach (1971) on the basis of their tenacious binding to the membrane, requiring detergents for solubilization. These are polypeptide III (mol. wt 100 000), polypeptide IV (mol. wt 77 000) and the major glycoprotein of the human erythrocyte membrane, PAS-1 (mol. wt 30 000), also known as *glycophorin* (Marchesi et al., 1972). The latter glycoprotein will be referred to in this chapter as MN-glycoprotein for reasons to be discussed.

All of the major polypeptide chains have been localized exclusively to the inside surface of the erythrocyte membrane with the exception of polypeptide III and MN-glycoprotein. Both of these latter appear, on the basis of labeling (Bretscher, 1971a, b; Segrest et al., 1973), enzymatic degradation (Bender, Garan and Berg, 1971; Kant and Steck, 1972) and amino acid sequence analysis (Segrest et al., 1972), to span the membrane.

Studies with various cross-linking reagents have suggested that polypeptide component III is present as a dimer (Yu and Steck, 1974) or even a tetramer (Wang and Richards, 1974) *in situ*, presumably forming a portion of the erythrocyte membrane intramembranous particles (seen by freeze-etch electron microscopy) as the multimer penetrates the bilayer. Further, there is evidence that one of the functions of component III is the transport of chloride ions across the membrane (Cabantchik and Rothstein, 1972).

Polypeptide VI (glyceraldehyde-3-phosphate dehydrogenase) appears to bind reversibly to the erythrocyte membrane (Kant and Steck, 1973). This protein has been shown to bind *in vitro* to isolated component III (Yu and Steck, 1974).

Polypeptides I and II (spectrin) have been shown to form actin-like filaments under certain conditions *in vitro* (Steers and Marchesi, 1969) and have been localized *in situ* to a heavy filamentous coat on the cytoplasmic surface of the human erythrocyte membrane (Nicolson, Marchesi and Singer, 1971).

1.2 MN-GLYCOPROTEIN

The details of the topomolecular anatomic relationship of MN-glycoprotein with the membrane are better characterized than for any other membrane protein, integral or peripheral, including rhodopsin (Hong and Hubbell, 1972) and cytochrome b_5 (Spatz and Strittmatter, 1971). Much of this characterization has been the result of work by Marchesi and coworkers (Segrest et al., 1972, 1973; Jackson et al., 1973). More recent work on this problem will be discussed later in this chapter.

1.2.1 Isolation

A glycoprotein containing the MN blood group activity was independently isolated by Winzler (1969) and Morawiecki (1964) from the human red cell membrane. These authors proposed that the glycoprotein was attached to the outside of the membrane by a hydrophobic segment. In addition to the phenol procedure utilized by these workers, MN-glycoprotein has been isolated by several other techniques (Zvilichovsky, Gallop and Blumenfeld, 1971; Kabylka *et al.*, 1972) including the lithium diiodosalicylate–phenol extraction utilized in the studies to be described (Marchesi and Andrews, 1971).

1.2.2   Characterization

The MN-glycoprotein comprises approximately 10 percent of the human erythrocyte membrane protein and contains external receptor sites for phytohemagglutinin, influenza virus and wheat-germ agglutinin (Segrest *et al.*, 1973; Jackson *et al.*, 1973). Initial chemical and structural characterizations of this glycoprotein have shown that it has a molecular weight of approximately 30 000 (Marton and Garvin, 1973; Grefrath and Reynolds, 1974) and is divided into a carbohydrate-rich N-terminal region, a hydrophilic C-terminus rich in proline residues but containing no carbohydrate, and an intervening nonpolar portion (Segrest *et al.*, 1972, 1973). Labeling studies, utilizing a lactoperoxidase-catalyzed iodination of the glycoprotein tyrosine residues *in situ* on erythrocytes and erythrocyte ghosts of varying degrees of permeability, suggest that the N-terminus and C-terminus of this molecule are on opposite sides of the membrane, the N-terminus being extracellular (as are the plant lectin sites) and the C-terminus intracellular (Segrest *et al.*, 1972, 1973).

Tillack, Scott and Marchesi (1972) used phytohemagglutinin (PHA) conjugated with ferritin to produce marker complexes capable of being visualized by freeze-etch electron microscopy. The PHA binds to MN-glycoprotein sites on the surface of the erythrocyte membrane. By a combination of fracture and etching, these workers demonstrated that MN-glycoprotein is in some manner associated with the 8-nm intramembranous particles of the erythrocyte membrane, since the MN-glycoprotein corresponds exactly to the distribution of intramembranous particles, even when the distribution of the latter is markedly altered by prior trypsin treatment.

Several studies seem to suggest that there is some form of interaction between the MN-glycoprotein and spectrin *in vivo*, in that certain physical and chemical changes which alter the state of spectrin concurrently alter the distribution of ferritin-conjugated antibodies and plant lectins directed against the glycoprotein (Ji and Nicolson, 1974). If this is true, then it seems most probable that spectrin is interacting with the C-terminal intracellular end of MN-glycoprotein.

The results of the labeling experiments support the presumption that the nonpolar portion of the MN-glycoprotein penetrates the hydrophobic core of the membrane. This portion of the molecule has a known amino acid sequence with a linear distribution of polar and nonpolar residues identical to the polar–nonpolar–polar cross section of a phospholipid bilayer (Segrest *et al.*, 1972). This nonpolar domain is contained intact within a hydrophobic tryptic peptide (35 residues and a residue weight of 3700) produced from

MN-glycoprotein (Segrest *et al.*, 1972, 1973; Jackson *et al.*, 1973). The amino acid sequence of this peptide is as follows:

Val	Gln	Leu	Pro	His	Pro	Phe	Ser	Glu	Ile	Glu	Ile
				5					10		
Thr	Leu	Ile	Val	Phe	Gly	Val	Met	Ala	Gly	Val	Ile
		15					20				
Gly	Thr	Ile	Leu	Leu	Ile	Ser	Tyr	Gly	Ile	Arg	
25					30					35	

The nonpolar domain extends from residues 12 to 34, a distance of 23 residues.

1.3   MN-GLYCOPROTEIN–LIPID INTERACTIONS

Clearly the forces involved in protein–lipid interactions in biological membranes are basic to any understanding of membrane structure and function. In these interactions, both electrostatic and hydrophobic forces have been implicated. One assumption has been that proteins which associate with the membrane would have special characteristics of amino acid composition or sequence that would differentiate them from those which do not.

Examination of the available amino acid composition of tightly associated (integral) membrane proteins seems to suggest that there might be an increase in hydrophobic amino acids. However, the surface MN-glycoprotein of the human erythrocyte has an overall amino acid composition which is more polar than most water-soluble proteins. It seems a reasonable hypothesis, as first suggested by Morawiecki (1964) and Winzler (1969), that the membrane-associative properties of this protein reside in its 23-residue hydrophobic domain (Segrest *et al.*, 1972).

1.3.1   Hydrophobicity of nonpolar domain

If the hydrophobic domain of MN-glycoprotein is involved in protein–lipid interactions in the membrane, it seems likely that its properties should be in some manner distinguishable from similar segments of proteins that are not associated with membranes. This hypothesis has been tested by computer analysis (Segrest and Feldmann, 1974). Continuous sequences of at least 10 amino acids devoid of charged residues (glutamic acid, aspartic acid, lysine and arginine) were selected from a file of all amino acid sequences known to occur in proteins. These noncharged sequences were compared with the 23-residue hydrophobic domain of the MN-glycoprotein. A total of 774 sequences were identified, having lengths ranging up to 64 amino acid residues. The average hydrophobicity per residue of each noncharged segment was calculated by assigning hydrophobic values for each noncharged amino acid residue. The results of this study, shown in *Figure 1.1*, show that the 23-residue hydrophobic domain of the MN-glycoprotein is distinguishable by informal cluster analysis from other segments of globular proteins when sequence length is plotted against hydrophobicity. This analysis suggests

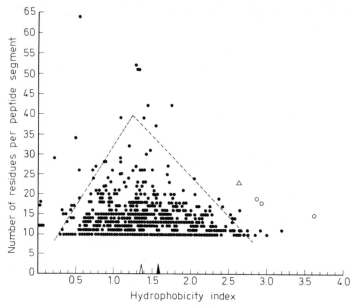

Figure 1.1   Hydrophobicity diagram. The number N (≥ 10) of amino acid residues of each of the 774 uncharged polypeptide segments selected by computer is plotted against its respective hydrophobicity index (HI). Since there is much overlap, especially where N is ≤ 15, not all points appear in the plot. The open elongated triangle on the baseline indicates the mean HI for the 774 segments, and the closed triangle, the mean HI (1.58) calculated from the mean distribution in 207 proteins of noncharged amino acid residues. Four hydrophobic peptide segments which cluster well outside the 'main sequence' of uncharged segments are designated as follows: △, 23-residue nonpolar domain of MN-glycoprotein; ○, 19- and 18-residue segments from two filamentous bacteriophage coat proteins and a 15-residue polypeptide antibiotic, gramicidin A

the possibility that other membrane-penetrating segments of proteins may be identified in the same way (Segrest and Feldmann, 1974).

1.3.2   Helical properties of nonpolar domain

Nonpolar media, such as the interior of a membrane or phospholipid bilayer, are helix-inducing (Fasman, 1967; Singer, 1971; Tanford, 1974). Fasman (1967) has shown that homopolymers of hydrophobic amino acids, particularly poly-L-methionine, poly-L-leucine and poly-L-alanine, form extremely stable helices in nonpolar media.

The nonpolar domain is the only portion of MN-glycoprotein with any reasonable possibility of being α-helical (*Figure 1.2*). The N-terminal domain, because of charge repulsion between sialic acid residues, and the C-terminal domain, because of the helix-breaking tendency of its prolines, are both unlikely to be helical (Segrest *et al.*, 1973).

Examination of the sequence of the 23-residue nonpolar domain (*Figure 1.2*) reveals that most of its residues are helix-inducing according to the classification of Lewis and Scheraga (1971). In particular, note first the high content of leucine, which has been suggested by Chou, Wells and Fasman

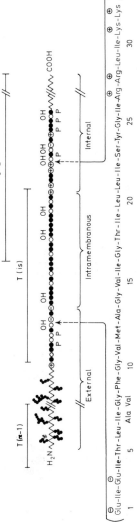

Figure 1.2 Molecular topography of MN-glycoprotein. Details of the nonpolar domain are shown. Each amino acid residue of the 51-residue sequence defined for the center of the polypeptide chain (Segrest et al., 1972) is represented by a circle, closed for hydrophobic, open for neutral and charged. The location of each hydroxyamino acid residue (threonyl and seryl) is indicated by the letters OH above the residue; the location of each prolyl residue is indicated by the letter P below the residue. The sections of the polypeptide chain represented by the N-terminal tryptic peptide T(α−1), the hydrophobic tryptic peptide, T(is), and the C-terminal cyanogen bromide fragment, C-2, are indicated. The sequence of the 23-residue nonpolar domain is shown below the diagrammatic polypeptide chain in bold-face lettering and the immediately adjacent charged residues by regular lettering

(1972) and Robson and Pain (1972) to be the key residue involved in helix induction, and, secondly, the absence of proline, an absolute helix breaker except when located at the N-terminal end of a helix (Lewis and Scheraga, 1971; Robson and Pain, 1972). The absence of proline from the nonpolar domain appears to be significant in a sequence 23 residues in length. This assumption is supported by the computer analysis described previously (Segrest and Feldmann, 1974).

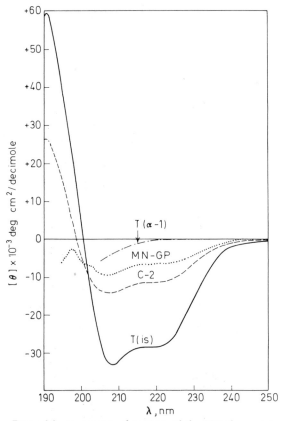

Figure 1.3 CD *spectra of portions of the MN-glycoprotein. Spectra were obtained on the intact MN-glycoprotein and T(α−1) in 0.1* M *Tris-*NaCl *and on the fragments C-2 and T(is) in 100% TFE*

A computer analysis by Robson and Pain (1972) suggests that certain residues have a tendency to promote helix formation in either an N- or C-terminal direction along the polypeptide chain from their location; glutamyl residues promote helix in a C-terminal direction and arginyl and lysyl residues in an N-terminal direction. Note from *Figure 1.2* that the non-polar domain lies immediately C-terminal to two glutamyl residues (residues 1 and 3) and N-terminal to two arginyl residues (residues 27 and 28) and two lysyl residues (residues 31 and 32).

The N-terminal tryptic peptide of MN-glycoprotein, $T(\alpha - 1)$, is 34 residues in length and contains approximately 80 percent of the carbohydrate (Segrest et al., 1973; Jackson et al., 1973). The C-terminal cyanogen bromide fragment of MN-glycoprotein forms a partial overlap with the nonpolar domain (*Figure 1.2*). The tryptic fragment T(is) contains the nonpolar domain intact (Segrest et al., 1972, 1973).

Figure 1.3 shows CD spectra for the intact glycoprotein and the major carbohydrate-containing tryptic peptide, $T(\alpha - 1)$, in aqueous solution, and the hydrophobic fragments T(is) and C-2, in trifluoroethanol (TFE).

Helix contents for the intact glycoprotein and $T(\alpha - 1)$ in 0.1 M Tris-NaCl and the fragments C-2 and T(is) in TFE were determined by a three-component curve-fitting procedure as follows.

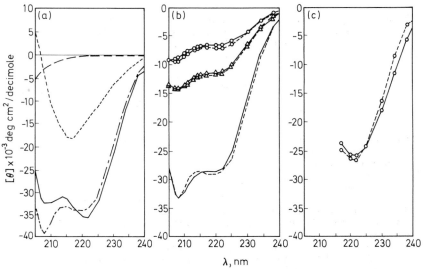

Figure 1.4   Curve fitting of a linear function of α, β and random conformations against experimental CD curves (Figures 1.4 and 1.7). (a) Standard CD curves: (— — — — —) 100 percent α-helix (PLA in TFE); (———) 100 percent α-helix (PLL in aqueous solvent); (— — — — —) 100 percent β-conformation (PLL in aqueous solvent); (— — —) 100 percent random coil (PLS in water). See text for details and references. (b) Experimental and fitted curves (205–240 nm) for samples in solvents: (———) experimental CD curve; (— — — — —) fitted curve; (○) MN-glycoprotein in 0.1 M Tris-NaCl (100 percent α-helix in aqueous solvent used as standard); (△) C-2 in TFE (100 percent α-helix in TFE used as standard); (null point) T(is) in TFE (100 percent α-helix in TFE used as standard). (c) Experimental and fitted curves (217–240 nm) for T(is) associated with egg lecithin liposomes: (———) experimental CD curve; (— — — — —) fitted CD curve (100 percent α-helix in aqueous solvent used as standard). Owing to a small amount of light scattering, the fit was not done below 217 nm

Reference CD curves for α, β and random conformations were selected and are shown in *Figure 1.4a*. The curve for 100 percent α-helix in aqueous solvents is that of poly-L-lysine (PLL) in the helical conformation as described by Greenfield and Fasman (1969); the curve for 100 percent α-helix in TFE is that of poly-L-alanine (PLA); the curve for 100 percent β-helix is that of PLL in the β conformation as described by Greenfield and Fasman (1969);

and the curve for 100 percent random coil* is that of poly-L-serine (PLS) in aqueous solutions as described by Quadrifoglio and Urry (1968). No attempt was made to correct the experimental curves for contributions other than these three basic peptide ellipticities because other possible contributions were felt to be relatively minor (as was exemplified by the fact that computed curves closely match the experimental ones).

Curve fitting was accomplished by use of the MLAB curve-manipulating computer program of Knott and Reece (1972). A linear function composed of percentages of α, β and random conformations was fitted to each experimentally derived CD curve.

The three-component conformational fits to these curves of *Figure 1.3* [excluding T($\alpha - 1$), which can be classified as random by inspection] are shown in *Figure 1.4b*, and the conformation parameters of these fits are given in *Table 1.1*.

Table 1.1 COMPARISON OF THE PERCENTAGE OF NONPOLAR DOMAIN AND PERCENTAGE OF HELIX FOR THE INTACT GLYCOPROTEIN AND FRAGMENTS C-2 AND T(is)

Polypeptide	Nonpolar domain*, %	Parameters determined by curve fitting	
		α-Helix, %	β Conformation, %
MN-Glycoprotein	20	19.4†	< 0.1
T (α − 1): N-terminal fragment	0	0†	
C-2: C-terminal fragment	28	31.2‡	5
T(is) in TFE	65	83.6‡	3.2
T(is) in liposomes	65	74.5§	< 0.1

* The nonpolar domain was assumed to contain exactly 23 residues. The molecular weight of the MN-glycoprotein is 30 000 (120 amino acid residues). The residue length of C-2 is 51 amino acid residues.
† In 0.1 M Tris-NaCl. C-2 and T(is) are not soluble in aqueous buffers.
‡ In 100% TFE. The intact glycoprotein and T($\alpha - 1$) are not soluble in TFE.
§ Liposomes in distilled water.

These data indicate that the MN-glycoprotein contains approximately 19 percent α-helix (*Table 1.1*) with the rest of its structure being random. As seen from *Figure 1.2*, the nonpolar domain of the MN-glycoprotein is composed of 23 amino acid residues, which is equal to 20 percent of the intact polypeptide chain (*Table 1.1*). It is further evident that the nonpolar domain represents 65 percent of the hydrophobic tryptic peptide T(is), 28 percent of the C-terminal cyanogen bromide fragment C-2 and 0 percent of the major carbohydrate-containing tryptic peptide, the N-terminal T($\alpha - 1$) (*Table 1.1*). On the basis of curve-fitting (*Figure 1.4*), the helix content of

* The resulting fits were quite sensitive to the CD parameters of the random-coil component. There is controversy as to what constitutes the CD pattern for a random coil (Tiffany and Krimm, 1973; Rao and Miller, 1973). There certainly is no question but that this pattern varies from protein to protein (Cortijo, Panijpan and Gratzer, 1973). The CD spectrum of PLS was chosen to represent the random conformation because: (a) the MN-glycoprotein is quite rich in hydroxyamino acids; (b) the CD curve of the N-terminal sialoglycopeptide, T($\alpha - 1$), closely resembles that of PLS; and (c) PLS gives the best quantitative and qualitative fit of all random CD patterns examined.

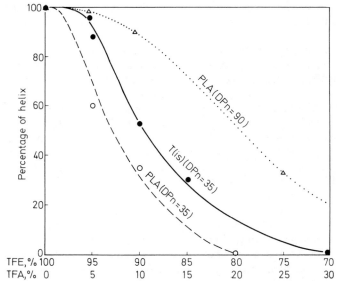

Figure 1.5 *Helix–coil transition for T(is) compared with poly-L-alanine (PLA) of the same degree of polymerization. Lots AL-19 and AL-52 obtained from Miles–Yeda Ltd (Rehovoth, Israel) were used. The former has a well characterized DPn of 35 (Ferretti and Paolillo, 1969) and the latter a DPn of 90 calculated from NMR measurements. Each sample in TFE (upper scale) was titrated with increasing amounts of TFA (lower scale) and its helix–coil transition was followed by CD; the percentage of helix was calculated from molar ellipticity at 222 nm*

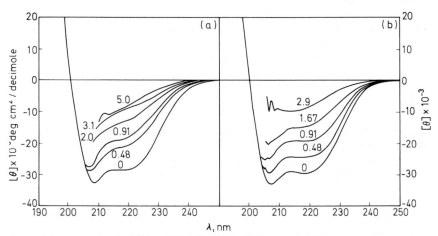

Figure 1.6 *CD spectra for T(is) and PLA (lot 19) in TFE titrated with increasing concentrations of lithium bromide, a known helix-breaker (Franzen, Bobik and Harry, 1966). The concentration of lithium bromide (g ml⁻¹ × 100) is indicated for each curve. (a) T(is); (b) PLA*

each of the latter three peptides is in close agreement with that portion of it represented by the nonpolar domain (*Table 1.1*); i.e. T(is) is approximately 85 percent helical and can account for all of the helix of the intact glycoprotein*.

These studies could not be done in the same solvent since there is a diversity in the chemical properties of the four samples. The MN-glycoprotein and the tryptic sialoglycopeptide T(α – 1) are soluble in aqueous solvents but not in organic solvents such as TFE, whereas C-2 and T(is) are soluble only in TFE. In truth, the author can see no advantage to be gained by using a single solvent in additive experiments such as these where fragments of an intact protein are examined. Logically one would desire to duplicate closely the dielectric environment in which the fragment is located in the intact protein (i.e. exposed to high-dielectric solvent or buried in a low-dielectric environment). The solubility properties of a fragment should in most cases be related to its native environment.

Nevertheless, since many peptides are helical in TFE one has some reservations about the conclusion suggested by the data of *Table 1.1*, namely that the intramembranous domain (IMD) is the unique helical portion of the MN-glycoprotein. To alleviate this objection, we sought to quantitate the stability of the IMD helix by measuring its helix–coil transition in two systems: TFE–TFA (trifluoroacetic acid) and TFE–LiBr. The helix–coil transition for T(is) was followed by CD and compared with PLA of a similar degree of polymerization (DPn = 35) since PLA has been suggested by Fasman (1967) to form the most stable helix of known homopolymers based on its stability to TFA. The results of the TFA titration are shown in *Figure 1.5*. The midpoint of the T(is) helix–coil transition is at 11% TFA, and for PLA the transition midpoint is at 7% TFA, suggesting that the IMD represented by T(is) is at least as stable as, and probably more stable than, PLA of the same degree of polymerization. The studies with another helix breaker, LiBr (Franzen, Bobik and Harry, 1966), tend to substantiate this conclusion (*Figure 1.6*). Based on this conclusion, it seems reasonable to postulate that, in the absence of unusual perturbation, the IMD will be helical in the low-dielectric environment of the membrane core, thereby comprising the major helical portion of the MN-glycoprotein.

1.3.3   Lipids associated with isolated MN-glycoprotein

As shown in *Table 1.2*, lipid represents approximately 7 percent of the mass of the isolated glycoprotein and 50 percent of the hydrophobic peptide T(is). Since T(is) before delipidation represents approximately 15 percent of the weight of the glycoprotein before trypsinization, the lipid associated with this peptide accounts quantitatively for the lipid bound to the solubilized glycoprotein. All of the major lipids of the red cell membrane are present in the chloroform–methanol extracts from the glycoprotein and T(is), as

---

* The assumption of conformation additivity of fragment peptides to account for conformation distribution in larger fragments and the whole glycoprotein is perhaps lessened in this specific case since this protein is not typically globular but appears to have its main segments in rather separate domains. This would thereby decrease the likelihood of strong intersegmental interactions in the manner which might preclude fragment conformation additivity.

Table 1.2 QUANTIFICATION OF BOUND LIPIDS

Sample	Lipid*, %	Major component†
MN-Glycoprotein (before ethanol wash)	10‡	Cholesterol, phospholipids
MN-Glycoprotein (after ethanol wash)	7.1	Cholesterol
T(is)	50	Cholesterol
T(is), delipidated	0	

* Measured by dry weight extracted into Folch–Lee lower phase (chloroform–methanol).
† Determined by thin-layer chromatography.
‡ The exact value is uncertain owing to the presence of bound lithium diiodosalicylate in this fraction.

shown by thin-layer chromatography. However, the level of phospholipids decreases following the ethanol wash to remove excess lithium diiodosalicylate (Marchesi and Andrews, 1971).

1.3.4   Recombination of nonpolar domain with lipids

Studies were undertaken to examine the interaction of T(is) with hydrated phospholipid bilayers (Segrest, Gulik-Krzywicki and Sardet, 1974) based on the working hypothesis that the membrane-associative properties of MN-glycoprotein reside in its 23-residue hydrophobic domain and therefore

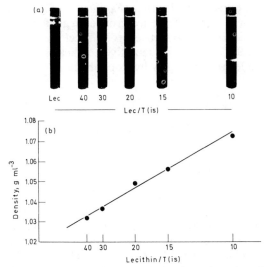

Figure 1.7   Sucrose density gradients of complexes formed between the membrane-penetrating peptide T(is) and lecithin with increasing T(is) concentration. Gradients are between the densities of 1.01 g ml^{-1} and 1.12 g ml^{-1}. Equilibrium conditions reached by centrifugation at $300\,000$g for 15 h. (a) Gradients. Samples are from left: lecithin alone and lecithin/T(is) 40, 30, 20, 15, 10 moles/mole. (b) Plot of equilibrium density of each T(is)/lecithin complex versus concentration of T(is)

in the tryptic peptide T(is); this hypothesis was supported by the computer and lipid analyses just described.

A technique which allows T(is) to be associated with phospholipid was developed. T(is) is dissolved in ethanol with egg lecithin, the ethanol evaporated away and the resulting T(is)–lecithin complex hydrated. Sucrose density gradients show that the complex forms a single band which increases in density with increasing peptide concentration in a linear fashion (*Figure 1.7*). These results show that a tight complex is formed between T(is) and egg lecithin under the conditions of the experiment.

1.3.5   Physical studies

Vesicles (proteoliposomes) can be prepared from these complexes either by brief sonication (Huang, 1969) or ethanol dispersion (Batzri and Korn, 1973) in water. These vesicles have proved suitable for spectroscopic studies, including PMR, CD and fluorescence.

It is indicated by CD studies (*Figure 1.8*) that the T(is) peptide is approximately 75 percent α-helical when associated with lecithin in the form of proteoliposomes* (*see Figure 1.4c and Table 1.1* for curve fitting). This finding supports the tentative conclusions of the CD studies described earlier, namely that in the absence of strong perturbations, the nonpolar domain of MN-glycoprotein will be helical in the membrane core.

It can be shown by PMR that association of T(is) with lecithin vesicles leads to broadening of the resonance peaks for the choline, methylene and terminal methyl groups of the lecithin (*Figure 1.9*). The broadening of the resonance peaks for the terminal methyl groups suggests that T(is) penetrates deeply into the bilayer. Fluorescence probe spectroscopy by the method described by Shechter *et al.* (1971) tentatively indicates that the interaction between T(is) and lecithin is largely hydrophobic.

Table 1.3   ASSOCIATION OF T(is) WITH BLACK FILMS

	Resistance, Ω cm^2	No. of films
Control black films	$(7.8 \pm 1.2) \times 10^8$	10
Black film with associated T(is)	$(6.3 \pm 0.9) \times 10^6$	11

In addition to vesicles, artificial bilayers or black films can be formed from the T(is)–lecithin complexes (Lea, Rich and Segrest, 1975). At the low concentration of 1 mole of T(is) per 400 moles lecithin, resistance of black films decreases by a factor of 100 (*Table 1.3*). This finding is compatible with a hydrophobic interaction with the black film in the form of bilayer penetration by T(is) (Kimelberg and Papahadjopoulos, 1971), as suggested by the PMR experiment (*Figure 1.9*).

* Since the peptide is presumably partially dissolved in the nonpolar hydrocarbon phase of the liposomes, the choice of solvent for the 100 percent helix standard is not a clear-cut decision. Use of TFE as the standard solvent, however, has little effect on the resulting fit.

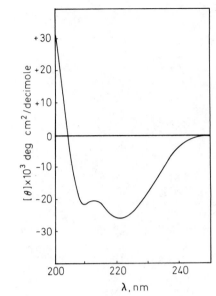

Figure 1.8 CD spectrum of the nonpolar domain, represented by T(is), associated with egg lecithin liposomes. T(is) in TFE and egg lecithin in 100% ethanol were mixed at a mole to mole ratio of 1:50 and all the solvent removed by evaporation. The complex was then used to make single bilayer liposomes by the technique of Batzri and Korn (1973)

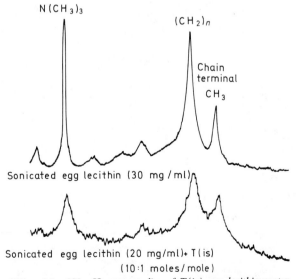

Figure 1.9 220 mHz PMR studies of T(is)–egg lecithin proteoliposomes. T(is)–egg lecithin complexes were formed as described in text and minimal size single-bilayer vesicles for PMR study were formed by sonication for 10 min under nitrogen

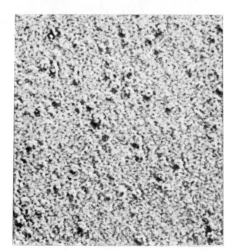

Figure 1.10   Freeze-etch electron micrograph at
an approximate magnification of ×300 000 of
T(is)–egg lecithin multilamellar proteoliposomes.
Note the well defined torus shape of the majority
of the particulate structures

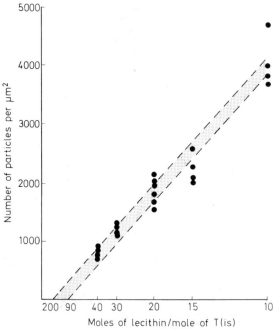

Figure 1.11   Particle count of T(is)–lecithin associations
plotted versus T(is) concentration. Each point represents a
particle count done over an area of 4 in² on each print extra-
polated to particles per square micrometer. Rather than
attempting to fit a unique straight line to the data points, a
range of possible straight lines has been indicated (shaded area
within dotted lines)

1.3.6   Freeze-etch studies

The technique that has proved most useful in studying the nature of the T(is)–egg lecithin complexes is freeze-etch electron microscopy. Freeze-etching of hydrated egg lecithin multilamellar vesicles alone produces smooth fracture faces, representing the hydrocarbon interior of a bilayer. Addition of T(is) results in the appearance of particulate structures on the fracture face (Segrest, Gulik-Krzywicki and Sardet, 1974). The best replicas (*Figure 1.10*) show that the particles are torus-shaped structures 8 nm in diameter. These particles extend a distance above the surrounding fracture face sufficient to suggest that they penetrate completely through the bilayer.

As the T(is) concentration is increased, the number of particles increases in a linear fashion and a plot of particle count versus T(is) concentration (*Figure 1.11*) extrapolates to zero particles at a finite T(is) concentration. This suggests a micellar-like phenomenon in which, at a critical concentration (the critical multimer concentration), T(is) preferentially associates with itself rather than with the surrounding phospholipids (Elworthy, Florence and MacFarlane, 1968). From the slope of the curve in *Figure 1.11*, knowing the surface area occupied per lecithin molecule (0.65 nm^2) in a bilayer, and estimating a surface area for each helical T(is) molecule, one can calculate that each particle is a multimer composed of from 4 to 10 T(is) molecules (Segrest, Gulik-Krzywicki and Sardet, 1974).

There is further evidence for bilayer penetration by T(is) multimers; as one goes below the liquid-crystalline phase transition temperature of an artificial lecithin, such as the saturated compound dipalmitoyllecithin, the multimeric particles visualized by freeze-etching clump (*Figure 1.12*). The

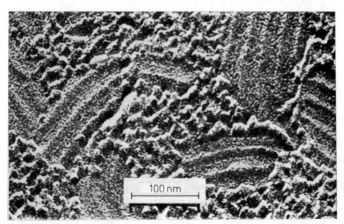

Figure 1.12   Freeze-etch electron micrograph of T(is)–dipalmitoyllecithin multilamellar proteoliposomes below the liquid crystalline temperature (40 °C) for dipalmitoyllecithin. Note the exclusion, and subsequent clumping, of T(is) multimers from the crystalline regions

likely reason for this clumping is that impurities, such as protein present in the hydrocarbon phase, are excluded into a separate phase as crystallization of the phospholipid hydrocarbon chains occurs.

The results summarized to this point suggest a model for T(is) association

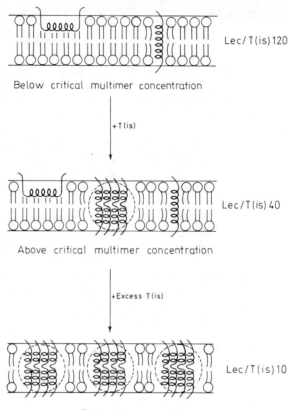

Lec/T(is)120

Below critical multimer concentration

+T(is)

Lec/T(is)40

Above critical multimer concentration

+Excess T(is)

Lec/T(is)10

Saturation

*Figure 1.13   Schematic diagram of a proposed model for inter-
actions of the membrane-penetrating peptide T(is) with hydrated
lecithin vesicles. The peptide, which is partially helical when
associated with lecithin, could be oriented either perpendicularly
or parallel to the paraffin chains of the phospholipid bilayer.
However, because the intramembranous particles appear to
penetrate the bilayer and because of the topographical distribution
of charge along the T(is) molecule, the orientation of the peptides
forming these structures appears likely to be parallel to the
hydrocarbon chains*

with lecithin bilayers (*Figure 1.13*). Below a critical concentration of T(is),
monomers of the peptide exist in the bilayer in two possible orientations.
As this critical concentration is exceeded, multimeric aggregation of T(is)
occurs to produce intramembranous structures with defined torus-like
shapes. The size of these structures is independent of T(is) concentration
(Segrest, Gulik-Krzywicki and Sardet, 1974).

1.4   MODEL FOR THE TOPOMOLECULAR ANATOMY OF MN-GLYCOPROTEIN

1.4.1   Molecular topography of nonpolar domain

The assumption is that T(is) forms multimeric helical structures across the plane of phospholipid bilayers, with the helices parallel to the hydrocarbon chain of the phospholipids (*Figure 1.13*). On this basis, a space-filling molecular model of T(is) was constructed assuming the peptide to be in an α-helical conformation. The topomolecular features of this model, although somewhat speculative, are extremely interesting in light of a recent paper by Inouye (1974) in which he proposes a topomolecular model for the structure *in situ* of a lipoprotein from the *E. coli* outer membrane. Inouye's model, a transmembrane superhelical multimer formed of intertwining α-helixes, has features similar to the model proposed in *Figure 1.13* for a T(is) multimer in phospholipid bilayers.

The amino acid sequence of T(is) has been plotted on an α-helical net in *Figure 1.14*. One can see that the helix is divided roughly into two faces, one

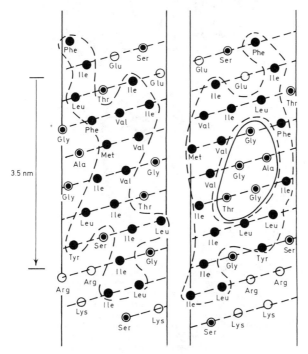

Figure 1.14   α-Helical net of the nonpolar domain of MN-glyco-protein. Each turn of the helix (3.6 residues) is represented by the diagonal dotted lines. This diagram has the advantage of allowing the surface topographical distribution of amino acid residues of an α-helix to be visualized in a two-dimensional diagram. The hydrophobic residues form a cluster which has been outlined with a dashed line in the diagram on the left. In the right-hand diagram, the central cluster of neutral residues has been outlined with a solid line. The center of projection of the right-hand diagram has been shifted slightly from that of the left-hand

having bulky, extremely hydrophobic residues and the opposite face having essentially neutral, nonbulky residues, at least in its middle portion. Rather than running parallel to the long axis of the helix, these faces run at an angle [*note* the two rows of hydrophobic residues forming a right-handed super-helix on the extremely hydrophobic face of T(is)], as do the nonpolar residues in Inouye's model.

This model for T(is) may be able to explain the tendency for the formation of multimers in phospholipids. It seems reasonable, assuming T(is) is helical in phospholipid bilayers as the CD data suggest, that T(is) might tend to aggregate in a micellar-like fashion in order to sequester its relatively neutral, nonbulky face away from the surrounding hydrocarbon phase. In this way the bulky, hydrophobic face would be exposed to the hydrocarbon phase.

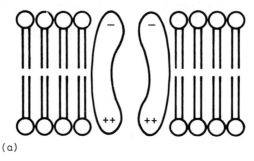

(a)

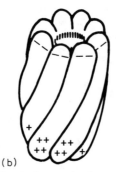

(b)

Figure 1.15   Proposed topomolecular anatomy for T(is) multimers associated with a phospholipid bilayer. (a) Cross section of T(is) multimer in a bilayer formed from α-helical monomers, in which the bulky, hydrophobic faces are oriented outward and the neutral faces inward. The result is a tubular transbilayer structure with a central cavity which narrows at both ends. (b) Proposed coiled-coil structure for the T(is) multimer. Note the rim of charged residues at either end

Further, assuming that the α-helical monomers of T(is) are supercoiled in a right-handed fashion to form a multimer with eight or more strands, a structure with several interesting features results. The helical rope formed is quite hydrophobic on the outside except at the ends where charged residues occur. Inside the helical rope a channel is formed which has two intriguing features. First, a relatively wide, essentially neutral cavity is formed in the central region of the channel because of the juxtaposition of the nonbulky, neutral residues. Secondly, the channel narrows considerably at both ends owing to the juxtaposition of bulky nonpolar residues. These features are diagrammatically illustrated in *Figure 1.15*.

1.4.2   State of self-association of MN-glycoprotein

The apparent similarity in size of the intramembranous particles produced by T(is) in phospholipid vesicles and those seen in native erythrocyte membranes (8 nm) suggests a possible relationship between the two phenomena. The apparent association of the MN-glycoprotein with intramembranous particles has already been discussed (Tillack, Scott and Marchesi, 1972). Two additional considerations, namely the torus-shaped, apparently symmetrical, almost subunit-like images of the T(is) particles (*Figure 1.10*), and the similarity in size of intramembranous particles from essentially all biological membranes (Branton, 1969), lend additional credibility to this possibility.

Assuming that each intramembranous particle contains at least one MN-glycoprotein molecule (there are two distinct distributions of particles, those on the outer and those on the inner fracture faces), one can calculate that each particle corresponds to approximately 2 to 3 molecules of glycoprotein (Segrest, Gulik-Krzywicki and Sardet, 1974).

However, there is an uneven distribution of particles on the two fracture faces. The outer fracture face contains one-third of the particles; the inner, two-thirds of the particles (Branton, 1969). This suggests two distinct kinds of particles. Further, reasoning *a priori*, it would seem likely that the MN-glycoprotein, owing to its outwardly oriented, bulky, carbohydrate-rich N-terminus, would stay with the outer fracture face. Finally, trypsin treatment of intact erythrocytes causes a subtle alteration in the size and distribution of the particles of the outer fracture face but not those of the inner; under these conditions, only MN-glycoprotein, of all the erythrocyte membrane proteins, is fragmented (Triplett and Carraway, 1972).

Therefore, if only the particles of the outer fracture face of the erythrocyte membrane represent MN-glycoprotein, each particle would contain *ca*. 6 MN-glycoprotein molecules (Segrest, Gulik-Krzywicki and Sardet, 1974).

One example of supportive evidence for this model is that the MN-glycoprotein as isolated by the lithium diiodosalicylate procedure (Marchesi and Andrews, 1971) exists as multimers in aqueous solution corresponding to from 6 to 12 molecules per multimer. Negative-stain electron-microscopic studies of these multimers reveal that they have a torus shape similar to that seen for T(is) multimers in phospholipid. Since the lithium diiodosalicylate extraction procedure does not disrupt protein–protein interactions, it seems likely that the procedure is extracting the MN-glycoprotein as it exists in the membrane. At the very least, the MN-glycoprotein seems to exist in the membrane in association with itself and no other protein. This conclusion has been supported by the cross-linking studies of Ji (1974), in which he finds cross-linked multimers on MN-glycoprotein of molecular weight approximately 200 000, corresponding to approximately 7 or so glycoprotein monomers.

1.4.3 Multimeric model for topomolecular anatomy of MN-glycoprotein

One can therefore arrive at a working model for the structure of the MN-glycoprotein *in situ*, that of a multimeric unit of about 6 MN-glycoprotein monomers oriented in parallel, exposing the N-terminal regions to the external aqueous environment. The helical nonpolar domains of these monomers are intertwined to form a membrane-spanning coiled-coil structure having a relatively neutral central channel. This intramembranous multimeric structure corresponds to the particles of the outer fracture face seen by freeze-etch electron microscopy of intact erythrocyte membranes. The combined C-terminal domains of the MN-glycoprotein multimer would then be free to interact with fibrils of spectrin on the cytoplasmic surface of the erythrocyte membrane (Ji and Nicolson, 1974). This speculative model has features suggestive of some transport role for MN-glycoprotein. Presently there is no evidence for or against such a possibility.

Acknowledgements

The author gratefully acknowledges the contributions of the following scientists: T. Gulik-Krzywicki, D. G. Dearborn, C. Sardet, R. J. Feldmann, R. Carter and L. D. Kohn.

REFERENCES

BATZRI, S. and KORN, E. D. (1973). *Biochim. biophys. Acta*, **298**:1015.
BENDER, W. W., GARAN, H. and BERG, H. C. (1971). *J. molec. Biol.*, **58**:783.
BLAUROCK, A. E. (1972). *Nature, Lond.*, **240**:556.
BLAUROCK, A. E. and STOECKENIUS, W. (1971). *Nature, New Biol.*, **233**: 152.
BRANTON, D. (1969). *A. Rev. Pl. Physiol.*, **20**:209.
BRETSCHER, M. S. (1971a). *J. molec. Biol.*, **59**:351.
BRETSCHER, M. S. (1971b). *Nature, New Biol.*, **231**:229.
BRETSCHER, M. S. (1972). *J. molec. Biol.*, **71**:523.
CABANTCHIK, Z. I. and ROTHSTEIN, A. (1972). *J. Membrane Biol.*, **10**:331.
CARRAWAY, K. L. and SHIN, B. C. (1972). *J. biol. Chem.*, **247**:2102.
CHAPMAN, D. (Ed.) (1968). *Biological Membranes, Physical Fact and Function*, pp. 125–202. London; Academic Press.
CHAPMAN, D., KAMAT, V. B., DE GIER, J. and PENKETT, S. A. (1968). *J. molec. Biol.*, **31**:101.
CHOU, P. Y., WELLS, M. and FASMAN, G. D. (1972). *Biochemistry*, **11**:3028.
CORTIJO, M., PANIJPAN, B. and GRATZER, W. B. (1973). *Int. J. Peptide Protein Res.*, **5**:179.
ELWORTHY, P. H., FLORENCE, A. T. and MACFARLANE, C. B. (1968). *Solubilization by Surface-Active Agents*, pp. 13–60. London; Chapman & Hall.
ENGELMAN, D. M. (1971). *J. molec. Biol.*, **58**:153.
FAIRBANKS, G., STECK, T. L. and WALLACH, D. F. H. (1971). *Biochemistry*, **10**:2606.
FASMAN, G. D. (1967). *Poly-α-Amino Acids: Protein Models for Conformational Studies*, pp. 499–508. Ed. G. D. FASMAN. New York; Marcel Dekker.
FERRETTI, J. A. and PAOLILLO, L. (1969). *Biopolymers*, **7**:155.
FRANZEN, J. S., BOBIK, C. and HARRY, J. B. (1966). *Biopolymers*, **4**:637.
FRYE, L. D. and EDIDIN, M. (1970). *J. Cell Sci.*, **7**:319.
GLASER, M., SIMKINS, H., SINGER, S. J., SHEETZ, M. and CHAN, S. I. (1970). *Proc. natn. Acad. Sci. U.S.A.*, **65**:721.
GREENFIELD, N. and FASMAN, G. D. (1969). *Biochemistry*, **8**:4108.
GREFRATH, S. and REYNOLDS, J. A. (1974). *Fedn Proc. Fedn Am. Socs exp. Biol.*, **33**:Abstr. 1738.
HOFF, H. F., MORRISETT, J. D. and GOTTO, A. M., JR. (1973). *Biochim. biophys. Acta*, **296**:653.

HONG, K. and HUBBELL, W. L. (1972). *Proc. natn. Acad. Sci. U.S.A.*, **69**:2617.

HUANG, C. (1969). *Biochemistry*, **8**:344.

INOUYE, M. (1974). *Proc. natn. Acad. Sci. U.S.A.*, **71**:2396.

JACKSON, R. L., SEGREST, J. P., KAHANE, I. and MARCHESI, V. T. (1973). *Biochemistry*, **12**:3131.

JI, T. H. (1974). *Proc. natn. Acad. Sci. U.S.A.*, **71**:93.

JI, T. H. and NICOLSON, G. L. (1974). *Proc. natn. Acad. Sci. U.S.A.*, **71**:2212.

KANT, J. A. and STECK, T. L. (1972). *Nature, New Biol.*, **240**:26.

KANT, J. A. and STECK, T. L. (1973). *J. biol. Chem.*, **248**:8457.

KIMELBERG, H. K. and PAPAHADJOPOULOS, D. (1971). *Biochim. biophys. Acta*, **233**:805.

KNOTT, G. D. and REECE, D. K. (1972). *Proceedings of the ONLINE 1972 International Conference*, p. 497. Brunel University, Uxbridge, England.

KOBYLKA, D., KHETTRY, A., SHIN, B. C. and CARRAWAY, K. L. (1972). *Archs Biochem. Biophys.*, **148**:475.

KORN, E. D. (1969). *Fedn Proc. Fedn Am. Socs exp. Biol.*, **28**:6.

KORNBERG, R. D. and MCCONNELL, H. M. (1971a). *Proc. natn. Acad. Sci. U.S.A.*, **68**:2564.

KORNBERG, R. D. and MCCONNELL, H. M. (1971b). *Biochemistry*, **10**: 1111.

LEA, E. J. A., RICH, G. J. and SEGREST, J. P. (1975). *Biochim. biophys. Acta*, **382**:41.

LENARD, J. and SINGER, S. J. (1966). *Proc. natn. Acad. Sci. U.S.A.*, **56**:1828.

LEVINE, Y. K., BAILEY, A. I. and WILKINS, M. H. F. (1968). *Nature, Lond.*, **220**:577.

LEWIS, P. N. and SCHERAGA, H. A. (1971). *Archs Biochem. Biophys.*, **144**:576.

LUX, S. E., JOHN, K. M., FLEISCHER, S., JACKSON, R. L. and GOTTO, A. M., JR. (1972). *Biochem. biophys. Res. Commun.*, **49**:23.

MARCHESI, V. T. and ANDREWS, E. P. (1971). *Science, N.Y.*, **174**:1247.

MARCHESI, V. T. and STEERS, E., JR. (1968). *Science, N.Y.*, **159**:203.

MARCHESI, V. T., TILLACK, T. W., JACKSON, R. L., SEGREST, J. P. and SCOTT, R. E. (1972). *Proc. natn. Acad. Sci. U.S.A.*, **69**:1445.

MARTON, L. S. G. and GARVIN, J. E. (1973). *Biochem. biophys. Res. Commun.*, **52**:1457.

MAZIA, D. and RUBY, A. (1968). *Proc. natn. Acad. Sci. U.S.A.*, **61**:1005.

MCCONNELL, H. M., WRIGHT, K. L. and MCFARLAND, B. G. (1972). *Biochem. biophys. Res. Commun.*, **47**:273.

METCALFE, S. M., METCALFE, J. C. and ENGELMAN, D. M. (1971). *Biochim. biophys. Acta*, **241**:422.

MORAWIECKI, A. (1964). *Biochim. biophys. Acta*, **83**:339.

NICOLSON, G. L., MARCHESI, V. T. and SINGER, S. J. (1971). *J. Cell Biol.*, **51**:265.

PAPAHADJOPOULOS, D. (1973). *BBA Library 3: Form and Function of Phospholipids*, pp. 143–169. Ed. G. B. ANSELL, R. M. C. DAWSON and J. N. HAWTHORNE. New York; Elsevier.

PHILLIPS, M. C., LADBROOKE, B. D. and CHAPMAN, D. (1970). *Biochim. biophys. Acta*, **196**:35.

QUADRIFOGLIO, F. and URRY, D. W. (1968). *J. Am. chem. Soc.*, **90**:2755.

RAO, S. P. and MILLER, W. G. (1973). *Biopolymers*, **12**:835.

ROBERTSON, J. D. (1964). *Cellular Membranes in Development*, pp. 1–81. Ed. M. LOCKE. New York; Academic Press.

ROBSON, B. and PAIN, R. H. (1972). *Nature, New Biol.*, **238**:107.

SEGREST, J. P. and FELDMANN, R. J. (1974). *J. molec. Biol.*, **87**:853.

SEGREST, J. P., GULIK-KRZYWICKI, T. and SARDET, C. (1974). *Proc. natn. Acad. Sci. U.S.A.*, **71**:3294.

SEGREST, J. P., JACKSON, R. L., MARCHESI, V. T., GUYER, R. B. and TERRY, W. (1972). *Biochem. biophys. Res. Commun.*, **49**:964.

SEGREST, J. P., KAHANE, J., JACKSON, R. L. and MARCHESI, V. T. (1973). *Archs Biochem. Biophys.*, **155**:167.

SHECHTER, E., GULIK-KRZYWICKI, T., AZERAD, R. and GUS, G. (1971). *Biochim. biophys. Acta*, **241**:431.

SINGER, S. J. (1971). *Structure and Function of Biological Membranes*, pp. 145–222. Ed. L. F. ROTHFIELD. New York; Academic Press.

SINGER, S. J. and NICOLSON, G. L. (1972). *Science, N.Y.*, **175**:720.

SPATZ, L. and STRITTMATTER, P. (1971). *Proc. natn. Acad. Sci. U.S.A.*, **68**:1042.

STEERS, E. and MARCHESI, V. T. (1969). *J. gen. Physiol.*, **54**:655.

STEIM, J. M., TOURTELLOTTE, M. E., REINERT, J. C., MCELHANEY, R. N. and RADER, R. L. (1969). *Proc. natn. Acad. Sci. U.S.A.*, **63**:104.

TANFORD, C. (1973). *The Hydrophobic Effect: Formation of Micelles and Biological Membranes*. New York; John Wiley.

TANFORD, C. (1974). *Proc. natn. Acad. Sci. U.S.A.*, **71**:1811.

TIFFANY, M. L. and KRIMM, S. (1973). *Biopolymers*, **12**:575.

TILLACK, T. W., SCOTT, R. E. and MARCHESI, V. T. (1972). *J. exp. Med.*, **135**:1209.

TOURTELLOTTE, M. E., BRANTON, D. and KEITH, A. (1970). *Proc. natn. Acad. Sci. U.S.A.*, **66**:909.

TRIPLETT, R. B. and CARRAWAY, K. L. (1972). *Biochemistry*, **11**:2897.

VANDERKOOI, G., SENIOR, A. E., CAPALDI, R. A. and HAYASHI, H. (1972). *Biochim. biophys. Acta*, **274**:38.

WANG, K. and RICHARDS, F. M. (1974). *Fedn Proc. Fedn Am. Socs exp. Biol.*, **33**:Abstr. 424.

WILKINS, M. H. F., BLAUROCK, A. E. and ENGELMAN, D. M. (1971). *Nature, New Biol.*, **230**:73.

WINZLER, R. J. (1969). *Red Cell Membranes: Structure and Function*, pp. 157–171. Ed. G. A. JAMIESON and T. J. GREENWALT. Philadelphia; J. B. Lippincott.

YU, I. and STECK, T. L. (1974). *Fedn Proc. Fedn Am. Socs exp. Biol.*, **33**:Abstr. 1740.

ZVILICHOVSKY, B., GALLOP, P. M. and BLUMENFELD, O. O. (1971). *Biochem. biophys. Res. Commun.*, **44**:1234.

ADDENDUM

Recent ^{22}Na-leakage experiments with phospholipid–T(is) liposomes demonstrate a cooperative change in ^{22}Na efflux at a T(is) concentration which corresponds closely to the critical multimer concentration predicted by the freeze-fracture experiments summarized in *Figure 1.11* (J. P. Segrest and J. B. Stone, unpublished work). Based on the equation $\mu^0_{\text{micelle}} - \mu^0_{\text{lipid}} = RT \ln \text{CMC}$ [where μ^0 is the chemical potential of T(is) as a multimer (micelle) or monomer in lipid, R is the gas constant, T is the temperature in degrees Kelvin and CMC is the critical multimer concentration in mole fraction], a free energy of multimerization of approximately -3 kcal mol^{-1} can be calculated for T(is) in phospholipid bilayers. This free energy is compatible with the model proposed for T(is) multimer formation (*Figure 1.15*).

Further evidence for complete penetration of phospholipid bilayers by T(is) multimers has been obtained from fluorescence experiments utilizing the fluidity probe pyrene. T(is) at a 15:1 molar ratio is found to decrease the fluidity of dimyristoyl phosphatidylcholine vesicles at 30 °C by a factor of approximately six (J. P. Segrest, A. Y. Romans and H. J. Pownall, unpublished work). The author is not aware of any other protein as yet examined which will produce such a spectacular fluidity change upon interaction in lipid bilayers. Studies by differential scanning calorimetry data also support the T(is) multimeric model (A. Y. Romans, D. Papahadjopoulos and J. P. Segrest, unpublished work).

2

The platelet *

G. A. Jamieson
*The American National Red Cross Blood Research Laboratory,
Bethesda, Maryland*

There are three major types of cell in the circulating blood: the red cell, which is concerned mainly with oxygen transport; the white cells, which are concerned with immunity and tissue repair, and the platelet or thrombocyte, which is concerned with coagulation and hemostasis. When a blood vessel is severed, or when the endothelial cells lining the lumen of the blood vessel are damaged, the platelets circulating in the blood are exposed to the subendothelial components of the vessel wall. Platelets immediately adhere to certain of these structures and this initial adhesion is followed by (a) the release reaction, in which certain constituents of the platelet are expelled from its interior, and (b) the aggregation of other platelets to those already adhering to the subendothelium. Finally, membrane fusion occurs between adjacent platelets leading to viscous metamorphosis and the retraction of the clot under the influence of contractile proteins in the platelet. This complex sequence of hemostatic events can be divided into six distinct stages (Stormorken and Owren, 1971) but Holmsen (1974) has emphasized that several of these may not be causally related and may be interdependent expressions of one basic cellular function.

These events are quite rapid. Under optimal conditions *in vitro* the adhesion of platelets to collagen is complete in a few seconds, aggregation and release in about three minutes and membrane fusion, viscous metamorphosis and clot retraction in about thirty minutes. Each of these obviously involves the platelet membrane; firstly, in the recognition by the platelet surface of specific subendothelial elements such as collagen and basement membrane; secondly, in the transfer of this information to the interior of the platelet leading to the release reaction; thirdly, in the interaction of complementary sites on adjacent platelets during aggregation; and fourthly, in the loss of membrane integrity leading to membrane fusion and viscous metamorphosis. Thus the platelet membrane is of fundamental importance in its rapid reaction to specific stimuli.

* Contribution No. 317 from The American National Red Cross.

A comment regarding nomenclature is mandatory since the terms employed to describe platelet function are used in a somewhat different sense than when used to describe other cell–cell or cell–surface interactions. The term *adhesion* describes the attachment of a platelet to a surface and is probably best restricted to the interaction of platelets with their natural substrates collagen and basement membrane; interaction with foreign surfaces such as glass beads and plastics is probably best described as 'adhesiveness'. *Aggregation* refers to the attachment of platelets to each other and carries with it the implication of normal physiological mechanisms; the terms 'agglutination' and 'adherence' have been less widely used to describe platelet–platelet attachment resulting from, for example, the effects of lectins or immune complexes. Thus, adhesion describes the process by which one platelet attaches itself to the vessel wall while aggregation describes the process by which the other platelets of the microthrombus attach themselves to it. The term *release reaction* describes a secretory phenomenon which is central to the platelet's role in hemostasis. Following the adhesion of platelets in plasma to collagen fibrils there is a brief lag period followed by the release of ATP, ADP, serotonin and calcium from the electron-dense granules of the platelet; this is termed Release I. Release II involves the release of lysomal enzymes from the α-granules of the platelet under the influence of strong aggregating agents such as thrombin, in the presence of calcium. Cytoplasmic enzymes, such as lactic dehydrogenase, do not appear in the medium during either of these phases of release, indicating the ultrastructural specificity of the reaction. However, Release I does cause changes in the plasma membrane that result in the manifestation of thromboplastic procoagulant activity.

For a fuller discussion of the role of the platelet in hemostasis, the reader is referred to several monographs in this area (Johnson, 1971; Brinkhous, Shermer and Mostofi, 1971; Sherry and Scriabine, 1974; Baldini and Ebbe, 1974).

2.1   PLATELET ULTRASTRUCTURE

The platelet is apparently unique among cells in being formed by the fragmentation of its precursor rather than by the maturation of a more primitive cell; the megakaryocyte divides within the bone marrow, each megakaryocyte producing a total of about 2000 platelets (Harker, 1971). During the maturation of megakaryocyte, demarcation membranes are formed by invagination of its plasma membrane (Behnke, 1968b), although an origin from internal structures has been suggested (Schulz, 1966). These demarcation membranes are rich in carbohydrate and define the outer membrane of the developing platelet. Once formed, the platelets are discoid in shape, 2–3 μm in diameter, with a lifespan of about 10 days and a turnover of about 5×10^{11} platelets per day in man.

The internal structure of the platelet is quite complex and our knowledge of it has been gained mainly from studies with the electron microscope (*Figure 2.1*). The platelet is devoid of a nucleus and, therefore, it should not strictly be termed a cell, although it is convenient to do so. The platelet contains a few primitive mitochondria with relatively few cristae, occasional

Golgi bodies and glycogen granules which are usually concentrated in particular areas but may also be seen spread throughout the cell. The granules appear to be lysosomal in nature and are rich in phospholipids and contain hydrolytic enzymes such as acid phosphatase, β-glucuronidase and cathepsin and may also be storage sites for platelet fibrinogen, serotonin

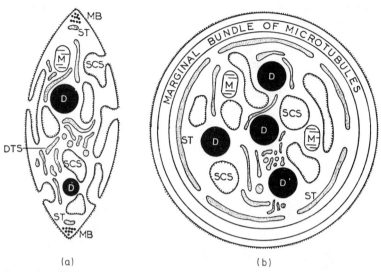

(a)   (b)

Figure 2.1   Diagrammatic representation of a rat blood platelet, (a) transverse and (b) equatorial section. The marginal bundle of microtubules, which encircles the platelet immediately beneath the plasma membrane, is seen in cross section in (a). Beneath is the submarginal tubule (ST) from the dense tubular system (DTS). Other components of this membrane are seen scattered in the cytoplasm. The surface coat also lines the surface-connected membrane system (SCS). D, dense platelet granule; G, glycogen; M, mitochondrion; MB, marginal bundle of microtubules. (From Behnke, 1967, courtesy of the author and The Wistar Press)

and mucopolysaccharides (White, 1971). The very dense granules, or dense bodies, are few in number but appear to be very important in hemostatic function. They are the primary secretory organelles of the platelet and contain serotonin, as well as nonmetabolic pools of ATP, ADP and catecholamines. There is some evidence that the dense bodies may be derived from the granules. The platelet also has a marked microtubular system which appears to provide a cytoskeleton that is active in platelet function.

The trilaminar structure of the platelet plasma membrane is morphologically indistinguishable from that of other cells. However, the surface membrane of the platelet is characterized by deep invaginations into the interior of the platelet leading to a complex network known as the 'surface-connected system' or 'canalicular system'. One may ask whether these invaginations represent uncompleted demarcation membranes remaining from the megakaryocyte. The canalicular system gives the platelet a 'sponge-like' quality which may assist in the transport of clotting factors, such as fibrinogen, and other plasma proteins within the circulation. It certainly gives to the platelet a high surface to volume ratio and brings the interior of the platelet and the exterior milieu into close proximity.

Outside the trilaminar membrane, both on the platelet surface and in the channels of the surface-connected system, is an extensive exterior coat or glycocalyx which is 20–50 nm in depth and is thought to be rich in carbohydrate on the basis of staining with specific cytochemical reagents (Behnke, 1967, 1968a; Rambourg and Leblond, 1967; Hovig, 1968). Emerging evidence, to be discussed later, suggests that components of this exterior coat may be important in platelet function. It can be removed by careful proteolysis without extensive changes in the ultrastructure of the membrane (Hovig, 1968). At the present time it is not clear whether the glycocalyx should be regarded as an integral component of the platelet membrane or as being more or less loosely associated with it.

In its normal environment the platelet, together with its exterior coat, is surrounded by an ill-defined nimbus of plasma proteins which has been called the 'plasmatic atmosphere'; differences between the reactions of platelets in plasma and washed platelets may arise from the loss of components in this compartment (Ardlie and Han, 1974).

Recent studies using freeze-fracture and freeze-etching techniques (Hoak, 1972; Reddick and Mason, 1973; Feagler *et al.*, 1974) have been particularly

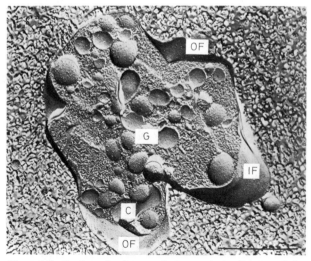

Figure 2.2   Replica of a freeze-etched normal platelet suspended in 25% glycerol. Inner faces (IF) and outer faces (OF) of the fractured platelet surface membrane contain scattered intramembranous particles. Fractured membranes of granules (G) and the canalicular system (C) also contain scattered intramembranous particles. The bar represents 1.0 µm; magnification × 34 000. (From Feagler et al., 1974, courtesy of the authors and The Rockefeller University Press)

valuable in allowing the visualization of surface details of the plasma membrane and its relationship to the canalicular or surface-connected system (*Figure 2.2*). Surface-associated particles have been found embedded in the outer surface of the membrane and extending into the canalicular system while the inner surface contained numerous fibrillar structures. These particles are absent from the surface of platelets which do not aggregate, either because they have been obtained from thrombasthenic patients or are normal platelets which have been treated with an anti-aggregating drug.

In addition to the plasma membrane, platelets contain membranes associated with their Golgi apparatus, their granules and dense bodies and with their dense tubular system; however, virtually nothing is known about the membranes of these last three compartments.

Thus, the platelet surface may be regarded as comprising three different zones: the plasmatic atmosphere, which is retained by platelets only under special conditions of isolation and is required for ADP-induced aggregation; the glycocalyx, which remains with the platelet during most isolation procedures and may be the site of thrombin action and lectin-induced agglutination, and the platelet membrane itself, which contains the structural elements associated with transport, release and the exchange of information between the interior and exterior of the platelet.

Studies on the nature of the platelet surface have been developed more from studies on intact platelets than on the isolated platelet membrane. The factors affecting platelet function have been extensively reviewed, particularly from the pharmacological viewpoint (Mustard and Packham, 1970) and theories of platelet aggregation have frequently focused on components of the plasmatic atmosphere or glycocalyx (Booyse and Rafelson, 1972; Ardlie and Han, 1974).

2.2 PLATELET CHARGE

It is only when charged groups are present in the outer nanometer or so of a cell surface that they contribute effectively to the electrophoretic charge, or zeta potential. Thus any change in the arrangement of charged groups at the cell surface, such as might arise following adhesion or aggregation, could lead to changes in electrophoretic mobility.

The platelet surface contains a range of ionic organic groups (Mehrishi, 1971) and platelets have a high net negative charge with an isoelectric point at pH 3.6 and an electrophoretic mobility of -0.85 μm s^{-1} V^{-1} cm^{-1} at pH 7 (Seaman and Vassar, 1966). This charge is at least partially due to sialic acid residues since treatment with neuraminidase removes about 10^5 molecules of sialic acid per μm^2 and results in a 50 percent reduction in electrophoretic mobility with a concomitant increase in the isoelectric point to pH 4.2. Calculations show that total surface sialic acid is about 1.9×10^6 molecules of sialic acid per μm^2, which is approximately 10 times the surface density on red cells. Although these values were obtained with intact platelets, similar values have been found for the isoelectric points of isolated platelet membranes by the techniques of isoelectric focusing, both before and after treatment with neuraminidase (Jamieson and Groh, 1971), suggesting that the charge distribution in the glycocalyx remains intact during the isolation of the membrane.

The electrophoretic mobility of platelets is affected by aggregating agents (Hampton and Mitchell, 1966; Seaman and Vassar, 1966), by aging *in vitro* (Mason and Shermer, 1971) and in certain pathological conditions. For example, the giant platelets of congenital macrothrombocytic thrombopenia (Bernard–Soulier syndrome) have a reduced electrophoretic mobility which may reflect a decreased density of sialic acid on the platelet surface since the total sialic acid content is the same as that of normal platelets (Gröttum and Solum, 1969). On the other hand, the electrophoretic mobility of platelets from patients suffering from Glanzmann's thrombasthenia is identical with that of normal platelets and they have been reported to have the same content of sialic acid (Zucker and Levine, 1963), although more recent studies indicate that these patients may lack a major glycoprotein on the platelet membrane (Nurden and Caen, 1974).

This bound sialic acid probably constitutes the viral receptor site of the platelet (Terada *et al.*, 1966) and is the source of its activity in inhibiting hemagglutination (Uhlenbruck, 1961); an active receptor fraction has been isolated from a crude preparation of platelet membranes by solubilization with aqueous pyridine (Pepper and Jamieson, 1968). Thrombocytopenia is a frequent concomitant of viral infections (Broun and Broun, 1962) and this condition is also induced in mice by injections of neuraminidase with, surprisingly, a concomitant decrease in the incidence of metastases (Gasic and Gasic, 1962). This may reflect a role for platelets in the spread of malignant disease (Gasic *et al.*, 1973) and high concentrations of Maloney mouse leukemia virus have been found in association with the demarcation membranes of mouse megakaryocyte (Schulz, 1968), reflecting, perhaps, the concentration of carbohydrate in this region.

2.3 ISOLATION OF PLATELET PLASMA MEMBRANES

Platelet disruption has been effected by sonication, the use of a tissue homogenizer or blender under various conditions, and osmotic lysis. The best-characterized membranes have been obtained by the use of the 'no clearance' tissue homogenizer (Marcus *et al.*, 1966) or by the glycerol lysis technique (Barber and Jamieson, 1970). Certain of these methods have been compared: blender and tissue homogenizer techniques (Day, Holmsen and Hovig, 1969), homogenizer, sonication, nitrogen cavitation, simple osmotic lysis and glycerol lysis (Barber, Pepper and Jamieson, 1971), and freeze-thawing, sonication, homogenization and glycerol lysis (Kaulen and Gross, 1973). In the second of these studies the use of membrane-hardening agents such as zinc chloride and fluorescein mercuric acetate was also evaluated. Taken together these studies showed the homogenizer and glycerol lysis techniques to be more satisfactory than the other methods tested in terms of simplicity, effectiveness and reproducibility, while the use of membrane-hardening agents reduced the degree of lysis and the activity of membrane enzymes without compensating advantages. The glycerol lysis technique appeared to have some advantage over the use of the tissue homogenizer in terms of a greater degree of lysis, a reduced release of lysosomal enzymes, and the presence of a greater number of intact granules in the debris fraction. A similar retention of subcellular structural elements has been found in the application of the glycerol lysis technique to nucleated cells (M. Jett, T. Seed and G. A. Jamieson, 1976, submitted for publication).

Membrane and granule fractions have been isolated by zonal centrifugation of homogenates prepared by the blender technique (Harris and Crawford, 1973) and the membrane fraction has been subfractionated into two components of which one probably arose from the surface membrane (Taylor and Crawford, 1974). Isolation of granule and mitochondrial fractions has also been obtained by the use of nitrogen cavitation (Broekman, Westmoreland and Cohen, 1974) and mitochondria have been isolated by the use of the French press (Fukami and Salgonicoff, 1973). Isolation of granule and mitochondrial membranes, as such, has not yet been reported.

The glycerol lysis technique involves the slow intracellular accumulation of glycerol into platelets by centrifugation through a continuous glycerol gradient (0–40%) followed by lysis in an isotonic sucrose solution. Platelet membranes have a very slow transit time for glycerol, of the order of minutes, whereas the transit time for water is measured in microseconds. Hence, the net effect of the above treatment is for a rapid influx of water and a slow efflux of glycerol, resulting in an 'explosive' lysis from the interior of the platelet. The plasma membrane, cellular cytoplasm and cellular debris are then separated by density-step (27% sucrose, d 1.106) centrifugation.

The membrane fraction isolated by the glycerol lysis technique and density-step centrifugation contains a high ratio of cholesterol to phospholipid (0.47), thought to be characteristic of plasma membranes in other cells, and shows an enrichment in plasma membrane enzymes such as phosphodiesterase (eightfold), acid phosphatase (fourfold) and ATPase (twofold) and a virtual absence of enzymes associated with subcellular elements.

The membrane fraction from the density-step procedure can be further separated into two equal subfractions of d 1.090 and 1.120 on a continuous sucrose gradient (15–40%, d 1.06–1.15). Both these subfractions appear to be derived from the platelet surface membrane. They differ in their protein to lipid ratios but are otherwise identical in chemical and enzymatic analysis, isoelectric points, and in protein distribution as determined by gel electrophoresis. They do not appear to be 'inside-out, outside-out' vesicles on the basis of their equal accessibility to neuraminidase and trypsin and to iodination catalyzed by lactoperoxidase (Barber and Jamieson, 1971a). However, they do differ in ultrastructure, the lighter (d 1.090) having an average diameter of 1.7 µm, close to that of the intact platelet, and with concentric double-membrane structures, while the heavier (d 1.120) consists of single membranes of average diameter 0.7 µm (*Figure 2.3*).

The two membrane subfractions have identical procoagulant activities but differences appear in the lipids extracted from each, the lipids extracted from the low-density membranes having a higher procoagulant activity than those from the high-density membranes (Barber *et al.*, 1972). The origin of these two subclasses of membrane is not known but they may reflect regions of anatomical specialization which have been proposed as existing in the platelet membrane (Nakao and Angrist, 1968) or the heterogeneity of platelet populations (*see* Karpatkin, 1972). However, this is now thought to arise from megakaryocytopoiesis, rather than from platelet senescence (Paulus and Kinet-DeNoel, 1973; Boneu *et al.*, 1973).

The platelet plasma membrane contains carbohydrate (7%), protein (36%) and lipid (52%). Cholesterol comprises 90 percent of the neutral lipid with approximately equimolar ratios of cholesterol to phospholipid, similar to values found for the plasma membranes of other cells. The sugar components identified were glucose, galactose, mannose, glucosamine and galactosamine, fucose and sialic acid.

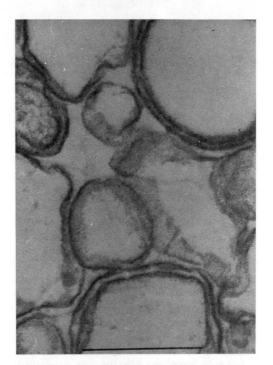

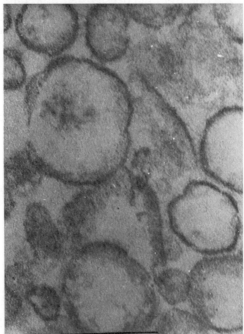

Figure 2.3   Electron micrographs of platelet membranes isolated by the glycerol lysis technique and continuous density centrifugation. Contaminating subcellular structures are absent. Magnification × 40 000. (a) Membrane band, d 1.090. Note the appearance of double membrane structures. The bar represents 2 μm. (b) Membrane band, d 1.120. Note the absence of double membrane structures. The bar represents 1 μm. (From Barber and Jamieson, 1970, courtesy of the American Society of Biological Chemists)

A number of enzymes known to be associated with the plasma membrane of other cells have been shown to be localized and concentrated in the platelet plasma membrane (*see* Jamieson and Barber, 1973). These include Mg^{2+},Ca^{2+}-dependent ATPase, Na^+,K^+-stimulated ATPase, adenyl cyclase, phosphodiesterase, cholinesterase, acid phosphatase, 5′-nucleotidase and several glycosyltransferases. On the other hand, the isolated platelet plasma membranes were deficient in those enzymes thought to be associated with subcellular elements: glycosidases, esterase and succinic dehydrogenase. These observations provide biochemical support for the platelet plasma membrane being derived from the plasma membrane of the megakaryocyte rather than from subcellular elements.

2.4 GLYCOPROTEINS OF THE PLATELET SURFACE

As mentioned above, indirect evidence for the presence of glycoproteins on the platelet surface arose from the use of specific cytochemical stains combined with electron microscopy and from studies on platelet electrophoresis and viral-induced agglutination.

Further indirect evidence for surface carbohydrate on the platelet has been provided by studies on the reaction of platelets with lectins having different carbohydrate specificities (Greenberg and Jamieson, 1974). These studies have shown that platelets are strongly agglutinated by wheat-germ agglutinin, which has *N*-acetylglucosamine or sialic acid as its haptenic determinants, by *Ricinus communis*, which has galactose as its haptenic determinant, and by the lectin of *Phaseolus coccineus*; this last lectin has no known monosaccharide determinant but is inhibited by an extended glycopeptide of the red cell surface (Kubanek, Entlicher and Kocourek, 1973). Lectins with specificities towards glucose and mannose (concanavalin A, and the lectins of *Lathyrus sativus, Lens culinaris* and *Pisum sativum*) do not cause platelet agglutination nor the release of ADP, but do cause the release of serotonin at high concentrations, suggesting that different receptor sites may exist on the platelet surface for the release of the two metabolites; another galactose-specific lectin (*Agaricus bisporus*) also falls into this class. Interestingly, although concanavalin A does not cause the agglutination of intact platelets it does cause the agglutination of isolated membranes. This is consistent with the suggestion that this lectin may be a marker for galactosyltransferase in certain cells (Lamont *et al.*, 1974) and with the fact that the galactosyltransferase of the platelet is manifest only in isolated membranes and not in intact platelets (Barber and Jamieson, 1971c), suggesting a location on the inner aspect of the membrane or buried within the intact membrane in a form which is revealed following the trauma of isolation.

Soybean lectin, with an *N*-acetylgalactosamine determinant, is virtually without effect on platelet agglutination or release.

Quantitative ultrastructural studies have also been carried out on the binding of lectins to platelets. The erythroagglutinating lectin (E-PHA) of *Phaseolus vulgaris* and the leukoagglutinating lectin (L-PHA) bind to intact platelets causing platelet aggregation and the inhibition of adenyl cyclase, changes which also occur during aggregation induced by thrombin

(Majerus and Brodie, 1972). About 600 000 molecules of E-PHA per platelet bind to the surface with an apparent dissociation constant of 0.5×10^{-7} M while about 300 000 molecules of L-PHA bind with an apparent dissociation constant of 4×10^{-7} M. Although the lectin of *Lens culinaris* does not cause aggregation it does bind tightly (400 000 molecules per platelet, 1.4×10^{-7} M) and there is an approximate doubling of receptor sites for this lectin on prior treatment of the platelet with thrombin or L-PHA. Subsequent studies showed differences in the rate and extent of serotonin release induced by thrombin and PHA and in resulting platelet morphology (Tollefsen, Feagler and Majerus, 1974a). The use of ferritin-conjugated PHA and the freeze-fracture technique demonstrated that the morphological changes induced by thrombin involve a decrease in the number of lectin receptor sites on the platelet surface and a marked increase in their number in the canalicular system (Feagler *et al.*, 1974). It was suggested that this may arise from the fusion of granule membranes with the plasma membranes of the canalicular system.

Direct evidence for the presence of surface glycoproteins on the platelet was obtained by the isolation of glycopeptides following treatment with trypsin, pronase and papain (Pepper and Jamieson, 1969, 1970), plasmin and thrombin (Jamieson *et al.*, 1971) and chymotrypsin (Barber and Jamieson, 1971a). In each case the glycopeptides obtained were generally similar to those obtained with trypsin, the most extensively studied system (Pepper and Jamieson, 1970). Three size classes of glycopeptides were obtained with molecular weights of 120 000 (glycopeptide I), 22 500 (glycopeptide II) and 5000 (glycopeptide III); the glycopeptide of highest molecular weight (glycopeptide I) has been termed a 'macroglycopeptide' since its molecular weight appears to be an order of magnitude higher than those of glycopeptides from plasma proteins or red cell stroma. The platelet glycopeptides showed several analytical differences between each class (*Table 2.1*); the macroglycopeptide (glycopeptide I) contained about 70 percent carbohydrate with galactose as essentially its sole neutral sugar with equimolar amounts of galactosamine and glucosamine, presumably as their *N*-acetyl derivatives, and with heterosaccharide residues joined to the polypeptide backbone in *O*- and *N*-glycosidic linkages. On the other hand, the glycopeptide of lowest molecular weight (glycopeptide III) contained equimolar amounts of galactose and mannose, had asparagine as its principal amino acid and appeared to contain only *N*-glycosidic linkages; the glycopeptide of intermediate molecular weight (glycopeptide II) was also intermediate in properties but was not studied extensively because it was present in relatively small amounts.

The presence of mucopolysaccharides on the outer surface of the platelet has frequently been suggested on the basis of ultrastructural studies with specific cytochemical reagents. In the proteolytic studies summarized above, it was found that treatment with enzymes which caused the release reaction (trypsin, pronase, papain, plasmin) resulted in the solubilization of a carbohydrate component (glycopeptide 0) which was eluted from Sephadex G200 at the void volume and had properties consistent with its being the chondromucopeptide derived from chondroitin sulfate (Jamieson *et al.*, 1971). On the other hand, this component was not obtained with chymotrypsin, which does not cause the release reaction, or among the tryptic

products of isolated membranes. These results suggest that mucopolysaccharides are not components of the platelet surface but are probably released
from the granules under appropriate circumstances.

Table 2.1   CARBOHYDRATE AND AMINO ACID ANALYSES OF
GLYCOCALICIN AND THE DERIVED MACROGLYCOPEPTIDE
AND PEPTIDE (mol %)

	Glycocalicin (mol. wt 148 000)	Macroglycopeptide (mol. wt 120 000)	Peptide (mol. wt 45 000)
NANA	12.5	21.1	1.5
Man	1.2	0.9	0.8
Fu	1.9	1.3	0.7
Gal	15.1	20.2	1.9
Glc	2.3	1.6	1.5
GlcNAc	7.2	10.9	0.3
GalNAc	5.9	10.0	0.2
Lys	3.3	2.1	4.2
His	1.1	0.6	1.7
Arg	1.1	trace	1.6
Asp	4.7	0.9	8.9
Thr	7.6	7.9	5.5
Ser	4.9	4.8	18.6
Glu	5.8	3.6	14.6
Pro	6.8	6.8	3.3
Gly	2.7	0.7	16.3
Ala	2.3	1.6	8.2
Cys	nd	nd	nd
Val	2.5	0.6	3.3
Met	0.8	trace	trace
Ile	1.3	1.0	2.2
Leu	6.7	2.3	4.0
Tyr	1.2	trace	1.7
Phe	1.6	1.1	1.6

Abbreviation—nd: none detected

Differences between the glycopeptides I, II and III were also shown in
their effect on the agglutination of platelets induced by plant lectins (Greenberg and Jamieson, 1974). Platelet agglutination induced by wheat-germ
agglutinin was inhibited by the macroglycopeptide but not by glycopeptide
II, while the agglutination induced by *Phaseolus coccineus* was inhibited by
glycopeptide II but not by the macroglycopeptide; glycopeptide III was
without effect in either case.

The arrangement of the parent glycoproteins, and other proteins, at the
platelet surface has been the object of several studies involving the lactoperoxidase-catalyzed iodination of intact platelets and their isolated
membranes.

Lactoperoxidase-catalyzed iodination of intact platelets and subsequent
gel electrophoresis in sodium dodecyl sulfate (SDS) leads to the labeling
of a heterogeneous group of polypeptides and of three glycoprotein bands
with molecular weights of approximately 150 000, 120 000 and 95 000
(Nachman and Ferris, 1972; Phillips, 1972; Nachman, Hubbard and

Ferris, 1973), all of which are accessible to trypsin (Phillips, 1972). Although the glycoprotein of lowest molecular weight (95 000) is present in the lowest amount, as judged from SDS staining, it was much more highly iodinated than the other two glycoproteins (Phillips, 1972), which may be solely a reflection of a higher content of aromatic amino acids rather than its quantitative significance in the membrane; on this basis the term 'major' surface glycoprotein for this component (Nachman, Hubbard and Ferris, 1973) is erroneous. This glycoprotein was partially purified following extraction of the membranes with lithium diiodosalicylate and was used to raise antibodies which could themselves induce platelet agglutination (Nachman, Hubbard and Ferris, 1973).

The glycoprotein component of highest molecular weight (*ca.* 150 000) is the one which stains most strongly with periodic acid–Schiff reagent. It has been found to occur in both membrane-bound and soluble forms in platelet homogenates and has been purified from the soluble fraction (Lombart, Okumura and Jamieson, 1974). The soluble form is thought to be present on the platelet surface, to judge from its labeling with $[^{14}C]$glycine ethyl ester by transglutaminase and on its accessibility to neuraminidase and to carboxypeptidase, which does not cause the release reaction of intracellular components (Okumura and Jamieson, 1975). The term *glycocalicin* has been suggested for this component on the basis of its origin as a component of the platelet glycocalyx.

Glycocalicin is cleaved by trypsin to yield a macroglycopeptide apparently identical with that obtained from intact platelets and containing 70 percent of carbohydrate, and a large peptide of molecular weight 48 000 which contains a small amount of carbohydrate (*ca.* 9 percent). Overall, this tryptic peptide lacks the preponderance of hydrophobic amino acids present in the tryptic peptide from the red cell glycoprotein (Winzler, 1969; Marchesi *et al.*, 1972) but may contain more limited hydrophobic domains which are responsible for membrane binding. Analytical data on glycocalicin and its derived peptide and macroglycopeptide are shown in *Table 2.1*.

Glycocalicin probably constitutes the thrombocyte-specific antigen previously reported as a soluble component of platelet homogenates (Milgrom, Campbell and Witebsky, 1968; Hanna and Nelken, 1971). Antiglycoprotein antibodies raised in chickens recognize two determinants in the glycoprotein, one against the macroglycopeptide portion and the other against the peptide portion (Okumura and Jamieson, 1975). These antibodies agglutinate intact platelets and isolated membranes, which suggests that the soluble and membrane-bound forms of the glycoprotein are immunologically cross-reactive. Glycocalicin also gives precipitin reactions with wheat-germ agglutinin and *Agaricus bisporus* and is presumably a receptor site for lectin (Greenberg and Jamieson, 1974).

2.5 LIPIDS OF THE PLATELET MEMBRANE

In addition to their role in primary hemostasis, platelets provide an active lipoprotein surface which interacts with plasma coagulation factors to activate prothrombin by both the intrinsic and, possibly, extrinsic pathways. Since exogenous phospholipids can perform the same function, the term *platelet factor 3* has been used although Marcus, Safier and Ullman (1971) have emphasized that it is not an identifiable biochemical substance. This activity resides in the platelet plasma membrane (Marcus, Ullman and Safier, 1969; Barber *et al.*, 1972). The mechanism of its conversion from an inactive to an active form during platelet aggregation is unknown but has been the stimulus for the isolation of well characterized platelet plasma membranes and the investigation of their lipid components.

Extensive analysis of the platelet membrane (Marcus, Ullman and Safier, 1969) has identified the individual fatty acids as well as the principal lipid classes. Phospholipids are the principal lipid class in the platelet membrane with lecithin as the predominant phosphatide, as in other cells, and with cholesterol comprising about 90 percent of the neutral lipids. The molar ratios of cholesterol and phospholipid are almost equal (Marcus, Ullman and Safier, 1969; Barber and Jamieson, 1970).

Phospholipids on the platelet surface may play a role in platelet function since the release of ADP and serotonin results from treatment with phospholipase C and morphological indications of degranulation have been found (Schick and Yu, 1974).

Gangliosides, which are glycosphingolipids containing sialic acid, have been suggested as being serotonin receptors in neuronal tissue (Woolley and Gommi, 1964), although this has been questioned (Fiszer and De Robertis, 1969). In view of the importance of serotonin in platelet function, several analytical studies have been carried on neutral and acid glycosphingolipids in intact platelets because of the difficulty of obtaining adequate amounts of membrane-derived material for analysis (Marcus, Ullman and Safier, 1972; Snyder, Desnick and Krivit, 1972; Tao, Sweeley and Jamieson, 1973), although it is generally agreed that glycosphingolipids are components of the plasma membrane of cells (Stoffel, 1971). Of the neutral glycosphingolipids, lactosylceramide and globoside were the most abundant with lesser amounts of glucosylceramide and trihexosylceramide. Gangliosides comprised 0.5 percent of the total platelet lipids and about 6 percent of the total sialic acid content with hematoside comprising 92 percent of the ganglioside fraction. Equilibrium dialysis studies showed that hematoside and disialolactosylceramide isolated from the platelet bind only weakly to serotonin and much less than do similar fractions from beef kidney (Marcus, Ullman and Safier, 1972). The content of hematoside in platelets is increased by treatment with trypsin (twofold), chymotrypsin (threefold) and thrombin (fivefold) (Tao, Sweeley and Jamieson, 1973). Time studies have shown that the increase in hematoside is complete in ten minutes and is accompanied by a concomitant decrease in ceramide and lactosylceramide (Chatterjee and Sweeley, 1973).

Recent studies have suggested that carbohydrate derivatives of dolichol and retinol may function as intermediates in glycoprotein biosynthesis. Surface glycoproteins appear to play an important role in certain aspects of platelet function. Dolichylphosphomannose and retinylphosphomannose are synthesized both by intact platelets and by isolated membranes although the ratio of the two components differs in each case (De Luca, De Luca and Jamieson, 1976). Although glycoproteins are apparently synthesized from mannose in intact platelets it has not been possible to show transfer from the purified mannolipid to endogenous receptors. Using isolated membranes and UDP–glucose, in order to avoid the multiplicity of pathways possible with glucose in intact platelets, it was found that only dolichylphosphoglucose was synthesized and this could not function as a donor to galactosylhydroxylysine, the characteristic glycopeptide collagen.

2.6 INTERACTION OF AGGREGATING REAGENTS WITH THE PLATELET SURFACE

This is probably one of the most important areas of investigation insofar as the hemostatic effectiveness of platelets in adhesion and aggregation is mediated through the platelet membrane. No adequate explanation of these phenomena has yet been advanced although ADP is considered to play a central role in platelet aggregation.

In the chronology of the events of hemostasis, the first event affecting the platelet membrane is probably the adhesion of platelets to collagen or other subendothelial structures of the vessel wall leading, subsequently, to the aggregation of platelets to those already adhering. A major limitation in the elucidation of these mechanisms has been the absence of adequate methods for their differential assay. Aggregation has been almost universally assayed by the decrease in the optical density of platelet-rich plasma following the addition of aggregating agents such as ADP or collagen (Born, 1962). This system has been modified by the addition of EDTA in order to measure adhesion to collagen (Spaet and Lejniecks, 1969), which has also been measured by optical (Lyman, Rosenberg and Karpatkin, 1971) or radioactive (Cazenave, Packham and Mustard, 1973) counting of platelets adhering to surfaces coated with collagen. However, the role of divalent cations in this phenomenon is not yet clear. In addition, the binding of radiolabeled collagen to platelets has been measured directly (Gordon and Dingle, 1974).

Probably the best assay is the 'ex vivo' system of Baumgartner (1973), which involves the morphometric analysis of electron photomicrographs of platelet adhesion or microaggregation on an everted segment of rabbit blood vessel from which endothelial cells have been removed. While this system gives an excellent representation of events in vivo, it is tedious and difficult in application since it requires extensive use of the electron microscope.

Blood platelets adhere both to the collagen of the basement membrane and to fibrillar collagen (Hugues and Mahieu, 1970), the latter causing a more extensive formation of platelet microaggregates while the former causes the adhesion of individual platelets (Baumgartner and Haudenschild, 1972; Huang, Lagunoff and Benditt, 1974).

The characteristic disaccharide of collagen and basement membrane is glucosylgalactosylhydroxylysine. Studies on the biosynthesis of glomerular basement membrane (Spiro and Spiro, 1971) have suggested that the glycosylation of collagen involves the transfer of galactose or glucose from the appropriate sugar nucleotide to hydroxylysine, or galactosylhydroxylysine, acceptors in collagen. The two enzymes involved in these transfer reactions, collagen:glucosyltransferase and collagen:galactosyltransferase, have been found in platelet membranes isolated by the glycerol lysis technique

(Barber and Jamieson, 1971a, b, c; Bosmann, 1971). However, it is probable that the collagen:galactosyltransferase is buried within the membrane or is present on its inner aspect and the role of this enzyme in platelet function is not known (Barber and Jamieson, 1971c).

A theory has been proposed that platelet to collagen adhesion is mediated by a lectin-like interaction between the collagen:glucosyltransferase of the platelet and incomplete heterosaccharide chains (galactosyl residues) present in collagen (Jamieson et al., 1971). This theory is based on the parallel requirements in adhesion and transferase activity for both free sulfhydryl groups and free ε-amino groups, and on the inhibition of adhesion by collagen glycopeptides. This theory is an extension of the original ideas of Roseman (1970) on the role of transferases in intercellular adhesion and those of Ashwell (see Ashwell and Morell, 1974) on the role of galactosyl residues in recognition processes.

Several studies have been directed towards determining the validity of this hypothesis although they have, without exception, addressed themselves to the study of collagen-induced aggregation rather than the adhesion of platelets to collagen, as originally proposed.

The ability of collagen to induce platelet aggregation is destroyed by oxidation of the terminal galactose residues to 6-aldehydogalactose with galactose oxidase, but is reestablished by their reduction back to galactose. This has been interpreted as indicating a role for the galactosyl residues of collagen in platelet aggregation (Chesney, Harper and Colman, 1972), but more recent data (Muggli and Baumgartner, 1973) suggest that the oxidation interferes with the formation of collagen fibrils and that the fibrillar structure is required for aggregation.

Galactosylhydroxylysine, the proposed receptor on collagen, binds to the platelet surface while the completed glycopeptide, glucosylgalactosyl-hydroxylysine, does not; on the other hand, collagen from which the carbohydrate has been removed by periodate oxidation is still able to support aggregation. These observations have led to the suggestion that the carbohydrate may function as a recognition site rather than an attachment site (Puett et al., 1973). Finally, the purified α1 chain of chick skin collagen has been reported to aggregate platelets and, of the peptides derived from it by cyanogen bromide treatment, only the glycopeptide α1-CB5 has been reported as active (Katzman, Kang and Beachey, 1973; Kang, Beachey and Katzman, 1974).

Thus the question of a role for the carbohydrate of collagen or basement membrane in platelet adhesion and aggregation remains open. An essential requirement of the glycosyltransferase theory is that the first step is the interaction of the enzyme on the platelet surface with the acceptor, galactosylhydroxylysine; the other substrates of the transferase reaction, namely UDP–glucose and Mn^{2+}, would then react subsequently but would not be required for the formation of an enzyme–acceptor complex. A soluble form of the enzyme has been purified approximately a hundredfold; preliminary studies (Smith, Kosow and Jamieson, 1975) show that the reaction mechanism is either ordered bi bi or a rapid equilibrium random bi bi with a dead-end EBQ complex. Thus, while UDP and, possibly, ADP inhibit the overall transferase reaction, these reagents may increase the affinity of the cell surface enzyme for its acceptor on collagen by either mechanism.

In view of these two possibilities, it is of interest that ADP has been found to potentiate collagen-induced aggregation (Packham *et al.*, 1973).

Other studies have been directed towards elucidating the mechanism of binding of ADP and of thrombin, both of which play a major role in platelet aggregation.

The binding of ADP to isolated platelet membranes was found to be maximal at 37 °C over 60 min with a binding constant of 6.5×10^6 M^{-1} and approximately 100 000 binding sites per platelet (Nachman and Ferris, 1974). Binding to the membrane was reduced by over 50 percent by modification of sulfhydryl groups. Inhibitors of platelet aggregation, such as AMP, ATP and chloroadenosine at 10^{-4} M reduced the binding of ADP by 50 percent or more whereas cAMP, α,β-methylene-ADP, adenosine and PGE were without effect, suggesting that they do not compete at the same site on the membrane. However, another report (Horak and Barton, 1974) suggests that α,β-methylene-ADP does compete with ADP in the intact platelet. This discrepancy remains to be resolved.

The possibility that other glycosyltransferases of the platelet surface may be involved in platelet–platelet aggregation has been suggested, based on the presence of a sialyltransferase which transfers sialic acid to exogenous glycopeptides from fetuin or prothrombin (Bosmann, 1972) although this has not been tested with endogenous glycopeptides of the platelet surface. However, some support for this hypothesis has been provided by the finding that bovine factor VIII, but not human factor VIII, can aggregate human platelets (Donati, de Gaetano and Vermylen, 1973; Forbes and Prentice, 1973). On the other hand, removal of terminal sialic acid from human factor VIII results in strong aggregating activity while treatment of this product, or bovine factor VIII, with galactose oxidase destroys the aggregating activity (Vermylen *et al.*, 1973). These results may be interpreted as indicating a role for a sialyltransferase of the platelet surface in this type of aggregation.

The exposure of platelets to thrombin results in a rapid increase in the heat production of the cell (Ross, Fletcher and Jamieson, 1973), reflecting stimulation of glycolysis and glycogenolysis (*see* Marcus, 1969a, b, c), increased synthesis of phosphatidylserine (Lewis and Majerus, 1969) and hematoside (Tao, Sweeley and Jamieson, 1973; Chatterjee and Sweeley, 1973), the release of calcium and ATP (Detwiler and Feinman, 1973a, b) and the modification of various 'thrombin-sensitive' proteins (Ganguly, 1971; Baenziger, Brodie and Majerus, 1971, 1972), among which may be platelet myosin (Booyse and Rafelson, 1972). These changes lead to modification in the pattern of labeling seen in platelets following iodination which suggests that the main effect of thrombin on the surface of intact platelets is on a glycoprotein of molecular weight about 120 000 (Steiner, 1973; Phillips and Agin, 1974); this glycoprotein appears to be identical to that which has been reported (Nurden and Caen, 1974) to be missing in Glanzmann's thrombasthenia.

The binding of thrombin to intact platelets appears to involve two classes of binding sites of different affinities, both located in the glycocalyx (Tollefsen, Feagler and Majerus, 1974b), and to result in an increase in the number of lectin receptor sites within the canalicular system, possibly from the fusion of granular membranes with it (Feagler *et al.*, 1974). The binding site of high affinity has about 500 receptors for platelet and that of low affinity, 50 000. Although diisopropylphosphorylthrombin competitively inhibits the binding of native thrombin it potentiates the release reaction at suboptimal concentrations of the latter, suggesting that thrombin effects the release reaction by a complex mechanism involving both binding and proteolytic activities. Once bound to the membrane thrombin appears to withstand solubilization in SDS and the solubilized receptor is thought to have a molecular weight of about 200 000 (Ganguly, 1974).

2.7 IMMUNOCHEMISTRY OF THE PLATELET SURFACE

Plasma proteins present in the plasmatic atmosphere, particularly fibrinogen, produce antibodies when intact washed platelets are used as an antigen. In addition, the platelet surface contains HL-A antigens identical to those found in lymphocytes and other tissues (Svejgaard, 1969) and these are of major importance in determining the therapeutic effectiveness of repeated platelet transfusions (Yankee, 1974).

At least three immunogenetic systems unique to platelets have been recognized. These are the PL^A, PL^E and Ko systems and appear to be products of genes located on three different chromosomes. The antigens PL^{A1}, PL^{E1} and Ko^b occur with a frequency of 97–99 percent in the population whereas the antigens produced by the alleles, PL^{A2}, PL^{E2} and Ko^a, have much lower frequencies, respectively 26 percent, 5 percent and 14 percent, and may be involved in neonatal thrombocytopenic purpura due to transplacental sensitization (Shulman et al., 1964).

Although nothing is known of the nature of the determinants of these three immunogenetic systems, some progress has been made regarding the structure of the so-called thrombocyte-specific antigen. An antibody specific for platelets is obtained if human platelets are injected into another species, preferably one phylogenetically distant such as chicken, and the resulting antiserum is carried through multiple adsorptions with plasma and various cells and tissues to remove nonspecific antibodies (Hanna and Nelken, 1971). Two antigens were recognized, a protein of molecular weight ca. 30 000 and a glycoprotein with a molecular weight of ca. 110 000. In view of its origin in the cell sap this antigen is probably identical to the glycoprotein of molecular weight 158 000 which has been isolated from the same source and is a component of the glycocalyx (Lombart, Okumura and Jamieson, 1974) since it is known to be readily cleaved to a glycopeptide of molecular weight 120 000. Antibodies to the purified glycoprotein contain determinants directed against both the protein and carbohydrate portions of the antigen (Okumura and Jamieson, 1975) and the glycopeptide inhibits the chicken antithrombocyte antisera prepared by Hanna and Nelken (Pepper and Jamieson, 1970). This antigen appears to occur in membrane-bound and soluble forms at the platelet surface.

Although it is not possible to compare the two antisera, a specific antibody to human platelets has also been prepared from intact washed platelets as the antigen and using immunosuppression to avoid the formation of nonspecific antibodies (Dzoga, Stoltzner and Wissler, 1972). This antiserum also gave two distinct precipitin lines and was used following conjugation with fluorescein or with ferritin to localize the antigen on the platelet surface and in the canalicular system (Stoltzner, Dzoga and Wissler, 1972).

A rabbit antibody to components of the cell sap of human platelets has been used to activate complement leading to complement-induced deletions in the pattern of polypeptide banding of the particulate (membrane) fraction in SDS gels (Zimmerman and Muller-Eberhard, 1973) which are reminiscent of, but distinct from, those found with thrombin but, like them, are obtained only when the platelets are intact and not when a platelet homogenate is exposed to complement. Platelets from patients with paroxysmal nocturnal hemoglobinuria are much more sensitive to complement than are normal platelets and a similar degree of sensitivity occurs following treatment of normal platelets with papain (Aster and Enright, 1969). Immunologic mechanisms of platelet damage and the principles of immune adherence of platelets, which are beyond the scope of this chapter, have been extensively reviewed elsewhere (Osler and Siraganian, 1972; Mylylla, 1973).

REFERENCES

ARDLIE, N. G. and HAN, P. (1974). *Br. J. Haemat.*, **26**:357.

ASHWELL, G. and MORELL, A. G. (1974). *Adv. Enzymol.*, **41**:99.

ASTER, R. H. and ENRIGHT, S. E. (1969). *J. clin. Invest.*, **48**:1199.

BAENZIGER, N. L., BRODIE, G. N. and MAJERUS, P. W. (1971). *Proc. natn. Acad. Sci. U.S.A.*, **68**:240.

BAENZIGER, N. L., BRODIE, G. N. and MAJERUS, P. W. (1972). *J. biol. Chem.*, **247**:2723.

BALDINI, M. G. and EBBE, S. (Eds) (1974). *Platelets: Production, Function, Transfusion and Storage.* New York; Grune & Stratton.

BARBER, A. J. and JAMIESON, G. A. (1970). *J. biol. Chem.*, **245**:6357.

BARBER, A. J. and JAMIESON, G. A. (1971a). *Biochemistry*, **10**:4711.

BARBER, A. J. and JAMIESON, G. A. (1971b). *Biochim. biophys. Acta*, **252**:533.

BARBER, A. J. and JAMIESON, G. A. (1971c). *Biochim. biophys. Acta*, **252**:546.

BARBER, A. J., KÄSER-GLANZMANN, JAKABOVA, M. and LÜSCHER, E. F. (1972). *Biochim. biophys. Acta*, **286**:312.

BARBER, A. J., PEPPER, D. S. and JAMIESON, G. A. (1971). *Thromb. Diath. haemorrh.*, **26**:38.

BAUMGARTNER, H. R. (1973). *Microvascular Res.*, **5**:167.

BAUMGARTNER, H. R. and HAUDENSCHILD, C. (1972). *Ann. N.Y. Acad. Sci.*, **201**:22.

BEHNKE, O. (1967). *Anat. Rec.*, **158**:121.

BEHNKE, O. (1968a). *J. Ultrastruct. Res.*, **24**:51.

BEHNKE, O. (1968b). *J. Ultrastruct. Res.*, **24**:412.

BONEU, B., BONEU, A., RAISSON, CL., GUIRAUD, R. and BIERME, R. (1973). *Thromb. Res.*, **3**:605.

BOOYSE, F. M. and RAFELSON, M. E. (1972). *Ann. N.Y. Acad. Sci.*, **201**:37.

BORN, G. V. R. (1962). *Nature, Lond.*, **194**:927.

BOSMANN, H. B. (1971). *Biochem. biophys. Res. Commun.*, **43**:1118.

BOSMANN, H. B. (1972). *Biochim. biophys. Acta*, **279**:456.

BRINKHOUS, K. M., SHERMER, R. W. and MOSTOFI, F. K. (Eds) (1971). *The Platelet.* Baltimore; Williams & Wilkins.

BROEKMAN, M. J., WESTMORELAND, N. P. and COHEN, P. (1974). *J. Cell Biol.*, **60**:507.

BROUN, G. D. and BROUN, G. O. (1962). *Proceedings 8th International Congress of Hematology*, pp. 1756–1760. Tokyo; Pan-Pacific Press.

CAZENAVE, J.-P., PACKHAM, M. A. and MUSTARD, J. F. (1973). *J. Lab. clin. Med.*, **82**:978.

CHATTERJEE, S. and SWEELEY, C. C. (1973). *Biochem. biophys. Res. Commun.*, **53**:1310.

CHESNEY, MCC., HARPER, E. and COLMAN, R. W. (1972). *J. clin. Invest.*, **51**:2693.

DAY, H. J., HOLMSEN, H. and HOVIG, T. (1969). *Scand. J. Haemat.*, Suppl. **7**:1.

DE LUCA, S., DE LUCA, L. and JAMIESON, G. A. (1976). Submitted for publication.

DETWILER, T. C. and FINEMAN, R. D. (1973a). *Biochemistry*, **12**:282.

DETWILER, T. C. and FEINMAN, R. D. (1973b). *Biochemistry*, **12**:2462.

DONATI, M. B., DE GAETANO, G. and VERMYLEN, J. (1973). *Thromb. Res.*, **2**:97.

DZOGA, K., STOLTZNER, G. and WISSLER, R. W. (1972). *Lab. Invest.*, **27**:351.

FEAGLER, J. R., TILLACK, T. W., CHAPLIN, D. D. and MAJERUS, P. W. (1974). *J. Cell Biol.*, **60**:541.
FISZER, S. and DE ROBERTIS, E. (1969). *J. Neurochem.*, **16**:1201.
FORBES, C. D. and PRENTICE, G. R. M. (1973). *Nature, New Biol.*, **241**:149.
FUKAMI, M. H. and SALGANICOFF, L. (1973). *Blood*, **42**:913.
GANGULY, P. (1971). *J. biol. Chem.*, **246**:4286.
GANGULY, P. (1974). *Nature, Lond.*, **247**:306.
GASIC, G. J. and GASIC, T. B. (1962). *Proc. natn. Acad. Sci. U.S.A.*, **48**:1172.
GASIC, G. J., GASIC, T. B., GALANTI, N., JOHNSON, T. and MURPHY, S. (1973). *Int. J. Cancer*, **11**:704.
GORDON, J. L. and DINGLE, J. T. (1974). *J. Cell Sci.*, **16**:157.
GREENBERG, J. H. and JAMIESON, G. A. (1974). *Biochim. biophys. Acta*, **345**:231.
GRÖTTUM, K. A. and SOLUM, N. O. (1969). *Br. J. Haemat.*, **16**:277.
HAMPTON, J. R. and MITCHELL, J. R. (1966). *Br. med. J.*, **1**:1074.
HANNA, N. and NELKEN, D. (1971). *Immunology*, **20**:533.
HARKER, L. A. (1971). *The Platelet*, pp. 13–25. Ed. K. M. BRINKHOUS, R. W. SHERMER and F. K.
    MOSTOFI. Baltimore; Williams & Wilkins.
HARRIS, G. L. A. and CRAWFORD, N. (1973). *Biochim. biophys. Acta*, **291**:701.
HOAK, J. C. (1972). *Blood*, **40**:514.
HOLMSEN, H. (1974). *Platelets: Production, Function, Transfusion and Storage*, pp. 207–220.
    Ed. M. G. BALDINI and S. EBBE. New York; Grune & Stratton.
HORAK, H. and BARTON, P. G. (1974). *Biochim. biophys. Acta*, **373**:471.
HOVIG, T. (1968). *Ser. Haemat.*, **1**:3.
HUANG, T. W., LAGUNOFF, D. and BENDITT, E. P. (1974). *Lab. Invest.*, **31**:156.
HUGUES, J. and MAHIEU, P. (1970). *Thromb. Diath. haemorrh.*, **24**:395.
JAMIESON, G. A. and BARBER, A. J. (1973). *Thrombosis: Mechanism and Control*, pp. 239–250.
    Ed. K. M. BRINKHOUS and S. HINNOM. Stuttgart; F. K. Schattauer Verlag.
JAMIESON, G. A., FULLER, N. A., BARBER, A. J. and LOMBART, C. (1971). *Ser. Haemat.*, **4**:125.
JAMIESON, G. A. and GROH, N. (1971). *Analyt. Biochem.*, **43**:259.
JOHNSON, S. A. (Ed.) (1971). *The Circulating Platelet.* New York; Academic Press.
KANG, A. H., BEACHEY, E. H. and KATZMAN, R. L. (1974). *J. biol. Chem.*, **249**:1054.
KARPATKIN, S. (1972). *A. Rev. Med.*, **23**:101.
KATZMAN, R. L., KANG, A. H. and BEACHEY, E. H. (1973). *Science, N.Y.*, **181**:670.
KAULEN, H. D. and GROSS, R. (1973). *Thromb. Diath. haemorrh.*, **30**:199.
KUBANEK, J., ENTLICHER, G. and KOCOUREK, J. (1973). *Biochim. biophys. Acta*, **304**:93.
LAMONT, J. T., PERROTTO, J. L., WEISER, M. M. and ISSELBACHER, K. J. (1974). *Proc. natn. Acad.
    Sci. U.S.A.*, **71**:3726.
LEWIS, N. and MAJERUS, P. W. (1969). *J. clin. Invest.*, **48**:2114.
LOMBART, C., OKUMURA, T. and JAMIESON, G. A. (1974). *FEBS Lett.*, **41**:31.
LYMAN, B., ROSENBERG, L. and KARPATKIN, S. (1971). *J. clin. Invest.*, **50**:1854.
MAJERUS, P. W. and BRODIE, G. N. (1972). *J. biol. Chem.*, **247**:4253.
MARCHESI, V. T., TILLACK, T. W., JACKSON, R. L., SEGREST, J. P. and SCOTT, R. E. (1972). *Proc.
    natn. Acad. Sci. U.S.A.*, **69**:1445.
MARCUS, A. J. (1969a). *New Engl. J. Med.*, **280**:1213.
MARCUS, A. J. (1969b). *New Engl. J. Med.*, **280**:1278.
MARCUS, A. J. (1969c). *New Engl. J. Med.*, **280**:1330.
MARCUS, A. J., SAFIER, L. B. and ULLMAN, H. L. (1971). *The Circulating Platelet*, pp. 241–258.
    Ed. S. A. JOHNSON. New York; Academic Press.
MARCUS, A. J., ULLMAN, H. L. and SAFIER, L. B. (1969). *J. Lipid Res.*, **10**:108.
MARCUS, A. J., ULLMAN, H. L. and SAFIER, L. B. (1972). *J. clin. Invest.*, **51**:2602.
MARCUS, A. J., ZUCKER-FRANKLIN, D., ULLMAN, H. L. and SAFIER, L. B. (1966). *J. clin. Invest.*,
    45:15.
MASON, R. G. and SHERMER, R. W. (1971). *The Platelet*, pp. 123–132. Ed. K. M. BRINKHOUS, R. W.
    SHERMER and F. K. MOSTOFI. Baltimore; Williams & Wilkins.
MEHRISHI, J. N. (1971). *Thromb. Diath. haemorrh.*, **26**:370.
MILGROM, F., CAMPBELL, W. A. and WITEBSKY, E. (1968). *Vox Sang.*, **15**:418.
MUGGLI, R. and BAUMGARTNER, H. R. (1973). *Thromb. Res.*, **3**:715.
MUSTARD, J. F. and PACKHAM, M. A. (1970). *Pharmac. Rev.*, **22**:97.
MYLLYLA, G. (1973). *Scand. J. Haemat.*, **19**, Suppl.:7.
NACHMAN, R. L. and FERRIS, B. (1972). *J. biol. Chem.*, **247**:4468.
NACHMAN, R. L. and FERRIS, B. (1974). *J. biol. Chem.*, **249**:704.
NACHMAN, R. L., HUBBARD, A. and FERRIS, B. (1973). *J. biol. Chem.*, **248**:2928.
NAKAO, K. and ANGRIST, A. (1968). *Nature, Lond.*, **217**:960.

NURDEN, A. T. and CAEN, J. P. (1974). *Br. J. Haemat.*, **28**:38.

OKUMURA, T. and JAMIESON, G. A. (1975). *J. biol. Chem.*, in press.

OSLER, A. G. and SIRAGANIAN, R. P. (1972). *Prog. Allergy*, **16**:450.

PACKHAM, M. A., GUCCIONE, M. A., CHANG, P.-L. and MUSTARD, J. F. (1973). *Am. J. Physiol.*, **225**:38.

PAULUS, J. M. and KINET-DENOEL, C. (1973). *Nouv. Revue fr. Hémat.*, **13**:494.

PHILLIPS, D. R. (1972). *Biochemistry*, **11**:4582.

PHILLIPS, D. R. and AGIN, P. A. (1974). *Biochim, biophys. Acta*, **352**:218.

PEPPER, D. S. and JAMIESON, G. A. (1968). *Nature, Lond.*, **219**:1252.

PEPPER, D. S. and JAMIESON, G. A. (1969). *Biochemistry*, **8**:3362.

PEPPER, D. S. and JAMIESON, G. A. (1970). *Biochemistry*, **9**:3706.

PUETT, D., WASSERMAN, B. K., FORD, J. D. and CUNNINGHAM, L. W. (1973). *J. clin. Invest.*, **52**:2495.

RAMBOURG, A. and LEBLOND, C. P. (1967). *J. Cell Biol.*, **32**:27.

REDDICK, R. L. and MASON, R. G. (1973). *Am. J. Path.*, **70**:473.

ROSEMAN, S. (1970). *J. Chem. Phys. Lipids*, **5**:270.

ROSS, P. D., FLETCHER, A. P. and JAMIESON, G. A. (1973). *Biochim. biophys. Acta*, **313**:106.

SCHICK, P. K. and YU, B. P. (1974). *J. clin. Invest.*, **54**:1032.

SCHULZ, H. (1966). *Verh. dt. Ges. Pathol.*, **50**:239.

SCHULZ, H. (1968). *Electron Microscopy of Blood Platelets and Thrombosis*. Berlin; Springer-Verlag.

SEAMAN, G. V. F. and VASSAR, P. S. (1966). *Archs Biochem. Biophys.*, **117**:10.

SHERRY, S. and SCRIABINE, A. (Eds) (1974). *Platelets and Thrombosis*. Baltimore; University Park Press.

SHULMAN, N. R., MARDER, V. J., HILLER, M. C. and COLLIER, E. M. (1964). *Prog. Hemat.*, **4**:222.

SMITH, D. F., KOSOW, D. P. and JAMIESON, G. A. (1975). *Fedn Proc. Fedn Am. Socs exp. Biol.*, **34**:241.

SNYDER, P. D., DESNICK, R. J. and KRIVIT, W. (1972). *Biochem. biophys. Res. Commun.*, **46**:1857.

SPAET, T. H. and LEJNIECKS, I. (1969). *Proc. exp. Biol. Med.*, **132**:1038.

SPIRO, R. G. and SPIRO, M. J. (1971). *J. biol. Chem.*, **246**:4899.

STEINER, M. (1973). *Biochim. biophys. Acta*, **323**:653.

STOFFEL, W. (1971). *A. Rev. Biochem.*, **40**:57.

STOLTZNER, G., DZOGA, K. and WISSLER, R. W. (1972). *Lab. Invest.*, **27**:357.

STORMORKEN, H. and OWREN, P. A. (1971). *Seminars Haemat.*, **8**:3.

SVEJGAARD, A. (1969). Iso-antigenic systems of human blood platelets—a survey, *Ser. Haemat.*, Vol. 2, No. 3.

TAO, R. V. P., SWEELEY, C. C. and JAMIESON, G. A. (1973). *J. Lipid Res.*, **14**:16.

TAYLOR, D. G. and CRAWFORD, N. (1974). *FEBS Lett.*, **41**:317.

TERADA, H., BALDINI, M., EBBE, S. and MADOFF, M. A. (1966). *Blood*, **28**:213.

TOLLEFSEN, D. M., FEAGLER, J. R. and MAJERUS, P. W. (1974a). *J. biol. Chem.*, **249**:2646.

TOLLEFSEN, D. M., FEAGLER, J. R. and MAJERUS, P. W. (1974b). *J. clin. Invest.*, **53**:211.

UHLENBRUCH, G. (1961). *Z. ImmunForsch. exp. Ther.*, **121**:420.

VERMYLEN, J., DONATI, M. B., DE GAETANO, G. and VERSTRAETE, M. (1973). *Nature, Lond.*, **244**:167.

WHITE, J. G. (1971). *The Circulating Platelet*, pp. 45–121. Ed. S. A. JOHNSON. New York; Academic Press.

WINZLER, R. J. (1969). *Red Cell Membrane—Structure and Function*, pp. 157–171. Ed. G. A. JAMIESON and T. J. GREENWALT. Philadelphia; J. B. Lippincott.

WOOLLEY, D. W. and GOMMI, B. W. (1964). *Nature, Lond.*, **202**:1074.

YANKEE, R. (1974). *Platelets: Production, Function, Transfusion and Storage*, pp. 313–326. Ed. M. G. BALDINI and S. EBBE. New York; Grune & Stratton.

ZIMMERMAN, T. S. and MULLER-EBERHARD, H. J. (1973). *Science, N.Y.*, **180**:1183.

ZUCKER, M. B. and LEVINE, R. U. (1963). *Thromb. Diath. haemorrh.*, **10**:1.

3

Lymphoid cells*

Marti Jett and Catherine Hickey-Williams

*The American National Red Cross Blood Research Laboratory,
Bethesda, Maryland*

3.1   INTRODUCTION

3.1.1   Scope of this chapter

The extent of the literature detailing current research on lymphocytes demands that this review be limited in its scope. Therefore, we have chosen to concentrate on those properties of the lymphocyte plasma membrane which relate to its antigenic nature and immunological functions. The structural properties of isolated membrane components will also be discussed. The role of the lymphocyte in the immune response has been elucidated mostly by using intact cells. Section 3.2 describes the properties of T and B cells, their separate and interdependent functions. Selective extraction of antigen material from intact cell surfaces by proteolytic cleavage or detergent solubilization has provided material on which to do preliminary chemical analysis. Section 3.3 details the progress made in characterizing cell surface antigens and receptor sites obtained by these methods.

It seems to us that for the elucidation of many lymphocyte properties, a simplified system is desirable or even necessary. To us, this means using isolated plasma membranes whenever possible. It is for this reason that we choose to include Section 3.4 as an extensive review of specific isolation procedures which have been used for lymphoid cells.

3.1.2   Function

The function of lymphocytes is the immunological protection of the host since these cells are involved in the defense against infections and invaders, both external and internal, such as transplants and malignancy (Miller and

* Contribution No. 325 from The American National Red Cross.

Mitchell, 1969; Muewissen, Stutman and Good, 1969; Miller, 1971; Claman, 1972; Katz and Benacerraf, 1972). This requires recognition of the foreign substance or antigen, processing so that the offending antigen is revealed to the appropriate lymphocytes, differentiation to produce either antibodies or kinins to combat the problem and then storage of the memory of the process so that future contact with the antigen will be met by more rapid resistance.

3.1.3　Morphology

The cytoplasmic structures are poorly developed in the small lymphocyte. There are few organelles present, other than mitochondria. There is very little rough endoplasmic reticulum, and only a very rudimentary Golgi apparatus, with few vesicles. The cell is nearly spherical and the surface may be slightly convoluted as viewed by phase contrast microscopy. Tightly packed chromatin is present throughout the nucleus. There are usually no nucleoli (Zucker-Franklin, 1969).

One model for observing cellular changes, which may parallel the antigen-stimulated changes, is the stimulation of peripheral blood lymphocytes by mitogens such as the phytohemagglutinin (PHA) from *Phaseolus vulgaris*. When a small lymphocyte is treated with PHA there is a gradual change from a small to medium and then large lymphocyte and finally to a large blast cell (Ling, 1968).

Morphologically, the most marked change is the decrease in the nuclear: cytoplasmic ratio, and the appearance of one or more nucleoli. The chromatin becomes loosely packed. The changes in cytoplasmic organelles reflect increased cellular activity. There is a marked increase in total cytoplasm; the endoplasmic reticular system expands and a highly developed Golgi apparatus appears. This large cell is not spherical, its surface appearing almost ragged and extremely convoluted when viewed by phase contrast microscopy (Pulvertaft, 1965).

3.2　T AND B LYMPHOCYTES

3.2.1　General characteristics

Chase and Landsteiner established the dual nature of the immune system by showing that certain types of immune reaction could be induced by a transfer of lymphoid cells, whereas others required only cell-free serum (*see* Chase, 1953). These are now known as *cellular* and *humoral* immunity. Cellular immunity involves recognition, the most important aspects of which are considered to involve the plasma membrane.

It has now been established that lymphocytes are derived from stem cells which migrate to certain organs, thereby determining the type of immunity their derivative cells will exhibit. Those stem cells which migrate to the thymus will produce lymphocytes responsible for cellular immunity, which includes combating fungi and viruses as well as tumor and transplant rejection (Cooper and Lawton, 1974). While the cells are maturing in the thymus they are called 'thymocytes' but as circulating cells they are designated 'T' lymphocytes. Thymocytes are capable of carrying out most but not all of the functions of mature T lymphocytes. Once they leave the thymus they seldom return, but thymic hormones may still influence circulating T lymphocytes. These are nondividing cells with a very long lifespan ranging from months in the rat to several years in man. Their path of circulation entails migration between epithelial cells which line the venules to the T-dependent region of the lymph nodes and spleen.

It is thought that passage through the postcapillary venules is controlled by carbohydrates in the lymphocyte plasma membrane and, presumably, corresponding specific receptor sites on the surface of the endothelial cells. The cells proceed through the lymphatic system into the thoracic duct where they reenter the bloodstream. This recirculation mechanism guarantees that in the course of a few days most competent lymphocytes will encounter any foreign substances or malignant cells; upon this contact the appropriate lymphocytes will divide as many as three times a day in the fulfillment of their ultimate function. During this differentiation process the cells produce lymphokinins which combat the foreign substance. T lymphocytes may or may not interact with macrophages in their cooperative effect with B lymphocytes and this will be discussed in greater detail later. Since 50–80 percent of circulating lymphocytes are T cells, athymic individuals show a marked decrease in peripheral blood lymphocytes and they lack cell-mediated immunity. Thymus grafts have been carried out successfully in newborn infants to restore cellular immunity (Cooper and Lawton, 1974).

There are certain surface markers (and functions) typical of T lymphocytes, such as the θ antigen in mice.

With sensitive radioiodinated anti-immunoglobulin (anti-Ig) sera, T cells were shown to have surface immunoglobulins (Roelants, 1972). The binding capacity was 100–400-fold less than for B lymphocytes and was about the same as was found on erythrocytes. Some argue that even a 3 percent contamination with B cells could show the same reactivity with anti-Ig sera. Others believe that immunoglobulins are present but more deeply buried in the surface and are therefore unreactive (Miller, 1972).

A function of T cells which serves as a means of their identification is the ability to form rosettes with sheep erythrocytes. (This is in contrast to B cells, which form rosettes only with sheep erythrocytes–antibody–complement.) About 60 percent of normal human peripheral lymphocytes and 99 percent of thymocytes form rosettes with sheep erythrocytes alone (Jondal, Holm and Wigzell, 1972), and these lymphocytes from either source do not carry immunoglobulins on their surface. Trypsin treatment of the intact cells abolishes their ability to form rosettes, but upon reculturing for less than 24 hours the ability is restored (Holmes, Morton and Schidlovsky, 1971).

The B lymphocytes confer humoral immunity toward bacterial infections and viral reinfections. They are responsible for the formation of antibody-producing or plasma cells. In the chick, the source of these cells has been clearly established as the bursa of Fabricius, a lymphoid body attached to the intestine near the cloaca. The organ for B cell development in mammals is not yet known. Stem cells migrate to the fetal liver, spleen and bone marrow. Owen, Cooper and Raff (1974) have shown that the fetal liver may be the primary site of B cell production. Indeed, transplantation of bone marrow and fetal liver seems to restore humoral immunity in patients with Bruton's disease, a sex-linked congenital absence of plasma cells and circulating antibodies (Cooper and Lawton, 1974).

B lymphocytes synthesize immunoglobulins but an individual cell makes only a single class, subclass, light-chain type, allelic type and specificity. The cell does not secrete the immunoglobulins but rather incorporates them into its plasma membrane. The antigen for which the surface-bound immunoglobulin is specific is recognized and bound by the cell. The sequence of events is not entirely clear, but the B cell requires a signal and then differentiates into a plasma cell which secretes large quantities of identical antibody molecules. Plasma cells live only a few days. There is a turnoff mechanism possible through antibody feedback and/or suppressor T cells.

In addition to surface-bound immunoglobulins, B cells can be recognized by the variety of their surface markers and receptors. In the mouse they display a specific B lymphocyte antigen. In all mammalian systems they bear antigen receptors, antigen–antibody–complement receptors and immunoglobulin receptors. These will each be discussed in detail in Section 3.2.4, 'B Cell Surface Receptors'.

3.2.2   Interdependence of T and B lymphocytes

Although the humoral immune system is considered to be independent of the thymus, there are certain occasions when B cells require some sort of stimulus from T cells in order to differentiate and produce antibody. Claman, Chaperon and Triplett (1966) showed that, in irradiated mice, the sum of the response upon reconstitution with T or B lymphocytes separately was less than the response when the two were given simultaneously. An example of the synergism is that both T and B cells are necessary to carry out the immune response with a hapten, such as dinitrophenol, attached to an immunogenic protein such as bovine serum albumin (BSA). In a mouse system, anti-θ serum prevented the anti-hapten response whereas athymic mice showed an unimpaired response to bovine serum albumin (Roelants, 1972; Miller, 1972).

It has been questioned whether the role of the T cells is active or passive. One theory suggested that the T cells concentrate monovalent antigens for presentation to the B cells (Miller, 1972); however, antigen-coated B cells cannot perform the same function. T cells treated with agents to prevent T cell activation (mitomycin C), RNA synthesis (actinomycin D) or protein synthesis (antimycin A) do not stimulate B cells (Miller et al., 1971; Sprent and Miller, 1971; Feldmann and Basten, 1972a). There seems little doubt that T cell cooperation is an active process.

The next question concerns the specificity of the synergism. Specific chemical mediators produced by T cells (Lawrence and Landy, 1969) and transferred by cell–cell contact would require a relatively high frequency of such T and B cells with matching specificities. The calculated frequency is 1 in 10^5 (Möller and Michael, 1971). When T and B cells are separated by a Nucleopore membrane (pore size 1 μm), the results show that stimulated T cells are required but that they produce a substance which can pass through the pores, therefore physical contact between T and B cells is not necessary.

This substance, postulated to be an immunoglobulin (Miller, 1972), can be conveyed *in vivo* by macrophages in the form of a polymeric antigenic structure which B cells recognize and to which they respond (Feldmann and Basten, 1972b). Also, T cells may influence the magnitude of the response in some cases, such as when exposed to sheep erythrocytes (Shearer and Cudkowicz, 1969).

3.2.3   Memory and tolerance

Memory resides in both T and B cells, since interaction between the cells is essential for an optimal secondary response (Miller, 1972). Tolerance can be achieved by both T and B cells, but T cells retain tolerance twice as long (100 days) and acquire it in one-fourth the time (2–5 days). B cells acquire tolerance slowly and lose it rapidly. In synergistic responses it is probable that only T cell tolerance is of importance (Miller, 1972).

3.2.4   B cell surface receptors

3.2.4.1   IMMUNOGLOBULIN RECEPTORS

Aggregated immunoglobulins or antigen–antibody complexes will bind specifically to B lymphocytes in the *absence* of complement through the Fc portion of the immunoglobulin. The binding is not affected by treating the lymphocytes with trypsin (Pernis *et al.*, 1971), or by preincubation of the cells with anti-Ig sera, eliminating the suggested possibility that the surface-bound immunoglobulins could be serving as the binding point for other immunoglobulins (Basten *et al.*, 1972). Receptors for immunoglobulins on macrophages are destroyed by phospholipase A (Davey and Asherson, 1967). In addition, aggregated immunoglobulin prevents antigen–antibody binding to lymphocytes but does not affect antigen–antibody–complement binding (Roelants, 1972).

3.2.4.2   ANTIGEN RECEPTORS

B lymphocytes have been shown to bind and concentrate antigens by using antigen-coated columns of glass or plastic beads (Wigzell and Andersson, 1969). Agammaglobulinemic patients, deficient in B but not T cells, have few lymphocytes which bind antigen (Naor, Bentwich and Cividalli, 1969), and several investigators have shown that a specific anti-Ig serum prevents the binding of that antigen to lymphocytes (Roelants, 1972; Miller, 1972).

It seems that the surface-bound immunoglobulins are serving as the recognition and receptor sites for the antigen specific for their own B cell. Some believe that it is this association which stimulates the T-independent B cell to undergo differentiation, plasma cell formation and antibody production, typical of the humoral response. The evidence is that anti-Ig sera can induce DNA synthesis in peripheral lymphocytes (only) in rabbits and chickens (Sell, 1972), although no corresponding increase in immunoglobulin synthesis occurs in the activated cell population.

Another theory is that an additional signal is required (Bretscher and Cohn, 1970) by polyclonal B cell activators such as pneumococcal polysaccharide, dextran, serum factors, synthetic polypeptides or lipopolysaccharides (Watson, Frenkner and Cohn, 1973). These induce polyclonal immunoglobulin synthesis, morphological transformation and increased DNA synthesis in B (but not T) lymphocytes. All immunoglobulin-bearing lymphocytes are stimulated by polyclonal B cell activators, thus displaying a nonspecific response (Coutinho *et al.*, 1974a, b). This may favor a third theory in which the cell with sufficient antigen bound to it is thus identified and will bind the polyclonal B cell activators until a sufficient number are bound to trigger proliferation and subsequent immunoglobulin production. This suggests a passive function for immunoglobulin receptors in concentrating antigens to a particular B cell. Complement (C-3) and immunoglobulin (Fc) receptors were also shown to play only passive roles in B cell activation by a T-independent antigen (Möller and Coutinho, 1975).

3.2.4.3   COMPLEMENT/IMMUNE COMPLEX RECEPTOR SITES

Antigen–antibody–complement (Ag–Ab–C) complexes bind to the B lymphocyte surface through the receptor site for complement. Cellular complexes, such as sheep erythrocytes, anti-Forssman antibody and guinea-pig C-3 (EAC), will bind to 60–70 percent of the B cells (separated by using anti-Ig-coated columns) and to only 1–2 percent of thymocytes (Wigzell, Sundquist and Yeshidu, 1972; Rutishauser and Edelman, 1972). Since 30–40 percent of the Ig-bearing cells do not appear to bind C-3 by rosette formation with EAC, it has been proposed that there may be a subpopulation of B cells which bear complement receptors (Nussenzweig, 1974). This may be due to the method of assay for C-3 receptors in contrast to the very specific antisera used against immunoglobulins. Indeed, if a soluble antigen–antibody–complement complex is used, a greater number of the B lymphocytes bind it (Eden, Bianco and Nussenzweig, 1973). Gelfand *et al.* (1974), using newborn mice, showed that at three days 23.8 percent of the spleen cells bore Ig receptors, whereas only a few showed C-3 receptor activity. At 6–12 weeks, 37 percent had Ig receptors and 25 percent had C-3 receptors, suggesting a gradual appearance of the C-3 receptors on Ig-bearing cells or perhaps a specific migratory effect into the spleen of cells bearing both receptors. Discussion of the characteristics of C-3 receptors and their role in the immune response requires a brief review of the complement system.

The complement proteins exist in an inactive form in serum. Sequential activation of the components is achieved by a series of enzyme–substrate interactions which exposes the active site on the molecule; this activates the next component in the reaction sequence (Cooper, 1973). Activation can occur by the classical (Nelson, 1963; Götze and Müller-Eberhard, 1971; Ruddy, 1974) or alternate (Götze and Müller-Eberhard, 1971; Cooper, 1973) pathways.

The complement component responsible for immune adherence, that is, the binding of antigen–antibody–complement to lymphoid cells, is C-3b.

This involves a two-stage binding, the first of which is a rather weak bond, while the second is so strong that it distorts the erythrocyte membranes in the EAC assay and cannot be disrupted by vigorous shaking (Nussenzweig, 1974).

The nature of the C-3b receptor has been explored by proteolysis. Complement receptor activity is abolished by briefly treating the lymphocytes with trypsin (Bianco, Patrick and Nussenzweig, 1970; Michlmayer and Huber, 1970). The fragments released do not inhibit complement binding (Dierich and Reisfeld, 1975). Chronic lymphocytic leukemia patients had fewer C-3b receptor lymphocytes and erythrocyte-binding lymphocytes (T cells); but a high percentage of the peripheral cells bore Ig receptors. As was mentioned earlier, the antigen–antibody receptor was not destroyed by trypsin treatment, nor did antigen–antibody or aggregated immunoglobulins interfere with EAC binding to lymphocytes (Nussenzweig, 1974).

Various probes have been used to learn more about the nature of the C-3b receptors. Dierich *et al.* (1974) showed that reducing agents destroyed the ability of the lymphocytes to undergo rosette formation, but this ability was recovered upon reculture for 8–10 hours. Free sulfhydryl groups are not required either for fixation or adherence to C-3b, since treatment of the cells with α-iodoacetamide (Dierich *et al.*, 1974) and *p*-chloromecuribenzoic acid (Bhakdi *et al.*, 1974) did not influence receptor function. The receptor was solubilized using 3 M KBr (Dierich *et al.*, 1974) and fractionated by gel filtration. The activity remained with a β-lipoprotein fraction and efforts to remove the lipid resulted in loss of activity.

Complexes of complement with iodinated BSA and its antibody remain on the surface of lymphocytes for several hours, when incubated in a tissue culture medium. Excess nonradioactive BSA quantitatively removes the label from the cells although simple exchange of BSA in the complex has been ruled out experimentally (Eden, Bianco and Nussenzweig, 1973). Addition of papain fragments of immunoglobulins to the incubation mixture removes the complex Ag–Ab–C from the cells. This may indicate that interaction between the complex and membrane is mediated by the antibody molecules or perhaps by other complement products which bind to Ig molecules.

The function of Ag–Ab–C in binding to lymphocytes is not entirely clear. It is thought to facilitate interaction between cells and the specific immune complex and it may assist in triggering antibody production. It is clearly involved in localization of antigen in lymphoid organs and in the migration of specific lymphoid cells to sites of inflammation (Bianco, Patrick and Nussenzweig, 1970). C-3-deficient patients, having one-thousandth the normal level of C-3, have normal levels of Ig (Alper *et al.*, 1972).

The interaction of C-3 with the plasma membrane may function as a nonspecific signal which also requires the binding of a specific antigen or mitogen to elicit antibody production (Dukor and Hartman, 1973). This theory is based on the finding that mitogenic stimulation and T-cell-mediated antibody formation *in vitro* both require the presence of C-3-sufficient serum (Pepys, 1972). Preincubation of B cells with C-3-sufficient serum but not C-3-deficient serum inhibits their interaction with immune complexes and also decreases their reactivity to T-cell-specific mitogens (Dukor *et al.*, 1974).

Another theory regarding the role of C-3 in the induction of the immune response is that T cell presentation of antigen to B cells may be enhanced by the interaction of the cross-linked antigen, fixed C-3 or T cell products (Pepys, 1972, 1974; Feldman and Pepys, 1974). Interaction of a T cell product with antigen could then be transferred to B cells, perhaps through macrophages, and the cell would differentiate and produce antibody to the specific antigen involved (Pepys, 1974). In this scheme, C-3 would act to increase the chance of the antigen making contact with the appropriate B cell. Evidence supporting this argument is that depletion of C-3 suppresses the T-dependent (not the T-independent) antibody response, and unbound C-3 fragments inhibit antigen-induced B lymphocyte transformation (Pepys, 1972).

Modulation of the immune response may also be a function of C-3 (Miller and Nussenzweig, 1974). Interaction of the Ag–Ab–C complex with the cell surface receptor first causes an increase in affinity of the cell for binding soluble complexes. This is followed by a decreased affinity leading to release of the complexes from the cell surface. There is some evidence that release is a result of activation of the alternate pathway (Miller, Saluk and Nussenzweig, 1973). Unbound immune complexes, when incubated with serum, become unable to attach to the appropriate lymphocytes. This activity may be a mechanism for modulation of the immune response (Nussenzweig, 1974). Alterations of this mechanism may play a role in certain immune complex diseases by allowing the complexes to accumulate on cell surfaces (Miller and Nussenzweig, 1974). The increased affinity of membranes for immune complexes containing C-3 is thought to increase the chance that the antigen will encounter B cells capable of reacting with it. C-3 is also thought to provide nonspecific signals for activation of the cell (Miller and Nussenzweig, 1974).

3.2.5   T and/or B cell surface receptors

3.2.5.1   HORMONE RECEPTORS

Certain hormones have been shown to bind to specific sites on lymphoid cells although it is unclear whether they are specific for either T or B cells, or if both are required. For an extended discussion of hormone receptors on various cells, *see* the review by Cuatrecasas (1974).

(a)   *Insulin receptors.* Normal human peripheral lymphocytes, whether intact or broken, do not bind insulin (Krug, Krug and Cuatrecasas, 1972). However, lectins such as concanavalin A (Con A) (Krug, Krug and Cuatrecasas, 1972) and phytohemagglutinin, as well as periodate treatment (Novogradsky and Katchalski, 1972), have been shown to cause these cells rapidly to develop the ability to bind insulin. Indeed, peripheral lymphocytes from patients with acute lymphocytic leukemia readily bind insulin, as do lymphoid cells (RPMI 6237) in long-term culture (Krug, Krug and Cuatrecasas, 1972), while chronic lymphocytic leukemia cells do not have this capacity. Insulin analogues have been shown to inhibit insulin binding but unrelated peptides such as glucagon or ACTH do not. The binding ability is destroyed by proteolysis with trypsin but is unaffected by digestion with ribonuclease, deoxyribonuclease or neuraminidase (Gavin *et al.*, 1973). Digestion with neuraminidase, followed by treatment with β-galactosidase, results in loss of receptor activity, suggesting the involvement of galactose in the receptor site. Binding is enhanced three- to sixfold by incubation of the cells with phospholipases C and A (Krug, Krug and Cuatrecasas, 1972) but unstimulated lymphocytes are still unable to bind insulin even after incubation with either phospholipase.

The insulin receptor on lymphoid cells is indistinguishable from those receptors found on the major insulin target tissues. Since the ability to bind insulin precedes the morphological change in the cell caused by mitogenic agents, it raises the question as to whether the receptors are synthesized *de novo* or are present in a modified form (Cuatrecasas, 1972; Roth, 1973). Cuatrecasas (1974) has postulated an interaction between the hormone–receptor complex and the enzymes which are affected, such as adenyl cyclase.

(b)   *Growth hormone receptors.* Growth hormone receptors are present on both stimulated and cultured lymphocytes (Lesniak *et al.*, 1973) and the receptors are specific for growth hormones since other polypeptide hormones do not cause the same responses (Hollenberg and Cuatrecasas, 1974). Only the human growth hormone component of molecular weight 45 000 shows binding to lymphoid cells; the derivative of molecular weight 22 000 is unable to bind (Gorden *et al.*, 1974).

The mechanism whereby growth hormone exerts its effect is poorly understood. It is postulated that it exerts its effects through the induction of a plasma factor, somatomedin, and has been reported to modify several membrane-bound enzymes (Rubin, Swislocki and Sonenberg, 1973).

(c)   *Glucocorticoids.* There seem to be at least two receptors on lymphocytes for the glucocorticoids, one of which may be a cytoplasmic receptor; the other is probably surface-associated (Kornel, 1973). Corticosteroids have an effect on lymphocytes which is the opposite of that on liver cells; thus, there is inhibition of synthesis of nucleic acids and proteins and in the rate of uptake of glucose and α-aminobutyric acid as well as an increase in the rate of release of amino acids (Kornel, 1973).

3.2.5.2   LECTIN RECEPTORS

Lectins obtained from plant and animal sources have been used as probes in studies of cell surfaces and have been used to stimulate resting lymphocytes to undergo mitosis so that the events which occur during transformation can be observed in a controlled system. This transformation may be a model for the events observed in malignancies or in the immunologic stimulation of cells by antigen which results in cell proliferation. Data regarding the carbohydrate specificity of lectins, molecular weights and other properties are tabulated in several reviews (Toms and Western, 1971; Lis and Sharon, 1973; Nicolson, 1974).

Although many lectins bind to surface structures of both T and B lymphocytes, there seems to be a cell specificity for a lectin to cause mitogenic stimulation. Soluble Con A or *Phaseolus vulgaris* PHA are thought specifically to stimulate T lymphocytes to undergo mitosis while pokeweed mitogen, wheat-germ agglutinin (WGA) and complexed Con A or PHA will stimulate mitosis in both T and B cells. It seems that T cells may be more easily stimulated to undergo lectin-induced mitosis than B cells (Boldt, MacDermott and Jorolan, 1975). However, there is no pattern in the ability of the mitogens to bind T or B cells selectively since this is a more general property than the induction of mitosis. The use of lectins labeled with fluorescent reagents, ferritin or radioiodine has allowed the measurement of this property.

Cells may bind large quantities of specific lectins but aggregation does not necessarily occur. Aggregation is dependent on many factors such as the density and fluidity of receptor sites, interference of cell surface structures, cell charge, repulsive forces as well as the concentration of the lectin, its binding constant and the number of binding sites per molecule.

Lectins modify certain functions of lymphocytes such as cell-mediated cytotoxicity (Perlmann and Holm, 1969). Con A stimulates the nonspecific release of a lymphotoxin which kills target cells *in vitro* (Schwartz and Wilson, 1971) and PHA stimulates release of lymphotoxins, which seem to act at the cell surface of the target cells (Kolb and Granger, 1968, 1970; Williams and Granger, 1973). In another system, the lectin-induced cytotoxicity to T cells involved cell–cell contact rather than mediation through lymphotoxins (Kirchner and Blaese, 1973) and B cells did not respond mitogenically to Con A but did exhibit the property of destroying target cells. This suggests that lymphotoxin release by lectins is probably a nonspecific reaction.

Lectin receptors on lymphocyte surfaces have been studied by a number of investigators. Virally transformed human peripheral lymphocytes were shown to be agglutinable with concentrations of Con A which would not agglutinate normal peripheral blood lymphocytes (De Salle *et al.*, 1972). The transformed cells bound more Con A per cell but the density of binding sites was essentially the same. The conclusions were that transformation had altered the 'agglutinability' factors rather than increasing binding ability.

Cells from patients with chronic lymphocytic leukemia do not agglutinate significantly more with Con A than do normal cells, but are substantially more agglutinable with WGA (Kapeller and Doljanski, 1972). In another study these cells bound approximately one-half as much PHA as did normal lymphocytes (Kornfeld, 1969). These results may relate to the fact that the ratio of T to B cells in peripheral blood is altered by proliferation of B cells in that disease.

More recently, Kornfeld and Siemers (1974) have shown that human peripheral blood lymphocytes contain six times as many mushroom PHA sites as do human thymocytes (derived from solid tissue) and twice as many binding sites for the erythroagglutinating PHA (E-PHA) and the leuko-agglutinating PHA (L-PHA). Since peripheral cell populations contain 60–80 percent T cells (equivalent to mature thymocytes), one can suggest that maturation from thymocyte to T cell results in an increase in the number of lectin-binding sites. Studies done with four populations of normal human lymphocytes confirm this. The number of binding sites per cell and the association constant for six different lectins have been determined using peripheral cells, isolated B cells, isolated Null cells and T plus Null cells

(Boldt, MacDermott and Jorolan, 1975). The number of sites per cell varies only slightly in each population for each lectin studied. However, differences do exist among the various lectins, ranging from 26×10^6 sites per cell for WGA to 0.58×10^6 sites per cell for L-PHA in each group. Human peripheral lymphocytes contain $2–3 \times 10^6$ sites per cell each for Con A, *Lens culinaris* (LcH), *Ricinus communis* and E-PHA. Isolated calf thymocyte membranes contain the same or slightly higher numbers of lectin receptors as do intact calf thymocytes (Kornfeld and Siemers, 1974).

Mouse L 1210 leukemia cells release a WGA-binding molecule, which is inhibitory in agglutination assays with intact cells upon hypotonic shock or sonication. Concomitantly, the residual cells lose their agglutinability although the cells are still viable by trypan blue exclusion (Burger, 1968). Janson and Burger (1973) described four different chemical extraction procedures for isolation of the WGA receptors from L 1210 cells using phenol, guanidine-HCl, pyridine and lithium diiodosalicylate, respectively. Four distinct bands, each displaying receptor activity, were found in each case by sodium dodecyl sulfate (SDS)–polyacrylamide gel electrophoresis. These receptor fractions did not inhibit agglutination of whole cells with Con A, although they did to some degree inhibit agglutination with PHA and LcH. Trypsinization of normal cells did not alter their inability to agglutinate with WGA, but pronase digestion did so to a minor degree. Antisera to the WGA receptor sites prevented WGA-induced but not Con-A-induced aggregation. WGA receptor antiserum was unreactive with normal lymphocytes (Janson, Sakamoto and Burger, 1973).

Allan, Auger and Crumpton (1972), using pig lymph-node cells, isolated the plasma membranes which they solubilized in sodium deoxycholate. The receptors for Con A were obtained from the solubilized membrane preparation by affinity chromatography using Sepharose-coupled Con A. Hayman and Crumpton (1972) obtained glycopeptides by the same technique from identically prepared membranes using LcH coupled to Sepharose (for a further discussion of this work, *see* Section 3.4.5).

3.3   ANTIGENS RELEASED FROM INTACT CELLS

3.3.1   Histocompatibility antigens

The mouse H-2 and human HL-A histocompatibility antigens are glycoprotein components of the plasma membrane especially prevalent in lymphocytes (Nathenson, 1970; Nathenson and Cullen, 1974).

On the basis of extensive genetic and serologic studies with inbred mouse strains, it has been concluded that H-2 antigens are controlled by two genes designated *H-2K* and *H-2D*. Some specificities or antibody combining sites called 'private' specificities are coded for exclusively by *D* or *K*, while others called 'public' map in either the *D* or *K* region (Klein and Shreffler, 1972; Klein, 1973). The *H-2* genes are part of a chromosomal region designated the *major histocompatibility complex*, whose other genes include: *Ir* or immune response genes (Lieberman *et al.*, 1972; McDevitt *et al.*, 1972; Günther, 1973); the *Ss-Slp* region, which controls the level of certain serum proteins (Passmore and Shreffler, 1968); and *Tla*, thymus leukemia antigen, which maps outside the *H-2* complex, but close to the *H-2D* region (Old *et al.*, 1968). Two or more co-dominant, closely linked genes specify the expression of HL-A antigens in man (Mayr *et al.*, 1973; Amos and Ward, 1975). The major histocompatibility complex in man also contains genes which control the immune response (Eijsvoogel *et al.*, 1972; Thorsby, Hirschberg and Helgesen, 1973).

H-2 and HL-A antigens are detected serologically with specific alloantisera, which are nonprecipitating but do lyse cells in the presence of complement. As transplantation antigens they elicit both a humoral and a cell-mediated immune response leading to rejection of allogeneic tissue grafts. H-2- and HL-A-active materials have been released from lymphoid tissue by extraction with 3 M KCl (Reisfeld, Pellegrino and Kahan, 1971) and with organic solvents (Manson and Palm, 1968), by proteolysis (Sanderson, 1969; Shimada and Nathenson, 1971; Cresswell, Turner and Strominger, 1973) and by detergent solubilization (Hilgert *et al.*, 1969; Schwartz and Nathenson, 1971; Snary *et al.*, 1974; Dawson *et al.*, 1974; Springer, Strominger and Mann, 1974).

Papain digestion of intact, radiolabeled human RPMI 4265 cells releases soluble radioactive HL-A antigens which can be separated on a Sephadex G150 column, concentrated and then reacted with specific alloantisera (Cresswell, Turner and Strominger, 1973). Further chromatography of specific immune complexes separates HL-A2 from H2-A7 antigen, each of which is specified by a different *HL-A* locus. Electrophoresis of the HL-A2 and A7 complexes under protein-denaturing conditions separates two fragments: a glycoprotein of molecular weight 34 000 and a protein of molecular weight 12 000. Identical results were obtained with six different HL-A specificities, released from two cultured and two peripheral lymphoid cell sources (Cresswell *et al.*, 1974).

Extraction of RPMI 4265 plasma membranes with Brij 99 detergent releases two subunits: a glycoprotein of molecular weight 44 000 and a protein of molecular weight 12 000. Papain digestion converts the glycoprotein of molecular weight 44 000 to a material of molecular weight 34 000 in several steps, leaving the protein of molecular weight 12 000 uncleaved (Springer, Strominger and Mann, 1974). This material of molecular weight 12 000 is identical to β-2 microglobulin, a protein found originally in urine, which is homologous with constant-region domains in immunoglobulins (Grey *et al.*, 1973; Nakamuro *et al.*, 1973; Peterson, Rusk and Lindblom, 1974).

HL-A antigens extracted from Im-1 lymphoid cells with 0.5% deoxycholate give two glycoprotein bands corresponding to molecular weights 82 000 and 43 000, plus the β-2 microglobulin of molecular weight 11 500 upon SDS gel electrophoresis in the absence of β-mercaptoethanol (Cresswell and Dawson, 1975). In the presence of reducing agent the band corresponding to molecular weight 83 000 was completely converted to molecular weight 43 000, suggesting that the HL-A antigens may occur as dimers.

H-2 antigens solubilized by NP-40 and subjected to SDS gel electrophoresis give a fragment of molecular weight 45 000 in the presence of reducing agents (Schwartz *et al.*, 1973). The H-2 material cleaved from cells by papain is smaller in molecular weight by 3000–6000. It is thought with both H-2 and HL-A antigens that the detergent-solubilized molecule is the intact antigen as it occurs in the membrane, while the papain-cleaved antigen is a water-soluble fragment consisting of approximately 80 percent of the molecule.

A protein of molecular weight 12 000, believed to be the murine counterpart of the β-2 microglobulin subunit of HL-A antigens, has been separated from H-2 and TL alloantigen complexes of mouse thymocytes (Vitetta, Uhr and Boyse, 1975). H-2 and TL antigens were individually precipitated from iodinated, lysed thymocytes by specific antisera. SDS gel electrophoresis of the immune complexes resulted in two ^{125}I-labeled peaks, one of molecular weight 45 000 and the other of molecular weight 12 000. Antisera directed against TL specificities did not contain antibodies to the low-molecular-weight protein, since anti-TL sera were unable to precipitate cells of genotype A/TL^-. This low-molecular-weight component appears to be noncovalently associated with both H-2 and TL molecules of mouse thymocytes.

H-2 and HL-A antigens are glycoproteins consisting of 6–10 percent carbohydrate. Antigenic activity is believed to reside in the protein as opposed to the carbohydrate portion (Nathenson and Muramatsu, 1971; Nathenson and Cullen, 1974). Glycopeptides, obtained by pronase digestion of detergent- or papain-solubilized antigen, neither inhibit target cell lysis in the ^{51}Cr-release cytotoxic assay, nor form specific immune complexes with anti-H-2 sera in an indirect precipitation test. Removal of 90 percent of the carbohydrate with specific exo- and endoglycosidases does not destroy antigenic activity. On the other hand, conditions which result in loss of protein conformation (6 M urea, pH extremes) or chemical modification of specific amino acid residues, do reduce antigenic activity. Alteration of lysine residues selectively destroys antibody-combining activity of some specificities, while retaining others (Pancake and Nathenson, 1973). Comparison of the ion-exchange chromatography pattern of tryptic peptides of *H-2K* and *H-2D* gene products from a single haplotype indicates 80 percent heterogeneity. When peptides from *H-2D* gene products from several different alleles are compared, 60–70 percent heterogeneity is found (Nathenson and Cullen, 1974). These differences in peptide patterns signal amino acid sequence differences, supporting the hypothesis that H-2 antigen specificities are related to protein structure.

Taking advantage of the glycoprotein nature of HL-A antigens, Snary *et al.* (1974) employed affinity chromatography using a column of LcH-Sepharose. Approximately 10 percent of the protein from BRI-8 lymphoblast plasma membranes solubilized with 1% deoxycholate adhered to the LcH-Sepharose and was eluted with methyl-α-D-mannopyranoside. The eluted fraction contained most of the HL-A activity (350-fold purification) and all the 5′-nucleotidase activity (Snary *et al.*, 1974), indicating that both activities reside on glycosylated proteins. Nonglycosylated forms of HL-A antigen may also occur on some cells since Dawson *et al.* (1974) found that up to 25 percent of the HL-A activity from Im-1 cells, solubilized by 0.5% deoxycholate, did not adhere to LcH-Sepharose.

The 'two gene' hypothesis discussed above for H-2 antigens has been confirmed serologically and biochemically by isolating two glycoproteins from cell extracts, one carrying specificities determined by the *H-2K* gene and another bearing *H-2D* specificities (Nathenson and Cullen, 1974). From heterozygous cells (F1 hydrids) solubilized with NP-40, four glycoprotein molecules were separated using monospecific anti-H-2 sera. Since all four antigens are expressed in these cells, the genes must be co-dominant.

3.3.1.1   CAPPING

In the plasma membrane of viable lymphocytes, H-2 and HL-A antigens controlled by different loci are expressed as separate molecules. The phenomenon of capping or clustering of antigen at one end of the cell after exposure to specific antiserum followed by anti-Ig serum occurs with a variety of lymphocyte antigens (Taylor *et al.*, 1971), including histocompatibility antigens (Neauport-Sautes *et al.*, 1973a).

Incubating mouse lymph node cells with fluorescein-labeled antisera to H-2K specificities resulted in the clustering of all the H-2K antigen molecules, leaving H-2D antigen widely distributed over the cell surface. A temperature-dependent phenomenon, capping occurs at 37 °C but not at 0 °C. Four independently migrating H-2 specificities have been demonstrated in heterozygous F1 cells (Neauport-Sautes *et al.*, 1973b).

The *HL-A* complex in man is currently believed to contain three loci, *SD-1*, *SD-2* and *SD-3*, which control serologically detectable surface HL-A antigens (Ceppellini, 1971; Mayr *et al.*, 1973; Amos and Ward, 1975). Antigen molecules specified by each of these loci migrate independently of each other in the presence of specific antiserum (Neauport-Sautes *et al.*, 1972; Svejgaard, Staub-Nielson and Ryder, 1972; Bernoco, Cullen and Scudeller, 1972), indicating that they are present on physically distinct molecules. All the β-2 microglobulin molecules on a lymphoid cell undergo redistribution or capping at 37 °C in the presence of anti-β-2 microglobulin sera plus anti-Ig sera (Neauport-Sautes *et al.*, 1974). To determine what effect capping of β-2 microglobulin molecules has on the distribution of HL-A antigens, experiments have been done using fluorescein-labeled specific antisera. Treating lymphocytes with anti-β-2 microglobulin antisera plus anti-Ig sera results in capping of all β-2 microglobulin molecules, plus all the HL-A antigens. No fluorescence with anti-HL-A sera can be detected outside the capped areas. Anti-HL-A sera are able to cap only some of the β-2 microglobulin molecules, leaving others distributed over the surface of the lymphocyte. These results indicate that all the HL-A serologically detectable antigens are associated with β-2 microglobulin on the cell surface.

3.3.1.2   MIXED LYMPHOCYTE REACTION

In many species, including mouse (Dutton, 1966) and man (Bach and Amos, 1967), unsensitized lymphocytes of different genotypes when mixed in culture undergo blastogenesis and mitotic activity, a phenomenon designated the *mixed lymphocyte reaction*. It was originally believed that serologically detectable major histocompatibility antigens (i.e. the *H-2K* and *H-2D* gene products) were responsible for stimulating this reaction in mice. Recent studies indicate that two genetic systems control the mixed lymphocyte reaction: the *M* locus, which is distinct from the *H-2* system and whose products have strong lymphocyte-activating properties (Festenstein *et al.*, 1972), and loci which map within the *H-2* major histocompatibility complex (Meo *et al.*, 1973; Widmer, Peck and Bach, 1973). Products of these loci are serologically undetectable by present methods.

In man similar results have been obtained; loci closely linked to the serologically detectable antigens—but genetically distinct—are responsible for the mixed lymphocyte reaction (Yunis and Amos, 1971). Lymphocytes stimulated by allogeneic cells in the mixed lymphocyte reaction are cytotoxic to target cells which share the same antigens as the stimulating cells, resulting in cell-mediated lympholysis. The structures against which the cytolysis is directed are the HL-A or serologically detected antigens (Eijsvoogel *et al.*, 1973).

3.3.1.3   ANTIGENIC ALTERATION BY PROTEOLYSIS

Trypsin treatment of peripheral lymphocytes increases the number of HL-A specificities present on the cells. Braun, Greclk and Murphy (1972) found that 50 percent of the peripheral lymphocytes tested had an increase of one to three additional specificities upon trypsinization. They also observed a case of renal transplant rejection in which lymphocytes of related donor and recipient, observed to be identical, displayed an additional specificity on donor lymphocytes following trypsinization. HL-A specificities can be masked by preincubation with a variety of substances such as normal human serum, calf serum, heat-inactivated serum, albumins, dextrans, etc., which may be attributable to nonspecific binding since this masking can be removed by repeated washing (Hirata and Terasaki, 1972). A decrease in cell viability is also frequently accompanied by an increase in additional HL-A specificities, which may be due to the unmasking of a hidden antigen by trypsin or by autolysis (Gibofsky and Terasaki, 1972). We believe that another controversial issue may be involved in the interpretation of these results; that is, the localization of transplantation antigens.

3.3.2   Tumor-associated antigens

Tumor-associated antigens represent the major differences in the cell surface known to exist between tumor and normal cells, but they are as yet un-characterized and it is not yet known whether they exist as cryptic entities in normal cell surfaces.

The distinction between a truly tumor-specific antigen and an isoantigen was not conclusively demonstrated until highly inbred mice became available. Among the early studies were those by Prehn and Main (1957), who induced tumors by using methylcholanthrene, transplanted them to syngeneic mice, allowed them to form palpable nodules and then removed them. These mice were subsequently challenged with the original tumor and found to be more resistant to its growth than were control mice. This led to the conclusion that these sarcomas possessed a tumor-specific transplantation antigen (TSTA) which could elicit immunity against the tumor cells transplanted into syngeneic recipients.

Since TSTA are detected by a single specific method of assay it may be more meaningful to describe them as tumor-associated antigens (TAA). The situation has been reviewed extensively (Haddow, 1965; Sjorgren, 1965; Klein, 1966; Smith, 1968a–c; Baldwin, 1971; Herberman, 1973; Herberman and Gaylord, 1973). Experiments have now shown that tumors induced by chemical or physical agents express individually distinct tumor antigens so that immunization against one tumor does not confer host resistance to other tumors even when induced by the same carcinogen (Baldwin, 1970).

On the other hand, virus-induced tumors contain cross-reacting antigen, suggesting virus-related specificities. Immunization against one tumor affords protection against others induced by the same virus, but not against those induced by different viruses (Habel, 1969; Pasternak, 1969). Virus-induced or associated tumor systems clearly offer an attractive model for the study of TAA in view of the presence of cross-reacting antigens. The apparent relationship between virus and TAA has raised the possibility that the TAA may be a direct viral product. In view of the fact that cross-reactivity between tumor cells is seen even though infectious virus is not released by the cells, an indirect role for the virus is probable. An example would be the induction of TAA resulting from a change in the synthesis of an already existing cellular antigen, as in lysogenic conversion in bacteria (Luria, 1962).

In view of the findings of membrane immunofluorescent assays (Herberman and Gaylord, 1973), TAA are assumed to be on the cell surface. The studies of Dickinson, Caspary and Field (1972) have supported this theory in the case of seven different cell types. There are parallels between TAA and transplantation antigens (some would say TAA are weak histocompatibility antigens). Histocompatibility antigens have been localized on the cell surface (Neauport-Sautes et al., 1972) although some believe them to be in different cellular compartments. Lewis et al. (1969), in studies involving 103 melanoma patients, found that two types of autoantibodies exist for the TAA. The one, directed against a cell surface antigen, was specific for the autochthonous tumor, whereas the antibody directed to the cytoplasmic antigen showed cross-reactivity with nearly all allogeneic tumor cells. Muramatsu et al. (1973), using murine leukemia cells, obtained a glycopeptide by papain digestion which exhibited TAA activity. Its molecular weight was 38 000. Exhaustive pronase digestion produced a glycopeptide of molecular weight 4500. In parallel experiments to define mouse transplantation antigens, the H-2 component released with papain was of molecular weight ca. 40 000, while its pronase digest resulted in a glycopeptide of molecular weight 3500. It was concluded that the TAA molecule contains a more complex carbohydrate structure than the H-2 antigen.

Another group (Baldwin, Bowen and Price, 1974), using rat hepatoma cells, obtained a tumor-associated antigen, also by papain digestion, or by digestion with β-glucosidase. It seems quite likely that the latter enzyme may have been contaminated with a protease although the authors suggested that removal of glucose may render the molecule susceptible to autolysis. This seems unlikely since glucose has never been shown to be a component of a complex glycoprotein which contains sialic acid. Antigens obtained by either enzyme exhibited a molecular weight of 50 000–60 000. Removal of sialic acid or galactose enzymatically did not affect the biological activity in an assay *in vitro*.

3.3.3   Surface-associated macromolecules

As has been discussed, proteolytic treatment of lymphoid cells has been used to obtain (glyco)peptides which exhibit such immunological activities as histocompatibility, antigenicity and tumor-associated antigenicity. It has also been used to observe the effect on immunological functions such as complement receptor activity, changes in histocompatibility types, etc., and to observe the ability of the cells to replace the proteolytically removed components. In addition, proteolysis of intact cells has been used to obtain surface components for analysis of the properties of the plasma membrane. This procedure, as will be discussed, is open to some question.

Trypsin and pronase release similar quantities (9–14 percent) of the dry weight of TA3 mammary adenocarcinoma cells (Codington, Sanford and Jeanloz, 1970), of which approximately one-quarter can be released upon incubation in the absence of any proteolytic enzyme; 40–60 percent of the hexose and 25 percent of the hexosamine and protein are also released in the absence of proteolysis.

In another study, Codington, Sanford and Jeanloz (1973) compared the products released by trypsin from TA3/ST and TA3/HA cells. TA3/ST cells are obtained from the ascites fluid of mice with a mammary adeno-carcinoma. TA3/HA cells arose from them by a spontaneous transformation during culturing *in vivo* (Hauschka *et al.*, 1971); TA3/HA cells are not rejected when injected into a mouse with an incompatible transplantation pheno-type, whereas TA3/ST are rejected. TA3/HA exhibit markedly less H-2 activity than the TA3/ST cells (Sanford *et al.*, 1973). Additional histocom-patibility antigens were not detected in either cell type following trypsin treatment or lyophilization by Sanford *et al.* (1973), who suggested that the equal H-2 levels found in lyophilized cells may be due to masking of the H-2 antigen by a sialoglycopeptide which they found only in the TA3/HA cells.

In view of the fact that a measure as drastic as lyophilization was necessary to reveal H-2 antigens, another explanation for the decrease in antigenicity may be a change in subcellular localization. This is because, as noted on p. 68, 'spontaneous transformation' can result in a change in subcellular localization of complement receptors.

TA3/HA cells release at least two high-molecular-weight components upon proteolysis that are present in only very low amounts in the trypsinates of TA3/ST cells, which have substantially higher levels of lower-molecular-weight components. It seems to us that this may suggest that a degradative enzyme could be absent in the HA subline since spontaneous transformation may be expected to lead to a de-differentiation and the deletion of synthetic capability rather than its enhancement as required for the synthesis of surface macromolecules.

The high-molecular-weight components have been analyzed for overall chemical content; only relative values were obtained. The high-molecular-weight material for TA3/HA cells gives a broad peak following chromato-graphy on BioGel A-5M (Codington, Sanford and Jeanloz, 1972) and Sepharose 4B (Slayter and Codington, 1973). In the latter case, this broad peak is divided into three parts and has been studied by shadow-casting electron microscopy. The conclusions were that the released material is present in rod-like structures with molecular weights 200 000–500 000 and 150 000–200 000; confirmation by more conventional techniques was not reported nor was the conclusion verified by ion-exchange chromatography.

In studies with RAJI lymphoid cells, surface products have been obtained by proteolysis from whole cells as well as from isolated plasma membranes (Jett and Jamieson, 1973). A 'trypsin barrier' had been proposed by Northrup (1926) and this has led many workers to proteolyse intact cells using trypsin. The assumption that trypsin does not enter the cell upon brief exposure is probably valid; however, it does not take into consideration the dynamic nature of living cells, which respond to external stimuli as simple as washing at 37 °C in physiologic buffers (Jett and Jamieson, 1973; van Blitterswijk et al., 1975) or at 4 °C (Lin, Wallach and Tsai, 1973) or relatively harsh measures such as brief proteolytic digestion. The data obtained with RAJI cells show that one of the ways cells respond is by 'exuding' an intracellular component which, one could speculate, may be a protective mechanism for the cell.

RAJI cells have been examined by light microscopy before and after brief (20 min) trypsin digestion; the viability remained at 95 percent based on uptake of trypan blue as well as morphological appearance. Examination of the trypsinized cells by electron microscopy (M. Jett, unpublished observa-tions) has shown that there is distortion of normal shape, some clumping and, more strikingly, the cells exhibit bald, smooth surfaces. Cells treated under the same conditions in the absence of trypsin were spherical in shape and had a rather uniform distribution of surface projections.

The products released by trypsin have been subjected to chromatography on Sephadex G-200 (Jett and Jamieson, 1973). Three size classes of glyco-peptides were obtained which had molecular weights of 15 000, 60 000 and $> 200 000$. When isolated and characterized plasma membranes were subjected to the same degree of trypsinization, only the lower two size classes were observed. Other subcellular fractions have been examined to see if a component of similar molecular weight and properties [such as 5% trichloroacetic acid (TCA) precipitation] could be observed. The cytoplasmic fraction was rich in this component, as was the culture medium. Following trypsin treatment of intact cells, Dounce homogenization of the residual cell released a component with these properties. The ratio of the high-molecular-weight material which was released by trypsin to that released by Dounce homogenization and to that which had accumulated in the culture fluid in 24 hours was 1:2.4:0.6.

Since none of the lower-molecular-weight components released from intact cells by trypsin was precipitable with 5% TCA, but only that of molecular weight $> 200 000$, a time course of the release of this component was followed, first in response to resuspending the cells in Tris-buffered saline, pH 7.4, at 37 °C (Figure 3.1). It was surprising that within 20 min large amounts of the component had been released upon merely resuspending the cells followed by gentle centrifugation at 800g for 2 min at 25 °C. After the release had reached a plateau (1 h), trypsin was added and again the release was very rapid, reaching a maximum after 20 min.

Ion exchange chromatography has been used to determine how many components make up the material of molecular weight > 200000 (M. Jett, unpublished observations). This material does not stain with periodate–Schiff reagent or Coomassie blue since it has a rather low sialic acid and aromatic amino acid content. It contains hexose and hexosamine in addition to sialic acid.

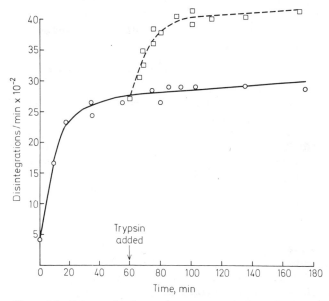

Figure 3.1   Release of radioactivity precipitable with 5% TCA from intact RAJI cells; as noted in the text, only the high-molecular-weight component (> 200000) is precipitated under these conditions. (○) Without, and (□) with trypsin. (From Jett and Jamieson, 1973, courtesy of Academic Press)

The conclusions from this study are that high-molecular-weight material may be only transiently associated with the surface membrane and is probably a secretion product. These experiments emphasize the importance of using isolated and characterized surface membranes when attempting to claim the localization of a component based on proteolytic experiments.

Released components have also been observed by Sakiyama and Burge (1972), who compared murine 3T3 cells with SV_{40}-transformed 3T3 cells. They found essentially no differences in the cell surface glycopeptides. However, a high-molecular-weight carbohydrate-containing component released into the culture medium was found in fivefold higher concentration in the normal as compared with the transformed cell line. A similar phenomenon was also observed by Lewis *et al.* (1969) using melanoma cells from 107 different patients; two types of TAA were found, one of which was specifically a secretion product. This has been observed with carcinoembryonic antigens (CEA) obtained from epithelial cells of the metastasized tumors (Denk *et al.*, 1972). CEA was thought to be pathognomonic for human colonic tumors. The molecule contains carbohydrate which may be necessary for antigen–antibody interaction since periodate oxidation results in loss of the biological activity (Banjo *et al.*, 1972), but more recent experiments, using synthetic oligosaccharides as inhibitors, suggest that the carbohydrate may not be involved in the antigenic determinant (Vrba *et al.*, 1975). CEA, while found as a cell surface component, has also been shown to be a secretion product and it has been suggested that it is wholly secreted and only transiently associated with the cell surface (Denk *et al.*, 1972).

That another lymphoid cell exhibits this property was shown by van Blitterswijk *et al.* (1975) using mouse ascites leukemia cells cultured *in vivo*. The cells were disrupted by hand homogenization and the subcellular fractions were separated by two methods:

1. The 'Ficoll' method (van Blitterswijk, Emmelot and Feltkamp, 1973), which involves hypotonic conditions.
2. A sucrose method (Molnar *et al.*, 1969), which maintained isotonic conditions throughout.

The two methods were used to confirm the results of the subcellular localization of the antigen induced by mammary tumor virus (MLr) and the normal differentiation antigen Thy. 1.2. The Ficoll method showed a twofold greater degree of purity based on the use of enzyme and chemical markers, whereas the sucrose method gave a twofold higher yield of membranes. The relative enrichment and recovery of the antigens were essentially the same by both methods.

Van Blitterswijk *et al.* (1975) found that the MLr TAA was not concentrated in the plasma membrane, as was the Thy. 1.2 antigen. The cytosol contained 69 percent of the MLr but only 3.8 percent of the Thy. 1.2 whereas the crude plasma membrane contained 63 percent of the Thy. 1.2 activity and 23 percent of the MLr activity. In addition, they found that the MLr activity occurred in the particle-free supernatant solution of the ascites fluid in which the cells had been grown. The authors proposed that the MLr antigen could be superficially located and is constantly being shed into the ascites fluid and/or that budding of the membrane, which removes clusters of antigen, is occurring (the Thy. 1.2 antigen does not accumulate in the extracellular fluid). They also found that the MLr could be easily eluted from the cells by simple washing with physiological buffers.

The MLr activity was present in the free state as well as in an immune complex (dissociable at pH 3.0) in culture fluid. Therefore van Blitterswijk *et al.* proposed that the ability to shed antigens which form complexes in the ascites fluid may be an example of the proposed mechanism (Alexander, 1974) whereby the tumor protects itself against the immunological defenses of the host.

In summary, studies of cell surfaces have been approached in two ways. The first is the release of components from intact cells using salt extraction or proteolytic digestion. Some investigators believe that salt extraction is too harsh for antigen isolation (Hollinshead, 1973; Harris, Price and Baldwin, 1973) or acts by stimulating cell autolysis (Mann, 1972), while the objection to the use of proteolytic enzymes involves the relatively low yields of antigen with possible destruction of antigenic specificities (Snary *et al.*, 1974). Also, it has been observed that digestion with trypsin, though an accepted method for obtaining surface components, in fact causes release of cytoplasmic components although the cells remain viable throughout the procedure (Jett and Jamieson, 1973; van Blitterswijk *et al.*, 1975).

The second approach involves the isolation of surface membranes with subsequent procedures to separate the appropriate structures. These procedures have been shown to concentrate surface-associated antigen. Procedures for membrane isolation have been reviewed by De Pierre and Karnovsky (1973). In the next section an updated review based on our experiences will be presented. We consider this approach to be particularly important as an essential step in differentiating components of the plasma membrane as such, from components of glycocalyx or intracellular origin.

3.4   MEMBRANE ISOLATION

3.4.1   Introduction

The study of membrane-related phenomena requires the isolation of plasma membranes of high purity. A variety of techniques for the disruption of cells and subsequent fractionation of subcellular particles has been described, each adapted to the particular type of starting material. Several critical reviews discuss the isolation of cell membranes in general (Warren and Glick, 1971; De Pierre and Karnovsky, 1973; Wallach and Lin, 1973; Wallach and Winzler, 1974; *see also* M. C. Glick, Vol. 1, Chapter 3 of this series) and lymphocyte membranes in particular (Crumpton and Snary, 1974) with regard to the merits of the various isolation and fractionation procedures, yield of membrane, contamination with subcellular organelles and criteria for characterization.

3.4.1.1   CELL SOURCES

Lymphoid cells used for plasma membrane isolation may be obtained from intact tissue such as lymph nodes, thymus, tonsils, bursa of Fabricius, from circulating peripheral lymphocytes, from cells maintained *in vivo* as solid or ascites tumors and from cells cultured *in vitro*. Solid tissue has the advantage that large quantities of cells may be obtained and processed, thereby increasing the overall yield of membranes, but the disadvantage that a single cell population is not present.

3.4.1.2   PASSAGE OF CELLS TO ENSURE REPRODUCIBILITY

Many studies of lymphocytes involve cells which are cultured *in vitro* or grown in mice and these allow a great deal of experimental flexibility. With human cells, culture systems permit the use of techniques which would be impossible for experimentation *in vivo*.

However, sufficient safeguards regarding the integrity of the cultures must be employed so that experimental data are both reliable and reproducible. Our experience with the human lymphoma cell line, RAJI, leads us to distrust undocumented continuous passage of malignant cells since spontaneous changes have been observed in culture (Hauschka *et al.*, 1971). Therefore, we propose the following guidelines:

1. The cells should be obtained from a source which is able to provide a full history of the cells including such details as the number of passages the cells have undergone, presence of viruses normally associated with the cells, absence of mycoplasma, etc.
2. There should be a strict limit to the number of passages the cells undergo. We have set 30 subcultures as the maximum, based on our experience. In addition, the cells are carefully examined biweekly for even slight signs of changes in characteristics. These include lack of 'background' material in the cultures, roundness of cells, and the fact that our cells, RAJI, grow in loosely associated clusters. These characteristics are immediately affected by 'external' agents.
3. Bacteriostatic agents should be cautiously employed. In our cultures, we use low levels of kanamycin, an antimycoplasma and bacteriostatic agent. Cultures showing any slight indication of variations are discarded and new cultures established from frozen cell stocks. The reason for this safeguard is that infections may go unnoticed when high levels of bacteriostatic agents are employed but, in their absence, will surely be obvious!

Sensitive enzymatic and chemical analyses have been reproducible after $2\frac{1}{2}$ years of cell culture using these guidelines. In addition, M. P. Dierich and R. A. Reisfeld (personal communication) found that 100 percent of the cells from our stock had complement receptor activity whereas less than 30 percent of the same type of cells from an undocumented source demonstrated this activity. The isolation of subcellular fractions which were able to inhibit complement receptor activity of intact cells, as demonstrated by rosette formation, was quite reproducible for up to approximately six months of continuous subculturing but began to decrease after about 60 passages to about 70 percent of the original activity, although rosette formation with intact cells showed that almost all cells still displayed receptor activity. There seemed to be a change in localization of the receptors from being membrane-associated to being in the soluble fraction, but no apparent signs of transformation or contamination existed.

3.4.2   Disruption methods

The first step in plasma membrane isolation involves cell disruption. In choosing a satisfactory method, a balance must be achieved between disruption of the cells and the preservation of intracellular organelles (primarily nuclei) and must take into consideration the effect of the method on biologic functions such as enzymatic or antigenic activity.

Ionic conditions should be tailored to the particular cell type, regardless of the method used. Increased cell fragility is brought about by using hypotonic conditions, and divalent cations are included to improve nuclear stability. In other cases, buffers at pH 8.5 have been thought to increase cell fragility (Neville, 1960; De Pierre and Karnovsky, 1973). *Table 3.1* shows a list of methods used for disruption of lymphoid cells.

The most popular methods with lymphocytes have employed motor- or hand-driven homogenizers in which large pieces of membrane are stripped from the cell. A hand-driven homogenizer used by Allan and Crumpton (1970, 1972) produced one of the best examples of maximum cell disruption (95 percent) with minimal nuclear damage (4 percent), and the isolated plasma membranes appear to have the least amount of contamination on the basis of enzymatic, chemical and optical methods. This method is useful only with solid tissue, not with single-cell suspensions. When a similar principle, but modified equipment, was used for single-cell suspensions (Crumpton and Snary, 1974), cell disruption was less than satisfactory (50 percent) although nuclear breakage was minimal.

An example of a large-scale membrane isolation method was described by Kornfeld and Siemers (1974), who used frozen organs homogenized in

Table 3.1 LYMPHOCYTE CELL-DISRUPTION METHODS

Cell type	Method of disruption	Vesicle size, μm	Nuclear breakage, % *	Reference
Hand-driven homogenizers				
Pig lymphocytes, pig or human thymus	Mince/press, isotonic saline–Tris	1–8	4	Allan and Crumpton (1970, 1972)
Human tonsil	Dounce, hypotonic Tris-DTT	0.3–0.6	nd	Lopes *et al.* (1973)
Mouse leukemia L 1210	Dounce, hypotonic NaHCO$_3$	nd	nd	Hourani, Chace and Pincus (1973)
Human chronic leukemia	Dounce, hypotonic NaHCO$_3$	nd	8	Marique and Hildebrand (1973)
Motor-driven homogenizers				
Calf thymus	Electric shredder	nd	2	Kornfeld and Siemers (1974)
Human tonsil	Potter–Elvehjem, isotonic saline–Tris	nd	30	Demus (1973)
Rat spleen and thymus	Potter–Elvehjem, hypotonic saline–Tris	nd	23	Ladoulis *et al.* (1973)
Chemical stabilization				
Mouse lymphoblast L 5178Y	Dounce, fluorescein mercuric acetate–hypotonic sucrose	9–11	nd	Manson, Hickey and Palm (1968)
Chick thymus, bursa and peripheral lymphocytes	Potter–Elvehjem, hypotonic ZnCl$_2$	0.1–0.2	All DNA sedimented with nuclear pellet	Ragland, Pace and Doak (1973)
Human WIL$_2$	Dounce, hypotonic ZnCl$_2$	0.6–1	nd	Kennel and Lerner (1973)

Table 3.1   LYMPHOCYTE CELL-DISRUPTION METHODS

Cell type	Method of disruption	Vesicle size, μm	Nuclear breakage, %*	Reference
N_2 cavitation				
Pig lymphocytes	N_2 cavitation, 700 lb in^{-2}, 15 min; 0.25 M sucrose	0.5–3	30	Ferber et al. (1972)
Calf thymocytes	N_2 cavitation, 500 lb in^{-2}, 15 min; Hanks balanced salts	1–2	4	van Blitterswijk et al. (1973)
Calf and rabbit thymocytes	N_2 cavitation, 400 lb in^{-2}, 20 min; HEPES buffer, salt solution	nd	nd	Schmidt-Ullrich et al. (1974)
Miscellaneous				
Human BRI-8	Mechanical shear, Eagle's balanced salts	1–3	9	Wright, Edwards and Jones (1974); Crumpton and Snary (1974)
Normal and leukemic mouse lymphocytes	EDTA–borate extraction	0.6–1.0	nd	Warley and Cook (1973)
Human RAJI	Glycerol loading, hypotonic lysis	1–8	10	Jett, Seed and Jamieson, manuscript in preparation

Abbreviations—DTT: dithiothreitol; nd: not determined.
* Percentage of DNA not recovered in the nuclear pellet.

an electric shredder. Plasma membranes of high purity, by enzymatic, chemical and optical criteria, were obtained. The degree of cell disruption was not reported, but nuclear breakage was minimal. Again, this method would probably be unsuitable with cell suspensions.

A difficulty with homogenization techniques is that the large nuclei of lymphoid cells (5–9 μm) are easily damaged before sufficient cell disruption has occurred. Ladoulis et al. (1974) employed hypotonic conditions to increase cell fragility and thereby to decrease the amount of shearing (motor-driven Potter–Elvehjem homogenizer) needed for cell disruption and 85 percent disruption was achieved. The degree of nuclear damage (23 percent) and increase in specific activity of a membrane marker enzyme were quite similar to that obtained by Demus (1973) using the same method for tonsil cells but employing isotonic buffers. Lopes et al. (1973) also used tonsil cells but disruption was accomplished by Dounce homogenization in hypotonic buffers. Enrichment of the plasma membrane marker enzyme was identical to that in the previous two methods. The recoveries of plasma membrane (based on enzymatic data) for these three examples were 18 percent, 12 percent and 7 percent respectively.

Chemical stabilization of cell surfaces with either metal ions (Zn^{2+}) or sulfhydryl reagents (fluorescein mercuric acetate), followed by homogenization in hypotonic buffer is among the methods described by Warren and Glick (1971) for isolating cell membranes as cell-size ghosts (*see also* Warren, Glick and Nass, 1961). The possible effects of the stabilizing agents on enzymatic and biologic activities must be considered when choosing this method of membrane isolation. Mouse lymphoblast cells, treated with fluorescein mercuric acetate did, however, retain their transplantation antigen activity when tested *in vivo* (Manson, Hickey and Palm, 1968). Another disadvantage is the degree of contamination one frequently seems to obtain using such harsh methods. For example, electron micrographs of the isolated WIL_2 membranes reveal some intracellular material still bound to the plasma membranes (Lerner, Meinke and Goldstein, 1971).

Borate–EDTA extraction is thought to cause an effect opposite to that described for chemical stabilization. According to procedures developed by McCollester (1970), inhibition of FAD-dependent enzymes by extraction with a solution of borate and EDTA leads to a preferential breakdown of the cytoskeleton. Intracellular contents are expelled, leaving the membranes as cell ghosts. In the application of this method to lymphoid cells, Warley and Cook (1972) reported very low recoveries of 5′-nucleotidase. Binding of borate to cell surface carbohydrates may be the cause of reduced enzyme activities. Rather small membrane vesicles resulted as compared with the chemical stabilization methods.

In nitrogen cavitation, cell suspensions equilibrated with nitrogen gas under pressure are released to atmospheric pressure by passing through a small orifice (Hunter and Commerford, 1961). The pressure at which the cells are equilibrated, the suspending medium, and the presence of divalent cations are important factors in determining the extent of cell disruption, the number of nuclei broken and aggregation of subcellular particles. The disadvantage of this method is that very small vesicles result, and are difficult to separate from intracellular organelles; one study (Ferber *et al.*, 1972) showed 30 percent nuclear breakage.

Glycerol loading was described for human platelets by Barber and Jamieson (1970). Its application to lymphoid cells required total modification of the method (M. Jett, T. Seed and G. A. Jamieson, manuscript in preparation). In order to preserve maximum cell viability until the final lysis step, the cells were kept at 37 °C and in balanced salt solutions during the washing steps. Eighty-five percent of the cells were disrupted by hypotonic lysis. The most striking feature of this method is that intact nuclei devoid of contaminating organelles were obtained within ten minutes of the lysis step. Based on enzymatic, chemical and optical data, the large fragments of plasma membranes were quite free of contaminating structures.

3.4.3   Isolation methods

Fractionation of the homogenate in the shortest possible time after cell disruption is most desirable to preserve enzyme activity and decrease cross-contamination with subcellular organelles. The swift removal of nuclei from the homogenate will reduce the possibilities of contamination with DNA. Most investigators have employed a combination of differential and isopycnic gradient centrifugation to separate the nuclei, lysosomes and mitochondria from the homogenate and recover the membranes (Wallach and Winzler, 1974). The size and shape of the membrane fragments obtained during cell disruption generally influence the choice of fractionation procedures. The general procedures used for subcellular fractionation of lymphocyte homogenates are shown in *Table 3.2*.

Table 3.2 LYMPHOCYTE MEMBRANE ISOLATION PROCEDURES

Fractionation procedure	Centrifugation time	Reference
1. *Polymer systems*		
(a) Dextran cushion	3 h	Ferber *et al.* (1972)
	Not reported	Schmidt-Ullrich *et al.* (1974)
(b) Discontinuous Ficoll gradient	15 h	van Blitterswijk *et al.* (1973)
(c) Polyethylene glycol–dextran system	2 h	Kennel and Lerner (1973) Hourani, Chace and Pincus (1973) Warley and Cook (1973)
2. *Sedimentation centrifugation*		
(a) Discontinuous glycerol gradient	2 × 20 min	Manson, Hickey and Palm (1968)
(b) Discontinuous sucrose gradient	12 h	Lopes *et al.* (1973)
3. *Flotation centrifugation*		
(a) Multistep flotation	22 h	Demus (1973)
	Not reported	Ladoulis *et al.* (1974)
(b) Step flotation	18 h	Allan and Crumpton (1970) Crumpton and Snary (1974)
(c) Step flotation (zonal)	4 h	Kornfeld and Siemers (1974)
4. *Simultaneous sedimentation–flotation (zonal)*	2 h	Jett, Seed and Jamieson, manuscript in preparation

* In every case, this major centrifugation step was preceded by one or more differential centrifugation steps.

In homogenates produced by nitrogen cavitation, plasma membrane vesicles have been found associated with the microsome fraction ($100\,000g$ pellet), owing to the size of the vesicles. Differences in ion binding capacity and permeability properties allow the plasma membrane vesicles to be separated from endoplasmic reticulum vesicles also present. The techniques used were first described by Wallach and Kamat (1966) for Ehrlich ascites cells.

Some specialized techniques such as the filtration of the homogenate over a column of glass beads (Warley and Cook, 1973) and its partition in a two-phase polymer system (Brunette and Till, 1971) have been employed to facilitate membrane isolation.

Continuous and discontinuous sucrose gradients have been used by a large number of investigators to isolate membranes from homogenates derived by hand- or motor-driven homogenization. One or more steps of differential centrifugation usually precede the sucrose gradients, and the fraction applied varies from the $4000g$ supernatant (Lopes *et al.*, 1973) to the $100\,000g$ pellet (Ladoulis *et al.*, 1974).

In the larger-scale method (Kornfeld and Siemers, 1974), nuclei were removed by low-speed centrifugation and the plasma membranes were aggregated by 0.1 M LiCl and then pelleted. A single-step (zonal) flotation gradient was employed.

Flotation in multistep, discontinuous gradients as used by Demus (1973) for tonsil cells and by Ladoulis *et al.* (1974) for rat spleen and thymus cells separates two bands of membranes. Fractionation of Zn^{2+}-hardened chick lymphocyte membranes also yields two bands of plasma membrane material on discontinuous sucrose gradients (Ragland, Pace and Doak, 1973). In each of the above cases, cells were disrupted by manual homogenization, a process which can be difficult to reproduce and regulate. The distribution of membrane material among several bands may reflect variation in homogenization conditions, such as the number of strokes required to disrupt cells, gentle versus hard homogenization.

Mouse lymphoblast membranes treated with fluorescein mercuric acetate were first localized in the 65% glycerol fraction of a multistep gradient then further purified on a continuous glycerol gradient. The very short centrifugation times allowed preparation and isolation of the L5178Y membranes within one day (Manson, Hickey and Palm, 1968).

In our system (M. Jett, T. Seed and G. A. Jamieson, manuscript in preparation), the first step is differential centrifugation to remove nuclei. Removal of the soluble and mitochondrial fractions, most of the endoplasmic reticulum and about 85 percent of the lysosomal material is accomplished by sedimentation of the plasma membrane at $100\,000g$ for 1 h onto a 38% sucrose step.

A batch method utilizing a two-hour simultaneous sedimentation and flotation centrifugation has been developed. This also permits the processing of sufficient quantities of the two types of membrane vesicles necessary for biochemical studies.

There are certain pitfalls which should be emphasized to ensure the reproducibility of these procedures for membrane isolation. First, sucrose solutions should be standardized by refractive index. In our experience, different batches of ribonuclease-free sucrose vary from 10–30 percent in water content. Secondly, we question the results obtained from cells stored at $-20\,°C$ in dimethyl sulfoxide or glycerol before disruption and fractionation. The assumption made by investigators who follow this practice is that the intracellular organelles will not be damaged if a cryoprotective agent is used. Simply suspending the cells in such agents, under those conditions, cannot provide protection sufficient to keep the intracellular organelles and plasma membrane intact. Freezing cells for reculture usually results in approximately 20 percent of the cells being viable; the procedure is carried out under special conditions by lowering the temperature in a controlled fashion to $-80\,°C$. Furthermore, storage at $-20\,°C$ is not sufficient to stop proteolytic cleavage. Baenziger and Majerus (1974) showed that lysosomal enzymes are active at $-20\,°C$. Thirdly, the use of deoxyribonuclease (Hourani, Chace and Pincus, 1974) to remove DNA seems to be a relatively crude procedure. These quantities of DNA indicate that extensive nuclear damage has occurred and the enzymatic data suggest that nuclear membranes are probably contaminating the plasma membrane fraction. Therefore, simply removing the DNA is treating the 'symptom' rather than the 'disease'. Furthermore, other investigators (Lerner, Meinke and Goldstein, 1971;

Lerner, 1972) have identified a type of DNA which is probably truly associated with plasma membranes of WIL2 cells and it was not present in the quantities found by Hourani, Chace and Pincus (1973). Fourthly, the two-phase system (Brunette and Till, 1971) may be questionable for use with lymphoid cells. Two separate groups have used this polyethylene glycol/dextran system and in both cases, the designated plasma membrane fraction was badly contaminated with intracellular particles as judged by electron micrographs. Neither group used sufficient enzymatic or chemical markers to allow a valid judgment as to purification. It may be unfair to attribute the poor results to the two-phase system, since in one of these cases (Hourani, Chace and Pincus, 1973) the problem may have resulted because the cells were stored at $-20\,^\circ$C before disruption and fractionation. In the other case (Kennel and Lerner, 1973) a chemical stabilization method was used for disruption and this may have contributed to the problem.

3.4.4 Membrane markers

There is no question that 'purified membranes' is a relative term. Careful biochemical techniques must be employed in membrane isolation in order to be able to draw any conclusions regarding surface phenomena. Indeed, this must be balanced by the fact that the isolation is not an end in itself but a stepping-stone to understanding the function of cell surfaces.

Isolated membranes of each cell type are generally characterized as to their purity by using chemical, enzymatic and electron microscopic criteria. Lack of a single chemical or enzyme marker which can be associated uniquely with the plasma membrane has led investigators to analyze those materials which are enriched in the fraction morphologically identifiable as membrane (De Pierre and Karnovsky, 1973; Wallach and Winzler, 1974). Obtaining a chemical and enzymatic profile of each subfraction of the homogenate is extremely important in determining the fractionation profile of subcellular organelles and in calculating the recoveries of enzymatic activities. We do not agree with those who use only optical methods for deciding the purity of a plasma membrane fraction since the vesicles obtained may well contain contaminating structures which can be shown by the determination of enzymatic activities.

5′-Nucleotidase, while a useful and popular plasma membrane marker, must be used in conjunction with other enzyme markers, and with chemical and optical methods, to obtain a broad picture of purification and recovery.

With RAJI lymphoid cells, we have found that the total activity of 5′-nucleotidase is 4–12 percent of that of other plasma membrane enzyme markers. This has been observed in other tumor cells as well and creates technical difficulties, since the large amounts of inorganic phosphate in the samples of nuclei and soluble fractions in particular result in small changes in large numbers in terms of the increase of enzymatic activity with time. Furthermore, in RAJI cells as well as other cells (Crumpton and Snary, 1974) this activity is also clearly associated with nuclear membranes, as shown in *Tables 3.3* and *3.4*.

Nuclear membrane marker enzymes such as 5′-nucleotidase, phosphodiesterase, ATPases, etc., have been tabulated by Wallach and Winzler

Table 3.3 LOCALIZATION OF ENZYMES IN SUBCELLULAR ORGANELLES

Enzyme	Localization	Localization reference	Assay reference
5′-Nucleotidase	1. Plasma membrane	Misra, Gill and Estes (1974), Wood (1967), Emmelot et al. (1964)	Mitchell and Hawthorne (1965)
	2. Nuclear membranes	Wallach and Winzler (1974), Crumpton and Snary (1974), Jett, Seed and Jamieson, manuscript in preparation	
Na$^+$,K$^+$-ATPase	1. Plasma membrane	Emmelot et al. (1964)	Wallach and Kamat (1966)
	2. Nuclear membrane	Crumpton and Snary (1974), Stern et al. (1952)	
Alkaline phosphatase	1. Plasma membrane	Bridgers and Kaufman (1962)	Bosmann, Hagopian and Eylar (1968)
	2. Nuclear membrane	Stern et al. (1952)	
Phospho-diesterase(s)	1. Plasma membrane	Erecinska, Sierakowska and Shugar (1969)	Erecinska, Sierakowska and Shugar (1969)
	2. Nuclear membrane	Erecinska, Sierakowska and Shugar (1969), Jett, Seed and Jamieson, manuscript in preparation	Koerner and Sinsheimer (1957)
NADH diaphorase	Microsomes	Williams, Gibbs and Kamin (1959)	Wallach and Kamat (1966)
Esterase	Microsomes	Weinhold and Rethy (1969)	Earl and Korner (1965)
Glucose-6-phosphatase	Microsomes	Hörtnagl, Winkler and Hörtnagl (1969)	Nordlie and Arion (1966)
Acid phosphatase	Lysosomes	Emmelot et al. (1964), de Duve et al. (1955)	Hubsch and West (1965)
β-N-Acetyl-glucosaminidase	Lysosomes	Sellinger et al. (1960)	Jett, Seed and Jamieson, manuscript in preparation
Succinate dehydrogenase	Mitochondria	Weiner (1970)	Earl and Korner (1965)
Cytochrome c oxidase	Mitochondria	Holtzman and Dominitz (1968)	Smith (1955)
Lactic dehydrogenase	Cytoplasm	Beattie, Sloan and Basford (1963)	Kornberg (1955)
DNA polymerase	Nuclei	Keir, Smellie and Siebert (1962), Behki and Schneider (1963)	Kaplan et al. (1973)

For additional marker enzymes see Schneider (1968), Shnitka and Seligman (1971), De Pierre and Karnovsky (1973) and Wallach and Lin (1973).

Table 3.4 CHEMICAL COMPOSITION AND ENZYME ACTIVITY OF ISOLATED PLASMA MEMBRANES*

	Human RAJI: glycerol loading—hypotonic lysis†	Human BRI-8: shear by mechanical pump‡	Human tonsil; Potter—Elvehjem§	Rat thymus; Potter—Elvehjem‖	Calf thymus; electric shredder¶	Pig lymph node; nitrogen cavitation**	Mouse leukemia; EDTA—borate extraction††
Protein in membrane fraction, %	1.7	0.6	0.9	1.4	0.5	1.7	nd
Protein, % dry wt	33	42	nd	nd	nd	nd	52
Lipid, % dry wt	35	51	nd	nd	nd	nd	48
Cholesterol:phospholipid molar ratio	0.69 and 0.97	1.02	0.69	0.98	nd	1.03	0.99
RNA, µg per mg protein	29.6	17	15	34	nd	25	nd
DNA, µg per mg protein	3.2	0	10	7.9	6.4	10	nd
5'-Nucleotidase	nd	5.4 (49×)	13.5 (13×)	6.8 (12×)	2.5 (33×)	7.5 (25×)	11.6 (1.9×)
Phosphodiesterase	0.475 (12.5×)	nd	nd	nd	nd	nd	nd
Succinic dehydrogenase	0	0	0.01	0.03	0	0	0
Glucose-6-phosphatase	0.4	0.18	0.98	0.24	nd	nd	nd
Acid phosphatase	nd	0	1.51	0.79	nd	0	nd
β-N-Acetylglucosaminidase	0.071	nd	nd	nd	1.87	nd	nd

* Enzyme activities are expressed as micromoles of product liberated per hour per milligram of protein. A value of 0 indicates no activity was detected, 'nd' means not determined. The additional data in parentheses for 5'-nucleotidase indicate the increase in specific activity compared with that of the total homogenate.

† Jett, Seed and Jamieson, manuscript in preparation; ‡ Crumpton and Snary (1974); § Demus (1973); ‖ Ladoulis et al. (1974); ¶ Kornfeld and Siemers (1974); ** Ferber et al. (1973); †† Warley and Cook (1973).

(1974), but these workers suggest that such enzymes can also be regarded as plasma membrane markers. For example, in our own studies we have found as much as 38 percent of plasma membrane markers [5'-nucleotidase and phosphodiesterase(s)] to be present in the nuclear fraction using p-nitrophenyl phosphate as the substrate but, in the same samples, using p-nitrophenyl-thymidine-5'-phosphate as the substrate, only 27 percent of the activity was associated with the nuclei. By electron microscopy, this fraction appears to be a pure population and does not contain either whole cells or plasma membrane fragments which could be the source of the enzyme.

Approximately 30 percent of the plasma membrane marker enzyme activities occur in the soluble fraction remaining after centrifugation at $100\,000g$ for one hour and in this case there is no significant difference between the two substrates used for phosphodiesterase assay. This soluble phosphodiesterase could be due to plasma membrane vesicles which do not sediment or it may represent enzyme which is dislodged from the plasma membrane during cell disruption. However, we believe that it is primarily a cytoplasmic form of the enzyme; phosphodiesterase, 5'-nucleotidase and ATPase have also been observed in cytoplasmic and nuclear fractions by other investigators (Erecinska, Sierakowska and Shugar, 1969; Warley and Cook, 1973; Wallach and Winzler, 1974; Misra, Gill and Estes, 1974).

The distribution of the total activity of plasma membrane enzymes in the cell fractions may not give a totally accurate picture of the percentage recovery. It seems to us that in studies where a claim for greater than 50 percent recovery of the plasma membrane is based on enzyme analyses (*see Table 3.4*), nuclear membranes are probably contaminating the plasma membrane fraction. This is based on the facts that nuclei have been shown to contain 37 percent of ATPase activity, 32 percent of 5'-nucleotidase activity (Crumpton and Snary, 1974) and 38 percent of phosphodiesterase activity (M. Jett, T. Seed and G. A. Jamieson, manuscript in preparation). In addition, all of these enzymes are found in the soluble form, as well (up to 30 percent). It is misleading to base percentage recovery of plasma membranes on the whole-cell homogenate for these reasons. Perhaps it would be more meaningful to base the recovery on the activity remaining in the homogenate after removal of the nuclei. This would require a very careful evaluation of the degree of nuclear damage. Certainly the distribution of activity is essential for evaluating various disruption or fractionation techniques on any one particular cell type. In combination with chemical data, the enzyme distribution may be more meaningful.

The degree of purity of the membrane fraction may be seen by the increase in the specific activity of the plasma membrane enzymes as well as the decrease in both total and specific activities of enzymes associated with other subcellular fractions. *Table 3.3* shows a list of such marker enzymes.

In most studies with lymphoid cells, cytoplasmic and mitochondrial components are readily removed from the plasma membrane fraction. Smooth endoplasmic reticulum and lysosomal membranes are the major contaminants, most of which are probably due to entrapment of these particles within plasma membrane vesicles. In our system, only 3.5 percent of the total lysosomal marker activity is coincidental with the peak of plasma membrane material and the specific activity is somewhat less than in whole cells. Crumpton and Snary (1974) have shown that washing plasma membranes decreases the contamination which had been observed enzymatically and by electron microscopy. In our system, pelleting and washing does remove contaminants, but also results in losses of plasma membrane.

Chemical data on the various cell fractions further clarify the information established by the enzymatic data. There are typical patterns which have been reported in membrane fractions for the distribution of DNA, sialic acid and carbohydrates, the lipid ratios, and certain amino acids such as the acidic amino acids, serine, threonine and proline, which seem to be higher in the plasma membrane fractions.

Although the chemical composition of isolated membranes will vary somewhat with the purity of the preparation, similar values have been obtained for a number of lymphocyte preparations (*see Table 3.4*). A cholesterol to phospholipid molar ratio approaching 1 is characteristic of plasma membrane from a variety of cell types (Coleman and Finean, 1966). With the exception of Demus (1973), who reported a ratio of 0.69 for tonsil cells, and Marique and Hildebrand (1973), with a value of 0.38 for chronic leukemia cells, most investigators have obtained a value close to 1.

Plasma membranes contain approximately equal amounts of protein and lipid by dry weight (Warley and Cook, 1973; Crumpton and Snary, 1974), with 6–10 percent carbohydrate. Present as neutral sugars, *N*-acetylhexosamines, fucose and sialic acid, the carbohydrate is found covalently bound to protein and lipid. The distribution of individual monosaccharides between glycoproteins and glycolipids has been discussed by Crumpton and Snary (1974). Analysis of the carbohydrate content of chick lymphocytes from blood, spleen, bursa and thymus by gas–liquid chromatography reveals a molar ratio of 1:1:1 for mannose:*N*-acetylglucosamine:galactose in each cell type (Droege *et al.*, 1975). Intact lymphocytes were subjected to papain digestion followed by isolation and analysis of the released glycopeptides. When chicks were either thymectomized or bursectomized and irradiated, the carbohydrate compositions of the surviving lymphocyte populations, especially spleen cells, were altered.

RNA composition varied from 15 to 35 μg per milligram of protein. Although the function of RNA in cell membranes has yet to be discerned, several investigators have postulated a synthetic role (*see* Crumpton and Snary, 1974). The range in DNA content of isolated membranes (from 10 μg per milligram of protein down to zero) may reflect to some extent the amount of nuclear breakage which occurred on cell disruption. Demus (1973) and Ferber *et al.* (1972), with approximately 30 percent nuclear breakage, have reported the highest specific content of DNA, while Crumpton and Snary (1974), with no detectable DNA, also found only 4–10 percent nuclear breakage. A unique species of cytoplasmic DNA associated with plasma membrane and comprising 0.5 percent of the total cell DNA was obtained by Lerner, Meinke and Goldstein (1971) from WIL2 cultured cells. Using NP-40 detergent to lyse plasma membrane selectively, the cytoplasmic DNA was isolated and characterized presumably in the absence of nuclear DNA. A role for membrane-associated DNA in synthesis of membrane immunoglobulin has been suggested (Lerner, 1972).

Measuring DNA and sialic acid in each other's presence has been a problem. The diphenylamine method for DNA (Dische, 1930) measures α-keto sugars and therefore sialic acid. However, development of the color reaction at 4 °C in the dark for 4 days yields a chromophore attributed to DNA (Croft and Lubran, 1965). The length of time and the frequency of spurious results are discouraging for even the most patient investigator. A more recent method for measuring DNA is based on helicity. This involves measuring fluorescence upon addition of ethidium bromide with and without prior incubation with 0.3 M KOH (to reduce RNA to its constituent nucleotides) (Donkersloot, Robrish and Krichevsky, 1972). In our hands, this assay for RNA gave 0.961 mg per 10 mg of cells (dry weight) as compared with 1.02 mg per 10 mg of cells using the Munro and Fleck (1966) modification for RNA analysis.

In the presence of small quantities of DNA, sialic acid can be determined by dichromatic readings at 532 and 549 nm (Warren, 1959), or by the method of Aminoff (1961) using extinction coefficients determined for this system (Jett and Jamieson, 1975) and substituted into the formula (Warren, 1959).

In the presence of higher concentrations of DNA, such as in whole cells and nuclear fractions, the absorbance of the aqueous phase *before* extraction is recorded at 532 and 549 nm. If A_{532} is more than twice as large as A_{549}, it is necessary to extract once with isoamyl alcohol, which preferentially extracts DNA (Warren, 1959; Kraemer, 1965, 1966). [1-^{14}C]N-Acetyl-neuraminic acid must be added as an internal standard to correct for the 20–30 percent of sialic acid which is also extracted by this procedure. This method has allowed the quantitative determination of the known sialic acid in the presence of large amounts of DNA, as in nuclear pellets and whole-cell homogenates (M. Jett and G. A. Jamieson, manuscript in preparation).

3.4.5 Solubilization of membrane components

Relating the structure and composition of membrane components to their biological functions requires the isolation of 'complete' molecules in a biologically active form. Proteolytic treatment of cell surfaces cleaves antigenically active, soluble pieces which generally do not contain that portion of the protein responsible for integration into the membrane (Nathenson and Cullen, 1974). Since most membrane proteins are not extractable by aqueous solutions owing to their strong hydrophobic interactions with membrane lipids, reagents such as detergents and organic solvents are necessary to achieve complete solubilization. Once solubilized, the components still require the presence of stabilizing agents to prevent aggregation. Further separation, characterization and biological assay then must be done in the presence of these detergents or solvents (Wallach and Winzler, 1974).

The most widely used technique for initial investigation of lymphocyte membrane components has been complete solubilization in SDS followed by electrophoresis in SDS–polyacrylamide gels (SDS–PAGE) (Maizel, 1969; Neville and Glossman, 1971). Conditions such as concentration of SDS, incubation temperature, and the presence or absence of disulfide-reducing agents have varied with each investigator. For this reason, caution should be employed in comparing electrophoresis patterns.

Lymphocyte plasma membrane from a variety of species displays a similar pattern overall after SDS–PAGE treatment: 20–30 polypeptide bands, ranging in molecular weight from 2×10^5 to 1×10^4. A small number of the higher-molecular-weight protein bands (7–10) also stain for carbohydrate with the periodate–Schiff reagent. Close examination of the gel patterns does reveal differences in the protein and glycoprotein compositions of lymphocytes from various tissues and species. For example, an intensely staining glycoprotein band with a mobility slightly less than that of the IgG marker appears very prominent in both pig thymocyte and lymphocyte membranes, but is hardly visible in human thymocyte membranes (Allan and Crumpton, 1971, 1972). Some quantitative and qualitative differences exist in individual protein bands when calf thymus and lymph node and rabbit thymus cell membranes are compared (Schmidt-Ullrich et al., 1974). These differences for the most part reflect intensity of staining, and difference in molecular weight of major components.

Electrophoresis of the microsome fraction from calf thymocytes was found to give 10 predominant polypeptide bands (Nos. 1–10) of molecular weight 1.5×10^4–2.8×10^5. Following dextran gradient centrifugation of the microsomes to separate plasma membrane from endoplasmic reticulum vesicles, the plasma membrane fraction was found to contain mostly high-molecular-weight components (bands 1–7) while the endoplasmic reticulum had small-size polypeptides (bands 8–10). In microsomes prepared from ^{125}I-labeled intact rabbit thymocytes, the radioactivity localized in bands Nos. 1–5, which also contain carbohydrate, suggesting that these glyco-proteins are located on the external cell surface (Schmidt-Ullrich et al., 1974).

Rat spleen-cell plasma membranes contain three glycoproteins of molecular weight 145 000 which are absent from rat thymocytes (Ladoulis et al., 1974). The two cell types share two major glycoproteins at molecular weights 50 000 and 65 000, while the thymus has an exclusive glycoprotein of molecular weight 27 000. ^{125}I label from intact thymocytes and lymphocytes iodinated before disruption localizes in a glycoprotein of molecular weight 117 000 present in both cells, and the glycoprotein of molecular weight 27 000 found in thymus cells. The spleen glycoproteins of molecular weight $> 200 000$ were also found to be labeled.

Tonsil-cell plasma membranes have an intensely staining polypeptide band of molecular weight 45 000–48 000 which may be related to a protein band found in liver, kidney and erythrocyte plasma membranes (Neville and Glossman, 1971; Lopes et al., 1973). This peptide was stained only faintly in tonsil-cell plasma membranes prepared by Demus (1973). On comparison of subcellular organelles in 10% polyacrylamide slab gels, it appeared to be limited to the plasma membrane fractions with a slight appearance in the lysosomal and smooth endoplasmic reticular membranes (Demus, 1973). Five glycoprotein bands, two at molecular weight 250 000 and one each at molecular weights 110 000, 33 000 and 28 000, are also present in tonsil plasma membranes (Lopes et al., 1973).

In a search for markers to differentiate T from B cells, Ragland, Pace and Doak (1973) compared the polypeptide patterns of chick thymus (T) cell, bursa (B) cell and circulating lymphocyte plasma membranes. Each cell type had approximately 30 polypeptide bands of which 26 were common to all three cells and 28 were shared by T and B cells. Circulating cells contained one unique protein of molecular weight 97 000, while B cells had two unique proteins of molecular weights 153 000 and 130 000. The T cells had a distinct band of molecular weight 75 000. All three cell types had the band at molecular weight 45 000–48 000 described above for tonsil cells (Lopes *et al.*, 1973).

There are several points which should be seriously considered when interpreting results obtained by SDS–PAGE (Wallach and Winzler, 1974). These involve the possibility that SDS may not entirely dissociate a protein into its subunits in the brief period of exposure to the detergent prior to electrophoresis; also that lipid–protein complexes may not be completely separated (Nelson, 1971; Katzman, 1972). The binding of SDS to glycoproteins is known to differ from that to nonglycosylated peptides, causing anomalies in molecular weight determinations (Segrest *et al.*, 1971). In addition, proteolysis of membranes by contaminating lysosomal or other enzymes during isolation or preparation of the purified membranes for PAGE would greatly alter the banding patterns. Age and storage conditions of membranes have an effect. Comparison of the polypeptide patterns of freshly prepared BRI-8 membranes with those stored at 2 °C for 7 days demonstrates a marked increase in the number of low-molecular-weight peptides (Crumpton and Snary, 1974).

Recent success in the isolation of biologically active macromolecules from lymphocyte membranes should promote advances in correlating structure with function. Certain detergents, especially nonionic Triton X-100, Brij 99, NP-40 and sodium deoxycholate dissolve membrane complexes but permit molecular interactions such as antigen–antibody reactions to occur (Crumpton and Parkhouse, 1972). These have been used, in preference to the strong dissociating agent SDS, to isolate histocompatibility antigens (*see* Section 3.3.1), enzymes, and lectin receptors from lymphocyte membranes. Ninety-five percent of the protein from pig lymphocyte membranes was solubilized with 2% deoxycholate at 23 °C (Allan and Crumpton, 1971). Fractionation of the solubilized membranes on sucrose density gradients or Sephadex G200 columns partially separated 5′-nucleotidase activity from leucine naphthylamidase activity; phospholipid was dissociated from protein. Certain enzymes such as Na^+,K^+-ATPase were inactivated by deoxycholate.

Glycoprotein fractions containing the receptor sites for Con A and LcH have been isolated from pig and human lymphocytes by affinity chromatography using lectins covalently bound to Sepharose (Allan, Auger and Crumpton, 1972; Hayman and Crumpton, 1972; Crumpton and Snary, 1974). Approximately 10 percent of the protein from solubilized membranes (pig and BRI-8 cells) adheres to the LcH-Sepharose and is specifically eluted with methyl-α-D-mannopyranoside in 1% deoxycholate. The eluate contains all the glycoproteins of the original membrane, most of the 5′-nucleotidase activity, and in the case of the BRI-8 cells, all the HL-A histocompatibility antigens (Snary *et al.*, 1974). Glycoprotein fractions eluted from Con A-Sepharose columns are 18 times as effective as whole membranes in reducing Con-A-induced transformation of lymphocytes (Allan, Auger and Crumpton, 1972). The eluates of LcH-Sepharose columns inhibit both LcH- and PHA-induced mitoses of pig lymphocytes (Hayman and Crumpton, 1972).

Glycoproteins extracted from mouse L1210 cells by lithium diiodosalicylate (Marchesi and Andrews, 1971) inhibit the binding to intact cells of wheatgerm agglutinin, LcH and Con A (Hourani, Chace and Pincus, 1973). The glycoprotein fraction, examined in 5% SDS–polyacrylamide gels, contains six of the protein bands found in whole membranes, and of these, four contain carbohydrate.

Immunoglobulins associated with the plasma membrane of lymphocytes have been demonstrated using fluorescent antibodies (Raff, Sternberg and Taylor, 1970). Believed to function as antigen receptors, these immunoglobulins have been isolated from lymphocytes of various tissues and species. ^{125}I-labeled human thymocytes incubated in tissue culture medium release ^{125}I-labeled surface proteins of molecular weight 200 000 (Marchalonis, Attwell and Cone, 1972) which co-precipitate with a purified human Waldenstrom surface macroglobulin. Reduction and alkylation of the precipitated material, followed by gel urea electrophoresis, resolves the labeled proteins into bands migrating with L chain and close to μ chain markers. M-like molecules were the only immunoglobulins detected. Similar results were obtained with murine thymocytes. Kennel and Lerner (1973) have characterized an immunoglobulin of molecular weight 265 000 associated with the plasma membrane of cultured WIL2 cells. The membrane immunoglobulin, solubilized by NP-40 extraction, was specifically precipitated with anti-human γ-globulin sera and anti-μ-chain sera. Other cells from which membrane immunoglobulins have been isolated include mouse spleen cells (Vitetta, Bauer and Uhr, 1971), mouse myeloma and human Burkitt's lymphoma cells (Bauer et al., 1971).

Identification of immunoglobulin molecules in the polypeptide pattern obtained upon SDS–polyacrylamide gel electrophoresis of solubilized, isolated membranes has been quite difficult. No H or L chains can be derived from reduction of large-molecular-weight (200 000 and above) polypeptides of calf or rabbit thymocytes (Schmidt-Ullrich et al., 1974). Schmidt-Ullrich et al. believe that an immature population of cells may be the reason for the lack of immunoglobulins. γ-Globulin was detected in one preparation of human tonsil plasma membranes and ribosomes by immunodiffusion (Demus, 1973) but was not found in a gel electrophoresis pattern of another tonsil plasma membrane preparation (Lopes et al., 1973).

Two labeled proteins of molecular weights 117 000 and 200 000 were precipitated from ^{125}I-labeled rat spleen and thymus membranes with anti-rat IgG sera (Ladoulis et al., 1974). From electrophoresis data obtained with rat myeloma IgM, the authors feel that the protein of molecular weight 117 000 may represent a half-molecule of membrane-bound IgM.

3.5   SUMMARY

Reasonably pure lymphocyte plasma membranes prepared by several different techniques have been chemically analyzed, solubilized and separated into components in the search to discover the structural bases of lymphocyte functions. Such techniques as affinity chromatography and gel electrophoresis offer an opportunity to isolate biologically active macromolecules whose structures can be readily determined. These techniques are being applied in the search for differences between tumor and normal cells and have been used in the isolation and characterization of cell surface antigens. This information will certainly be important as the more complex roles of lymphocytes are elucidated.

Differences definitely exist in the plasma membranes of T and B lymphocytes. These are reflected in the unique polypeptide bands found for bursal cells, thymus cells and circulating lymphocytes in chickens, and the altered carbohydrate composition occurring after thymectomy or bursectomy and irradiation. These structural differences may relate to the functional differences between T and B cells.

REFERENCES

ALEXANDER, P. (1974). *Cancer Res.*, **34**:2077.
ALLAN, D., AUGER, J. and CRUMPTON, M. J. (1972). *Nature, New Biol.*, **236**:23.
ALLAN, D. and CRUMPTON, M. J. (1970). *Biochem. J.*, **120**:133.
ALLAN, D. and CRUMPTON, M. J. (1971). *Biochem. J.*, **123**:967.
ALLAN, D. and CRUMPTON, M. J. (1972). *Biochim. biophys. Acta*, **274**:22.
ALPER, C. A., COLTEN, H. R., ROSEN, F. S., RABSON, A. R., MACNAB, G. W. and GEAR, J. S. S. (1972). *Lancet*, **ii**:1179.
AMINOFF, D. (1961). *Biochem. J.*, **81**:384.
AMOS, D. B. and WARD, F. E. (1975). *Physiol. Rev.*, **55**:206.
BACH, F. H. and AMOS, D. B. (1967). *Science, N.Y.*, **156**:1506.
BAENZIGER, N. L. and MAJERUS, P. W. (1974). *Meth. Enzym.*, **31**:149.
BALDWIN, R. W. (1970). *Revue fr. Étud. clin. biol.*, **15**:593.
BALDWIN, R. W. (1971). *Proc. R. Soc. Med.*, **64**:1039.
BALDWIN, R. W., BOWEN, J. G. and PRICE, M. R. (1974). *Biochim. biophys. Acta*, **367**:47.
BANJO, C., GOLD, P., FREEDMAN, S. O. and KRUPEY, J. (1972). *Nature, New Biol.*, **238**:183.
BARBER, A. J. and JAMIESON, G. A. (1971). *Biochemistry*, **10**:4711.
BASTEN, A., MILLER, J. F. A. P., SPRENT, J. and PYE, J. (1972). *J. exp. Med.*, **135**:610.
BAUER, S., VITETTA, E. S., SHERR, C. J., SCHENKEIN, I. and UHR, J. W. (1971). *J. Immun.*, **106**:1133.
BEATTIE, D. S., SLOAN, H. R. and BASFORD, R. E. (1963). *J. Cell Biol.*, **19**:309.
BEHKI, R. M. and SCHNEIDER, W. C. (1963). *Biochim. biophys. Acta*, **68**:34.
BERNOCO, D., CULLEN, S. and SCUDELLER, G. (1972). *Histocompatibility Testing*, pp. 527–537. Ed. J. DAUSSET and J. COLOMBANI. Copenhagen; Munksgaard.
BHAKDI, S., KNUFERMANN, H., SCHMIDT-ULLRICH, R., FISCHER, H. and WALLACH, D. F. H. (1974). *Biochim. biophys. Acta*, **363**:39.
BIANCO, C., PATRICK R. and NUSSENZWEIG, V. (1970). *J. exp. Med.*, **132**:702.
BOLDT, D. H., MACDERMOTT, R. F. and JOROLAN, E. P. (1975). *J. Immun.*, **114**:1532.
BOSMANN, H. B., HAGOPIAN, A. and EYLAR, E. H. (1968). *Archs Biochem. Biophys.*, **128**:51.
BRAUN, W. E., GRECLK, D. R. and MURPHY, J. J. (1972). *Transplantation*, **13**:337.
BRETSCHER, P. and COHN, M. (1970). *Science, N.Y.*, **169**:1042.
BRIDGERS, W. F. and KAUFMAN, S. (1962). *J. biol. Chem.*, **237**:526.
BRUNETTE, D. M. and TILL, J. E. (1971). *J. Membrane Biol.*, **5**:215.
BURGER, M. (1968). *Nature, Lond.*, **219**:499.
CEPPELLINI, R. (1971). *Progress in Immunology*, p. 973. Ed. B. AMOS. New York; Academic Press.
CHASE, M. W. (1953). *The Nature and Significance of the Antibody Response*, pp. 156–169. Ed. A. M. PAPPENHEIMER, JR. New York; Columbia University Press.
CLAMAN, H. N. (1972). *Prog. Allergy*, **16**:40.
CLAMAN, H. N., CHAPERON, E. A. and TRIPLETT, R. F. (1966). *J. Immun.*, **97**:828.
CODINGTON, J. F., SANFORD, B. H. and JEANLOZ, R. W. (1970). *J. natn. Cancer Inst.*, **45**:637.
CODINGTON, J. F., SANFORD, B. H. and JEANLOZ, R. W. (1972). *Biochemistry*, **11**:2559.
CODINGTON, J. F., SANFORD, B. H. and JEANLOZ, R. W. (1973). *J. natn. Cancer Inst.*, **51**:585.
COLEMAN, R. B. and FINEAN, J. B. (1966). *Biochim. biophys. Acta*, **125**:197.
COOPER, M. D. and LAWTON, A. E., III (1974). *Scient. Am.*, **231**:58.
COOPER, N. R. (1973). *Contemporary Topics in Molecular Immunology*, Vol. 2, pp. 155–183. Ed. R. A. REISFELD and W. J. MANDY. New York; Plenum Press.
COUTINHO, A., GRONOWICZ, E. and MÖLLER, G. (1974a). *Progress in Immunology*, Vol. 2, p. 167–175. Amsterdam: North-Holland.

COUTINHO, A., GRONOWICZ, E., BULLOCK, W. and MÖLLER, G. (1974b). *J. exp. Med.*, **139**:74.
CRESSWELL, P. and DAWSON, J. R. (1975). *J. Immun.*, **114**:523.
CRESSWELL, P., TURNER, M. J. and STROMINGER, J. L. (1973). *Proc. natn. Acad. Sci. U.S.A.*, **70**:1603.
CRESSWELL, P., ROBB, R. J., TURNER, M. J. and STROMINGER, J. L. (1974). *J. biol. Chem.*, **249**:2828.
CROFT, D. N. and LUBRAN, M. (1965). *Biochem. J.*, **95**:612.
CRUMPTON, M. J. and PARKHOUSE, R. M. E. (1972). *FEBS Lett.*, **22**:210.
CRUMPTON, M. J. and SNARY, D. (1974). *Contemporary Topics in Molecular Immunology*, Vol. 3, pp. 27–55. Ed. G. L. AUDA. New York; Plenum Press.
CUATRECASAS, P. (1972). *Insulin Action*, pp. 137–169. Ed. I. B. FRITZ. New York; Academic Press.
CUATRECASAS, P. (1974). *A. Rev. Biochem.*, **43**:169.
DAVEY, M. J. and ASHERSON, G. L. (1967). *Immunology*, **12**:13.
DAWSON, J. R., SILVER, J., SHEPPARD, L. B. and AMOS, D. B. (1974). *J. Immun.*, **112**:1190.
DE DUVE, C., PRESSMAN, B. C., GIANETTO, R., WATTIAUX, R. and APPLEMANS, F. (1955). *Biochem. J.*, **60**:604.
DEMUS, H. (1973). *Biochim. biophys. Acta*, **291**:93.
DENK, H., TAPPEINER, G., ECKERSTORFER, R. and HOLZNER, J. H. (1972). *Int. J. Cancer*, **10**:262.
DE PIERRE, J. W. and KARNOVSKY, M. L. (1973). *J. Cell Biol.*, **56**:275.
DE SALLE, L., MUNAKATA, N., PAULI, R. M. and STRAUSS, B. S. (1972). *Cancer Res.*, **32**:2463.
DICKINSON, J. P., CASPARY, E. A. and FIELD, E. J. (1972). *Nature, New Biol.*, **239**:181.
DIERICH, M. P. and REISFELD, R. A. (1975). *Fedn Proc. Fedn Am. Socs exp. Biol.*, **34**:4176.
DIERICH, M. P., FERRONE, S., PELLEGRINO, M. A. and REISFELD, R. A. (1974). *J. Immun.*, **113**:940.
DISCHE, Z. (1930). *Mikrochemie*, **8**:4.
DONKERSLOOT, J. A., ROBRISH, S. A. and KRICHEVSKY, M. I. (1972). *Appl. Microbiol.*, **24**:179.
DROEGE, W., STROMINGER, J. L., SINGH, P. P. and LUDERITZ, O. (1975). *Eur. J. Biochem.*, **54**:301.
DUKOR, P. and HARTMAN, K. U. (1973). *Cell. Immun.*, **7**:349.
DUKOR, P., SCHUMANN, G., GISLER, R., DIERICH, M., KONIG, W., HADDING, U. and BITTER-SUERMANN, D. (1974). *J. exp. Med.*, **139**:337.
DUTTON, R. W. (1966). *J. exp. Med.*, **123**:665.
EARL, D. C. N. and KORNER, A. (1965). *Biochem. J.*, **94**:721.
EDEN, A., BIANCO, C. and NUSSENZWEIG, V. (1973). *Cell. Immun.*, **1**:459.
EIJSVOOGEL, V. P., VAN ROOD, J. J., DU TOIT, E. D. and SCHELLEKENS, P. T. A. (1972). *Eur. J. Immun.*, **2**:413.
EIJSVOOGEL, V. P., DU BOIS, R., MELIEF, C. J. M., ZEYLEMAKER, W. P., RAAT-KONING, L. and DE GROOT-KOOY, L. (1973). *Transplantn Proc.*, **5**:1301.
EMMELOT, P., BOS, C. J., BENEDETTI, E. L. and RÜMKE, P. (1964). *Biochim. biophys. Acta*, **90**:126.
ERECINSKA, M., SIERAKOWSKA, H. and SHUGAR, D. (1969). *Eur. J. Biochem.*, **11**:465.
FELDMANN, M. and BASTEN, A. (1972a). *Eur. J. Immun.*, **2**:213.
FELDMANN, M. and BASTEN, A., (1972b). *Nature, New Biol.*, **237**:13.
FELDMANN, M. and PEPYS, M. B. (1974). *Nature, Lond.*, **249**:159.
FERBER, E., RESCH, K., WALLACH, D. F. H. and IMM, W. (1972). *Biochim. biophys. Acta*, **266**:494.
FESTENSTEIN, H., SACHS, J. A., ABBASI, K. and OLIVER, R. T. D. (1972). *Transplantn Proc.*, **4**:219.
GAVIN, J. R., III, GORDEN, P., ROTH, J., ARCHER, J. A. and BUELL, D. N. (1973). *J. biol. Chem.*, **248**:2202.
GELFAND, M. C., ELFENBEEN, G. E., FRANK, M. M. and PAUL, W. E. (1974). *J. exp. Med.*, **139**:1125.
GIBOFSKY, A. and TERASAKI, P. I. (1972). *Transplantation*, **13**:192.
GORDEN, P., LESNIAK, M. A., HENDRICKS, C. M., ROTH, J., MCGUFFIN, W. and GAVIN, J. R., III (1974). *Israel J. med. Sci.*, **10**:1239.
GÖTZE, O. and MÜLLER-EBERHARD, H. J. (1971). *J. exp. Med.*, **134**:905.
GREY, H. M., KUBO, R. T., COLON, S. M., POULIK, M. D., CRESSWELL, P., SPRINGER, T., TURNER, M. and STROMINGER, J. L. (1973). *J. exp. Med.*, **138**:1608.
GÜNTHER, E. (1973). *Transplantn Proc.*, **5**:1315.
HABEL, K. (1969). *Adv. Immun.*, **10**:229.
HADDOW, A. (1965). *Br. med. Bull.*, **21**:133.
HARRIS, J. R., PRICE, M. R. and BALDWIN, R. W. (1973). *Biochim. biophys. Acta*, **311**:600.
HAUSCHKA, T. S., WEISS, L., HALDRIDGE, B. A., CUDNEY, T. L., ZAMPFT, M. and PLANINSEK, J. A. (1971). *J. natn. Cancer Inst.*, **47**:343.
HAYMAN, M. J. and CRUMPTON, M. J. (1972). *Biochem. biophys. Res. Commun.*, **47**:923.
HERBERMAN, R. B. (1973). *Israel J. med. Sci.*, **9**:300.
HERBERMAN, R. B. and GAYLORD, C. E. (1973). *Natn. Cancer Inst. Monogr.*, **35**:1.

HILGERT, I., KANDUTSCH, A. A., CHERRY, M. and SNELL, G. D. (1969). *Transplantation*, **8**:451.
HIRATA, A. and TERASAKI, P. I. (1972). *Immunology*, **108**:1542.
HOLLENBERG, M. D. and CUATRECASAS, P. (1974). *Control of Proliferation in Animal Cells*, pp. 423–434. Ed. B. CLARKSON and R. BASERGA. New York; Cold Spring Harbor Laboratory Press.
HOLLINSHEAD, A. (1973). *Natn. Cancer Inst. Monogr.*, **37**:209.
HOLMES, E. C., MORTON, D. L. and SCHIDLOVSKY, G. (1971). *J. natn. Cancer Inst.*, **46**:693.
HOLTZMAN, E. and DOMINITZ, R. (1968). *J. Histochem. Cytochem.*, **16**:320.
HÖRTNAGL, H., WINKLER, H. and HÖRTNAGL, H. (1969). *Eur. J. Biochem.*, **10**:243.
HOURANI, B. T., CHACE, N. M. and PINCUS, J. H. (1973). *Biochim. biophys. Acta*, **328**:520.
HUBSCHER, G. and WEST, G. R. (1965). *Nature, Lond.*, **205**:799.
HUNTER, M. J. and COMMERFORD, S. L. (1961). *Biochim. biophys. Acta*, **47**:580.
JANSON, V. K. and BURGER, M. M. (1973). *Biochim. biophys. Acta*, **291**:127.
JANSON, V. K., SAKAMOTO, C. K. and BURGER, M. M. (1973). *Biochim. biophys. Acta*, **291**:136
JETT, M. and JAMIESON, G. A. (1973). *Biochem. biophys. Res. Commun.*, **55**:1225.
JONDAL, M., HOLM, C. and WIGZELL, H. (1972). *J. exp. Med.*, **136**:207.
KAPELLER, M. and DOLJANSKI, F. (1972). *Nature, New Biol.*, **235**:184.
KAPLAN, P. M., GREENMAN, R. L., GERIN, J. L., PURCELL, R. H. and ROBINSON, W. S. (1973). *J. Virol.*, **12**:995.
KATZ, D. H. and BENACERRAF, B. (1972). *Adv. Immun.*, **15**:1.
KATZMAN, R. L. (1972). *Biochim. biophys. Acta*, **266**:269.
KEIR, H. M., SMELLIE, R. M. S. and SIEBERT, G. (1962). *Nature, Lond.*, **196**:752.
KENNEL, S. J. and LERNER, R. A. (1973). *J. molec. Biol.*, **76**:485.
KIRCHNER, H. and BLAESE, R. M. (1973). *J. exp. Med.*, **138**:812.
KLEIN, G. (1966). *A. Rev. Microbiol.*, **20**:223.
KLEIN, J. (1973). *Transplantn Proc.*, **5**:11.
KLEIN, J. and SHREFFLER, D. C. (1972). *J. exp. Med.*, **135**:924.
KOERNER, J. F. and SINSHEIMER, R. L. (1957). *J. biol. Chem.*, **228**:1049.
KOLB, W. R. and GRANGER, G. A. (1968). *Proc. natn. Acad. Sci. U.S.A.*, **61**:1250.
KOLB, W. R. and GRANGER, G. A. (1970). *Cell. Immun.*, **1**:122.
KORNBERG, A. (1955). *Meth. Enzym.*, **1**:441.
KORNEL, L. (1973). *Acta endocr., Copenh.*, Suppl. **178**:1.
KORNFELD, R. and SIEMERS, C. (1974). *J. biol. Chem.*, **249**:1295.
KORNFELD, S. (1969). *Biochim. biophys. Acta*, **192**:542.
KRAEMER, P. M. (1965). *J. cell. Physiol.*, **67**:23.
KRAEMER, P. M. (1966). *J. cell. Physiol.*, **68**:85.
KRUG, U., KRUG, F. and CUATRECASAS, P. (1972). *Proc. natn. Acad. Sci. U.S.A.*, **69**:2604.
LADOULIS, C. T., MISRA, D. N., ESTES, L. W. and GILL, T. J., III (1974). *Biochim. biophys. Acta*, **356**:27.
LAWRENCE, H. S. and LANDY, M. (1969). *Mediators of Cellular Immunity*. New York; Academic Press.
LERNER, R. A. (1972). *Contemporary Topics in Immunochemistry*, Vol. 1, pp. 111–143. Ed. F. P. INMAN. New York; Plenum Press.
LERNER, R. A., MEINKE, W. and GOLDSTEIN, D. A. (1971). *Proc. natn. Acad. Sci. U.S.A.*, **68**:1212.
LESNIAK, M. A., ROTH, J., GORDEN, P. and GAVIN, J. R., III (1973). *Nature, New Biol.*, **241**:20.
LEWIS, M. G., IKONOPISOV, R. L., NAIRN, R. C., PHILLIPS, T. M., FAIRLEY, G. H., BODENHAM, D. C. and ALEXANDER, P. (1969). *Br. med J.*, **3**:547.
LIEBERMAN, R., PAUL, W. E., HUMPHREY, W. and STIMPFLING, J. H. (1972). *J. exp. Med.*, **136**:1231.
LIN, P. S., WALLACH, D. F. H. and TSAI, S. (1973). *Proc. natn. Acad. Sci. U.S.A.*, **70**:2492.
LING, N. R. (1968). *Lymphocyte Stimulation*. Amsterdam; North-Holland. New York; John Wiley.
LIS, H. and SHARON, N. (1973). *A. Rev. Biochem.*, **43**:541.
LOPES, J., NACHBAR, M., ZUCKER-FRANKLIN, D. and SILBER, R. (1973). *Blood*, **41**:131.
LURIA, S. E. (1962). *Science, N.Y.*, **136**:685.
MAIZEL, J. V., JR. (1969). *Fundamental Techniques In Virology*, pp. 334–362. Ed. K. HABEL and N. P. SALZMAN. New York; Academic Press.
MANN, D. L. (1972). *Transplantation*, **14**:398.
MANSON, L. A., HICKEY, C. A. and PALM, J. (1968). *Biological Properties of the Mammalian Surface Membrane*, pp. 93–103. Ed. L. A. MANSON. Philadelphia; Wistar Press.
MANSON, L. A. and PALM, J. (1968). *Advances in Transplantation*, pp. 301–304. Ed. J. DAUSSET. Copenhagen; Munksgaard.

MARCHALONIS, J. J., ATWELL, J. L and CONE, R. E. (1972). *Nature, New Biol.*, **235**:240.
MARCHESI, V. T. and ANDREWS, E. P. (1971). *Science, N.Y.*, **174**:1247.
MARIQUE, D. and HILDEBRAND, J. (1973). *Cancer Res.*, **33**:2761.
MAYR, W. R., BERNOCO, D., DEMARCHI, M. and CEPPELLINI, R. (1973). *Transplantn Proc.*, **5**:1581.
MCCOLLESTER, D. L. (1970). *Cancer Res.*, **30**:2832.
MCDEVITT, H. D., DEAK, B. D., SHREFFLER, D. C., KLEIN, J., STIMPFLING, J. H. and SNELL, G. D. (1972). *J. exp. Med.*, **135**:1259.
MEO, T., VIVES, J., MIGGIANO, V. and SHREFFLER, D. (1973). *Transplantn Proc.*, **5**:377.
MICHLMAYER, G. and HUBER, H. (1970). *J. Immun.*, **105**:670.
MILLER, G. W. and NUSSENZWEIG, V. (1974). *J. Immun.*, **113**:464.
MILLER, G. W., SALUK, P. H. and NUSSENZWEIG, V. (1973). *J. exp. Med.*, **138**:495.
MILLER, J. F. (1971). *Morphological and Functional Aspects of Immunity*, pp. 93–102. Ed. K. LINDAHL-KIESSLING, G. ALM and M. G. HANNA, JR. New York; Plenum Press.
MILLER, J. F. (1972). *Int. Rev. Cytol.*, **33**:77.
MILLER, J. F. and MITCHELL, G. F. (1969). *Transplantn Rev.*, **1**:3.
MILLER, J. F., BASTEN, A., SPRENT, J. and CHEERS, C. (1971). *Cell. Immun.*, **2**:469.
MISRA, D. N., GILL, T. J., III and ESTES, L. W. (1974). *Biochim. biophys. Acta*, **352**:455.
MITCHELL, R. H. and HAWTHORNE, J. N. (1965). *Biochem. biophys. Res. Commun.*, **21**:333.
MÖLLER, G. and COUTINHO, A. (1975). *J. exp. Med.*, **141**:647.
MÖLLER, G. and MICHAEL, G. (1971). *Cell. Immun.*, **2**:309.
MOLNAR, J., MARKOVIC, G., CHAO, H. and MOLNAR, Z. (1969). *Archs Biochem. Biophys.*, **134**:524.
MUEWISSEN, H. J., STUTMAN, O. and GOOD, R. A. (1969). *Semin. Hemat.*, **6**:67.
MUNRO, H. N. and FLECK, A. (1966). *Meth. biochem. Analysis*, **14**:158.
MURAMATSU, T., NATHENSON, S. G., BOYSE, E. A. and OLD, L. J. (1973). *J. exp. Med.*, **137**:1256.
NAKAMURO, K., TANAGAKI, N., KREITER, V. P., MOORE, G. E. and PRESSMAN, D. (1973). *Proc. natn. Acad. Sci. U.S.A.*, **70**:2863.
NAOR, D., BENTWICH, Z. and CIVIDALLI, G. (1969). *Aust. J. exp. Biol. med. Sci.*, **47**:759.
NATHENSON, S. G. (1970). *A. Rev. Genet.*, **4**:69.
NATHENSON, S. G. and CULLEN, S. E. (1974). *Biochim. biophys. Acta*, **344**:1.
NATHENSON, S. G. and MURAMATSU, T. (1971). *Glycoproteins of Red Cells and Plasma*, pp. 254–262. Ed. G. A. JAMIESON and T. J. GREENWALT. Philadelphia; J. B. Lippincott.
NEAUPORT-SAUTES, C., SILVESTRE, D., NICCOLAE, M. G., KOURILSKY, F. M. and LEVY, J. P. (1972). *Immunology*, **22**:833.
NEAUPORT-SAUTES, C., LILLY, F., SILVESTRE, D. and KOURILSKY, F. M. (1973a). *J. exp. Med.*, **137**:511.
NEAUPORT-SAUTES, C., SILVESTRE, D., LILLY, F. and KOURILSKY, F. M. (1973b). *Transplantn Proc.*, **5**:443.
NEAUPORT-SAUTES, C., BISMUTH, A., KOURILSKY, F. M. and MANUEL, Y. (1974). *J. exp. Med.*, **139**:957.
NELSON, C. A. (1971). *J. biol. Chem.*, **246**:3895.
NELSON, D. S. (1963). *Adv. Immun.*, **3**:131.
NEVILLE, D. M. (1960). *J. biophys. biochem. Cytol.*, **8**:413.
NEVILLE, D. M. and GLOSSMAN, H. (1971). *J. biol Chem.*, **246**:6335.
NICOLSON, G. L. (1974). *Int. Rev. Cytol.*, **39**:89.
NORDLIE, R. C. and ARION, W. J. (1966). *Meth. Enzym.*, **9**:619.
NORTHRUP, J. H. (1926). *J. gen. Physiol.*, **9**:497.
NOVOGRADSKY, A. and KATCHALSKI, E. (1972). *Proc. natn. Acad. Sci. U.S.A.*, **69**:3207.
NUSSENZWEIG, V. (1974). *Adv. Immun.*, **19**:217.
OLD, L. J., STOCKERT, E., BOYSE, E. A. and KIM, J. H. (1968). *J. exp. Med.*, **127**:523.
OWEN, J. J., COOPER, M. D. and RAFF, M. C. (1974). *Nature, Lond.*, **249**:361.
PANCAKE, S. and NATHENSON, S. G. (1973). *J. Immun.*, **11**:1086.
PASSMORE, H. C. and SHREFFLER, D. C. (1968). *Genetics, N.Y.*, **60**:210.
PASTERNAK, C. (1969). *Cancer Res.*, **12**:1.
PEPYS, M. B. (1972). *Nature, New Biol.*, **237**:157.
PEPYS, M. B. (1974). *J. exp. Med.*, **140**:126.
PERLMANN, P. and HOLM, G. (1969). *Adv. Immun.*, **11**:117.
PERNIS, B., FERRARINI, M., FORNI, L. and AMANTI, L. (1971). *Progress in Immunology*, pp. 95–106. Ed. B. AMOS. New York; Academic Press.
PETERSON, P. A., RUSK, L. and LINDBLOM, J. B. (1974). *Proc. natn. Acad. Sci. U.S.A.*, **71**:35.
PREHN, R. T. and MAIN, J. M. (1957). *J. natn. Cancer Inst.*, **18**:769.
PULVERTAFT, R. J. V. (1965). *J. clin. Path.*, **18**:261.

RAFF, M. C., STERNBERG, M. and TAYLOR, R. B. (1970). *Nature, Lond.*, **225**:553.

RAGLAND, W. L., PACE, J. L. and DOAK, R. L. (1973). *Biochem. biophys. Res. Commun.*, **50**:118.

REISFELD, R. A., PELLEGRINO, M. A. and KAHAN, B. D. (1971). *Science, N.Y.*, **172**:1134.

ROELANTS, G. (1972). *Curr. Topics Microbiol. Immun.*, **59**:135.

ROTH, J. (1973). *Metabolism*, **22**:1059.

RUBIN, M. S., SWISLOCKI, N. I. and SONENBERG, M. (1973). *Archs Biochem. Biophys.*, **157**:243.

RUDDY, S. (1974). *Transplantn Proc.*, **6**:1.

RUTISHAUSER, U. and EDELMAN, G. M. (1972). *Proc. natn. Acad. Sci. U.S.A.*, **69**:3774.

SAKIYAMA, H. and BURGE, B. W. (1972). *Biochemistry*, **11**:1366.

SANDERSON, A. R. (1969). *Nature, Lond.*, **220**:192.

SANFORD, B. H., CODINGTON, J. F., JEANLOZ, R. W. and PALMER, P. D. (1973). *J. Immun.*, **110**:1233.

SCHMIDT-ULLRICH, R., FERBER, E., KNÜFERMANN, H., FISCHER, H. and WALLACH, D. F. H. (1974). *Biochim. biophys. Acta*, **332**:175.

SCHNEIDER, W. C. (1968). *Handbook of Biochemistry. Selected Data for Molecular Biology*, p. K-3. Ed. H. A. SOBER. Cleveland, Ohio; Chemical Rubber Co.

SCHWARTZ, B. D., KATO, K., CULLEN, S. and NATHENSON, S. G. (1973). *Biochemistry*, **12**:2157.

SCHWARTZ, B. D. and NATHENSON, S. G. (1971). *J. Immun.*, **107**:1363.

SCHWARTZ, H. J. and WILSON, F. (1971). *Am. J. Path.*, **64**:295.

SEGREST, J. P., JACKSON, R. L., ANDREWS, E. P. and MARCHESI, V. T. (1971). *Biochem. biophys. Res. Commun.*, **44**:390.

SELL, S. (1972). *Transplantn Rev.*, **5**:19.

SELLINGER, O. Z., BEAUFAY, H., JAQUES, P., DOYEN, A. and DE DUVE, C. (1960). *Biochem. J.*, **74**:450.

SHEARER, G. M. and CUDKOWICZ, G. (1969). *J. exp. Med.*, **130**:1243.

SHIMADA, A. and NATHENSON, S. G. (1971). *J. Immun.*, **107**:1197.

SHNITKA, T. K. and SELIGMAN, A. M. (1971). *A. Rev. Biochem.*, **40**:375.

SJORGREN, H. O. (1965). *Prog. exp. Tumor Res.*, **6**:289.

SLAYTER, H. S. and CODINGTON, J. F. (1973). *J. biol. Chem.*, **248**:3405.

SMITH, L. (1955). *Meth. Enzym.*, **2**:735.

SMITH, R. T. (1968a). *New Engl. J. Med.*, **278**:1207.

SMITH, R. T. (1968b). *New Engl. J. Med.*, **278**:1268.

SMITH, R. T. (1968c). *New Engl. J. Med.*, **278**:1326.

SNARY, D., GOODFELLOW, P., HAYMAN, M. J., BODMER, W. F. and CRUMPTON, M. J. (1974). *Nature, Lond.*, **247**:457.

SPRENT, J. and MILLER, J. F. (1971). *Nature, New Biol.*, **234**:195.

SPRINGER, T. A., STROMINGER, J. L. and MANN, D. (1974). *Proc. natn. Acad. Sci. U.S.A.*, **71**:1539.

STERN, H., ALLFREY, V., MIRSKY, A. E. and SAETREN, H. (1952). *J. gen. Physiol.*, **35**:559.

SVEJGAARD, A., STAUB-NIELSON, L. and RYDER, L. P. (1972). *Histocompatibility Testing*, pp. 465–473. Ed. J. DAUSSET and J. COLOMBANI. Copenhagen; Munksgaard.

TAYLOR, R. B., DUFFUS, W. P. H., RAFF, M. C. and DE PETRIS, S. (1971). *Nature, New Biol.*, **233**:225.

THORSBY, E., HIRSCHBERG, H. and HELGESEN, A. (1973). *Transplantn Proc.*, **5**:1523.

TOMS, G. C. and WESTERN, A. (1971). *Chemotaxonomy of the Leguminosae*, pp. 367–462. Ed. J. B. HARBORNE, D. BOULTER and B. L. TURNER. New York; Academic Press.

VAN BLITTERSWIJK, W. J., EMMELOT, P. and FELTKAMP, C. A. (1973). *Biochim. biophys. Acta*, **298**:577.

VAN BLITTERSWIJK, W. J., EMMELOT, P., HILGERS, J., KAMLAG, D., NUSSE, R. and FELTKAMP, C. A. (1975). *Cancer Res.*, **35**:2743.

VITETTA, E. S., BAUER, S. and UHR, J. W. (1971). *J. exp. Med.*, **134**:242.

VITETTA, E. S., UHR, J. W. and BOYSE, E. A. (1975). *J. Immun.*, **114**:252.

VRBA, R., ALPERT, E., ISSLBACHER, K. J. and JEANLOZ, R. W. (1975). *Fedn Proc. Fedn Am. Socs exp. Biol.*, **34**:1871.

WALLACH, D. F. H. and KAMAT, V. B. (1966). *Meth. Enzym.*, **8**:164.

WALLACH, D. F. H. and LIN, P. S. (1973). *Biochim. biophys. Acta*, **300**:211.

WALLACH, D. F. H. and WINZLER, R. J. (1974). *Evolving Strategies and Tactics in Membrane Research*. New York; Springer-Verlag.

WARLEY, A. and COOK, G. M. W. (1973). *Biochim. biophys. Acta*, **323**:55.

WARREN, L. (1959). *J. biol. Chem.*, **234**:1971.

WARREN, L. and GLICK, M. C. (1971). *Biomembranes*, Vol. 1, pp. 257–288. Ed. L. A. MANSON. New York; Plenum Press.

WARREN, L., GLICK, M. C. and NASS, M. K. (1966). *J. cell. Physiol.*, **68**:269.

WATSON, J., FRENKNER, E. and COHN, M. (1973). *J. exp. Med.*, **138**:699.

WEINER, N. (1970). *A. Rev. Pharmac.*, **10**:273.

WEINHOLD, P. A. and RETHY, V. B. (1969). *Biochem. Pharmac.*, **18**:677.
WIDMER, M. B., PECK, A. B. and BACH, F. H. (1973). *Transplantn Proc.*, **5**:1501.
WIGZELL, H. and ANDERSSON, B. (1969). *J. exp. Med.*, **129**:23.
WIGZELL, H., SUNDQUIST, K. G. and YESHIDU, T. O. (1972). *Scand. J. Immun.*, **1**:75.
WILLIAMS, C. H., GIBBS, R. H. and KAMIN, H. (1959). *Biochim. biophys. Acta*, **32**:568.
WILLIAMS, T. W. and GRANGER, G. A. (1973). *Cell. Immun.*, **6**:171.
WOOD, J. G. (1967). *Am. J. Anat.*, **121**:671.
WRIGHT, B. M., EDWARDS, A. J. and JONES, V. E. (1974). *J. Immun. Meth.*, **4**:281.
YUNIS, E. J. and AMOS, D. B. (1971). *Proc. natn. Acad. Sci. U.S.A.*, **68**:3031.
ZUCKER-FRANKLIN, D. (1969). *Semin. Hemat. VI*, **4**:15.

4

Normal and transformed cells *in vitro*

C. A. Pasternak
Department of Biochemistry, University of Oxford *

4.1   INTRODUCTION

The development of culture techniques over the past few decades has enabled investigators to study different cell lines under identical conditions. Since malignant cells *in vivo* are generally exposed to the same environmental milieu as their normal counterparts, a comparison of normal and malignant cells *in vitro* should provide useful information concerning the nature of cancer. In order to try to pinpoint the basic lesion, cells that are as similar as possible in other respects are obviously desirable. Transformed cells have proved to be just that.

The key demonstration was that viral infection of cultured cells can lead to a permanent change such that the extent of growth *in vitro* is no longer subject to the same restrictions as before (Enders, 1965; Dulbecco, 1969; *note* Freeman and Huebner, 1973). As a result, transformed cells do not cease growing when they reach confluency, but pile upon, or rather below (Guelstein *et al.*, 1973), each other in layers.

Cells transformed with carcinogenic chemicals behave in a similar manner with respect to growth (Berwald and Sachs, 1963) and have also been investigated, together with spontaneously transformed cells selected for the property of unrestricted growth. The use of transformed cells to study malignancy is further justified by the fact that many transformed lines actually give rise to tumours when injected into an appropriate host (Fenner, 1968). The genetic implications of viral (Levine and Burger, 1972) or chemical transformation fall outside the scope of this review.

Suggestions that the cell periphery plays a role in the altered growth of transformed cells are the result of an accumulation of data implicating the

* Present address: Department of Biochemistry, St George's Hospital Medical School, University of London.

89

cell surface in the control of growth *in vitro* (Pardee, 1971). Indeed, a major difference between diverse cell types is often displayed at their surface. Not only, for example, is (a) the morphology of the brush border of epithelia, (b) the insulin sensitivity of muscle, or (c) the presence of antigen receptors on lymphoid cells, typical of that particular cell type but each accounts for much of the specialized function. The internal membranes, on the other hand, are more uniform. This is not surprising, for the major functions of mitochondria or endoplasmic reticulum are, after all, the same in epithelia, muscle or lymph.

The aim of this review, then, is to compare the plasma membrane of normal and transformed cells *in vitro*, and to assess the extent to which differences can account for the behaviour of transformed cells.

Two reservations with respect to malignancy must be borne in mind. First, most 'normal' cell lines grown in culture are not really normal, since they are selected for relatively rapid growth in rather artificial media. In fact several normal cell lines are potentially malignant in the proper host. In many cases then, one is comparing *degrees* of malignancy. Secondly, the properties of cells in culture do not necessarily give an indication of their behaviour *in vivo*. Although unrestricted growth *in vitro* is generally reflected in tumorigenicity *in vivo*, this correlation does not always hold (Koprowski *et al.*, 1966; Ponten, 1971).

The following discussion is divided for ease of presentation into morphological, functional and compositional differences but, of course, morphology, function and composition are intimately related; moreover, apparent compositional differences have proved in certain instances to be due merely to changes in topographical localization (Section 4.3.2). Although an attempt will be made to mention most of the surface properties that have been studied, the review is not a comprehensive one, but rather reflects particular areas of interest. In any case, other chapters in this series cover specific topics in greater detail. Of particular relevance is Vol. 5, Chapter 7, 'The Role of the Cell Membrane during the Resumption of Growth *In Vitro*', since many of the changes resulting from transformation *in vitro* are also seen when confluent normal cells are stimulated to further growth by serum or other agents. That being so, the reader might conclude that any differences between normal and transformed cells should disappear if both are in the exponential phase of growth. Up to a point this is true, and many of the differences described in the literature are between *static, normal* and *growing, transformed* cells. This is often unintentional, for the following reason. When normal cells reach confluency, they stop growing. In the case of transformed cells, growth may not cease in the same way. Instead, cell death begins to accompany further cell division, so that cell number remains constant despite the presence of growing cells (Stoker, 1972). Obviously it is important to specify the conditions as it is comparisons between normal, growing and transformed, growing cells that are the most pertinent to this discussion.

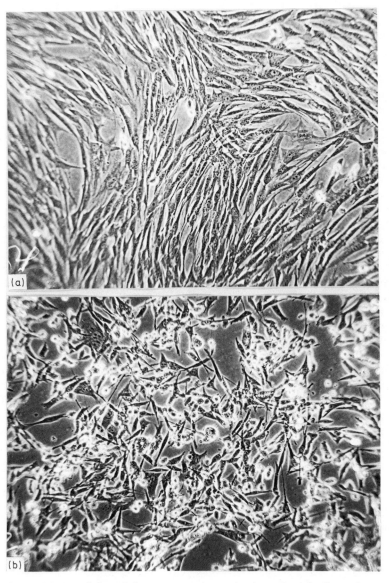

Figure 4.1   Morphological characteristics of normal and transformed cells in culture. (a) BHK C13 cells; (b) polyoma-virus-transformed BHK C13 cells. × 170. (Courtesy of Dr M. G. P. Stoker, Imperial Cancer Research Fund, London)

4.2   DIFFERENCES IN MORPHOLOGY

4.2.1   **Cell shape**

The most obvious difference between normal and transformed cells growing in culture is their appearance under the microscope (*Figure 4.1*). Many cell lines are fibroblast-like and retain the typical elongated shape and ability to grow in rows. This is generally lost on transformation, which often results in more rounded cells with random patterns of growth.

The reason for the morphological difference is not clear. One suggestion is that the system of microtubules and microfilaments is involved. This is based on the demonstration that demecolcine and vinblastine, which inhibit the proper aggregation of microtubular subunits, rapidly cause fibroblastic-like cells to lose their characteristic appearance, and to become rounded or 'epithelial-like' (Hsie and Puck, 1971). Microtubules have certainly been implicated in determining cell shape (Margulis, 1973), although how this is achieved is not known.

One of the major roles of microtubules is to form the mitotic spindle, by which chromosome pairs are separated during cell division (Bajer and Mole-Bajer, 1972). In addition, microtubules participate in secretion, pino-cytosis and phagocytosis and other more specialized functions concerned with motion. Whether the microtubular elements involved are the same in every case is not known. But if spindle formation and the maintenance of fibroblastic shape each require a large number of microtubules, the two processes may not be able to occur simultaneously. Certainly mitosis is the very time at which fibroblastic and other cells growing in monolayer become more rounded and temporarily loosen their contact with the substratum. Microtubules may therefore be part of the link between surface properties and cell division. Their availability is unlikely to be the actual trigger for the start of a new cell cycle (Fox, Sheppard and Burger, 1971) since DNA synthesis and other pre-mitotic events take place before assembly of the mitotic spindle. Their correct reassembly at the cell surface after mitosis, on the other hand, might be important for fibroblastic-like cells (*Figure 4.2*).

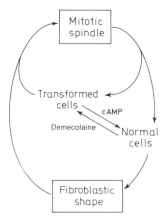

Figure 4.2 Involvement of micro-tubules in cellular functions

This would be in keeping with the fact that a defect in transformed cells is a failure to 'opt out' of the cell cycle and to enter G_0 (Stoker, 1972; Pasternak, 1974).

The rounding of fibroblastic cells is also brought about by cytochalasin B (Puck, Waldren and Hsie, 1972), which affects microfilaments (Wessels *et al.*, 1971) and other contractile processes (Estensen, Rosenberg and Sheridan, 1971). A possible site of action is on actin-like proteins found in the surface membrane (Pollard and Korn, 1973). At present, the exact interplay between the microtubular and microfilamentous systems is not clear. Morphological alterations of the type described—and the reverse may be achieved by the addition of dibutyryl cyclic AMP (diBcAMP) and testosterone (Puck, Waldren and Hsie, 1972)—have to be reconciled with the fact that trans-formation is heritable. Whatever the genetic change (Levine and Burger, 1972), its effect on the organization of the microtubular–microfilamentous system can presumably be overridden by diBcAMP and testosterone, or mimicked by demecolcine and cytochalasin B (*Figure 4.2*).

The difference between normal and transformed cells is not always as obvious as suggested in *Figure 4.1*. Many normal lines are of epithelial origin, and more subtle alterations, such as in the number of microvilli (O'Neill and Follett, 1970) or other processes affecting cell contact (Edwards, Campbell and Williams, 1971) or attachment to the substrate (Sanford *et al.*, 1967) may be involved.

4.2.2   Cell contacts

Another feature that may be lost by transformed cells is the existence of membrane junctions. These allow the passage of inorganic and other ions between adjacent cells and may therefore be revealed by establishment of an electric current or a movement of dye. In mixed populations of normal and malignant cells, junctional activity has in some instances been observed only between normal cells and not between a normal and a transformed cell (Azarnia and Loewenstein, 1971; Loewenstein, 1973). Lack of intercellular communication through loss of junctions is certainly an attractive hypothesis. Unfortunately, it does not appear to be true of transformed cells in general (Bennett, 1973) and, in any case, transformed cells *do* communicate with neighbouring normal cells, provided the latter are in an active state of growth (Stoker, 1972).

4.3   DIFFERENCES IN BIOCHEMICAL FUNCTIONS

4.3.1   Enzymes

One activity that presumably differs is that responsible for the organization of the microtubular–microfilamentous system (Section 4.2), but this has not yet been identified. Much attention has been paid to changes in transport of metabolites such as amino acids and sugars, for which an enzyme system of one sort or another is clearly required. Discussion of that topic will be followed by a consideration of adenyl cyclase and some other enzymes, the function of which is not always clear. Finally, the turnover of membrane components will be reviewed, since it has been suggested that this is different in normal and transformed cells (Warren, 1969).

4.3.1.1   TRANSPORT

Based partly on the high metabolic activity of many tumours, it has been suggested that an increased uptake of nutrients might be a characteristic feature of malignant cells. Certainly the intracellular concentration of some amino acids is greatly increased in rapidly growing cells (Johnstone and Scholefield, 1965) and a consequential 'swamping' of potential feedback mechanisms has been offered as an explanation for the insensitivity of many tumours to external stimuli (Pitot, 1963). However, the growth rate and metabolic activity of normal and transformed cells *in vitro* are rather similar —it is the *extent* of growth that is different—and it is therefore not surprising that little difference in transport was observed initially (Eagle, Piez and Levy, 1961). More recently, however, the suggestion of altered transport has been revived (Pardee, 1964), and an increased uptake has been reported for certain amino acids in virally transformed mouse cells (Foster and Pardee, 1969) and for sugars in virally transformed rat (Hatanaka, Huebner and Gilden, 1969), mouse (Hatanaka, Augl and Gilden, 1970; Isselbacher, 1972), chick (Hatanaka and Hanafusa, 1970; Martin *et al.*, 1971) and hamster (Isselbacher, 1972; Kalckar *et al.*, 1973) cells. While Foster and Pardee (1969) and Isselbacher (1972) found an altered V_{max} for the process, Hatanaka and Hanafusa (1970) reported an alteration in K_m. The methodology employed for measuring uptake, the state of the cells (whether growing or confluent) and other factors are probably responsible for some of these differences. Even when uptake is the same, there may be a differential susceptibility to inhibition by lectin binding (Inbar, Ben-Bassat and Sachs, 1971a) (*see* Section 4.3.2.2); this difference has not been confirmed in other studies (Isselbacher, 1972) though, and is more likely to reflect a general change in the fluidity of the membrane (Section 4.3.2.2) than in the intrinsic properties of the transport system.

What, then, is the relevance of an increase in transport? The observation (reviewed by Knox and Pasternak in Vol. 5 of this series) that transport of metabolites by confluent normal cells is rapidly increased after addition of serum implies that transport changes may have more to do with the physiological state of the cells than with transformation *per se*. Certainly insulin, which has clear-cut transport-stimulating properties, triggers growth of confluent normal cells. On the other hand, altered transport in virally transformed cells *is* dependent on the viral genome in so far as temperature-sensitive mutants do not show increased transport at the non-permissive temperature (Martin *et al.*, 1971); moreover, transformed cells show a decrease in nutritional requirements whether growing or static (Stoker, 1972). It is therefore likely that transport changes are the consequence of certain surface alterations in transformed cells, as a result of which the intracellular concentration of key metabolites alters. This may, in some situations, lead directly to an increased rate of growth and in others may itself alter the surface properties so as to affect the extent of growth. In fact

Holley (1972) has re-emphasized Pardee's (1964) suggestion and has postulated that all malignant cells have in common a defective transport system of one sort or another.

In so far as the transport of certain cations is dependent on plasma membrane ATPases, measurements of Na^+,K^+-ATPase are relevant. No consistent changes have emerged (*see* Section 4.3.1.2). Of course the enzyme is involved only with *active* transport of cations, and passive changes due to increased 'leakiness' (Wallach, 1973) might modify intracellular concentration. Although most reports on 'leakiness' have dealt with the loss of degradative enzymes and their relation to the invasiveness of malignant cells, it may be noted that a failure to maintain permeability accompanies modification of the cell surface by paramyxovirus (Pasternak and Micklem, 1973; 1974a, b).

4.3.1.2 ADENYL CYCLASE AND OTHER ENZYMES

The demonstration that cAMP or its dibutyryl derivative is able to reverse the morphology and several other properties of transformed cells (Bürk, 1968; Hsie and Puck, 1971; Johnson, Friedman and Pastan, 1971) has led to extensive research on the possible role of cAMP in controlling growth. Bürk (1968) originally suggested that the level of adenyl cyclase, the bulk of which is located in plasma membrane, is reduced in virally transformed cells compared with their normal counterparts (Heidrick and Ryan, 1971; Otten, Johnson and Pastan, 1971; Sheppard, 1972). But other studies on the activity of adenyl cyclase in a variety of normal and transformed cells (Makman, 1971; Peery, Johnson and Pastan, 1971) have failed to confirm a consistent decrease in the enzyme level, despite the fact that temperature-sensitive mutants show cAMP levels to be controlled by viral genes (Pastan *et al.*, 1974). Any changes that are seen are probably due more to the physiological state of the cells and to the concentration of hormones in the growth medium (Schorr *et al.*, 1971). In other words, the intracellular concentration of cAMP is regulated by factors such as phosphodiesterase, cell leakage or cAMP-binding protein (Daniel, Litwack and Tomkins, 1973) rather than by adenyl cyclase (Burger, 1973). It is the influence of the cell surface on these factors that is important in transformed cells.

Sheinin and Onodera (1972) have measured Na^+,K^+-ATPase in 3T3 cells and virally transformed derivatives. No significant difference was detected, the values being in the same range as those of other normal or transformed cells in culture. Nor does the activity of the enzyme show any consistent change in the liver of rats fed with carcinogens (Emmelot and Bos, 1969a) or in various mouse hepatomas compared with normal liver (Emmelot and Bos, 1969b). The reported difference in Na^+,K^+-ATPase between two cell lines, one of which is tumorigenic and is not contact-inhibited (Lelievre, Prigent and Paraf, 1971), appears to be related to the position of the cells in the growth cycle (cf. Pasternak and Graham, 1973).

Several other enzymes have been examined, but no obvious trends emerge (e.g. Graham, 1976). Lelievre, Prigent and Paraf (1971), for example, found a hundredfold increase in 5′-nucleotidase in the tumorigenic, non-contact-inhibited cell line referred to above; wheras of two tumorigenic mastocytoma cell lines, one (P815Y) has <1 percent of the 5′-nucleotidase activity of the other (HTC), without any obvious differences in growth characteristics (A. M. H. Warmsley, M. C. B. Sumner and C. A. Pasternak, unpublished observations). Whatever the significance of some of these plasma membrane enzymes (De Pierre and Karnovsky, 1973), it does not appear to lie with the crucial difference between normal and transformed cells.

4.3.1.3 MEMBRANE TURNOVER

In the case of membrane components, turnover, i.e. the synthesis and degradation of a particular molecular species, is superimposed upon insertion and removal of intact molecules. The latter process is the predominant one in surface membranes, since neither phospholipids, proteins, glycolipids nor glycoproteins can be synthesized *de novo*, or entirely degraded, by purified plasma membrane. However, partial synthesis and degradation of phospholipids is possible and in the case of protein- and carbohydrate-containing molecules, the reports of endogenous proteolytic (Burger, 1973) and glycosyltransferase (Bosmann, 1972a) activities may be significant.

To what extent, then, are the components—or portions thereof—of the surface membrane renewed in normal and transformed cells? Since the components do not appear to be covalently linked to each other, there is no reason why their turnover should be linked either, and this has proved to be the case. Thus phospholipids in general turn over faster than other components (Pasternak, 1973a), despite the fact that during normal assembly of plasma membrane in growing cells, protein and phospholipids are inserted in concert (Graham *et al.*, 1973).

The turnover of surface phospholipids is difficult to assess, since isolation of plasma membrane is generally a prerequisite. In studies on whole cells, little difference has been detected between normal and transformed fibroblasts (Cunningham, 1972; Kulas *et al.*, 1972) in the exponential phase of growth. Stimulation by serum of phospholipid turnover in normal (Pasternak, 1972) but not in transformed (Cunningham, 1972) cells at confluency, may be a consequence of other changes at the surface membrane (P. Knox and C. A. Pasternak, Vol. 5 of this series). Small changes in turnover of surface phospholipids, of course, would be masked by the general background of total cellular turnover. A particular reason for examining phospholipid turnover (Pasternak, 1973b) is that the fluidity of surface membranes may be greater in transformed than in normal cells (Section 4.3.2.2), and since fluidity is dependent partly on the nature of the apolar side-chains of phospholipids, enzymes which modify the acyl side-chains may be an important factor. In the studies just mentioned, the *polar group* of the molecules was labelled (by [^{32}P]phosphate, [^{3}H]choline or [^{3}H]inositol) and this is probably a measure of complete degradation and resynthesis, involving the removal and insertion of intact molecules; turnover of the acyl side-chains is not only much faster and occurs *in situ* (Ferber, 1971), it also allows chemical modification of the acyl side-chains (Hill and Lands, 1970; Numa and Yamashita, 1974). It is therefore pertinent to note that no difference in the turnover of acyl side-chains from isolated plasma membrane of normal and transformed cells has in fact been detected (Micklem *et al.*, 1976).

The turnover of glycoproteins and glycolipids is easier to assess than that of phospholipids since they are located predominantly at the cell surface, and the same is true of specific proteins such as surface antigens. On the other hand, the very specificity of such components limits any generalizations that may be attempted, as illustrated by the difference in turnover of immunoglobulin and histocompatibility antigens on lymphocytes (Uhr and Vitetta, 1973).

As a result of studies with isotopic tracers, Warren (1969) has suggested that protein and glycoprotein do not turn over at all in growing cells; that is, the rate of synthesis is just that required for cell proliferation. In the case of static cells, the synthetic rate is maintained but is matched by an equivalent rate of degradation. In transformed cells, however, Warren found the 'rate of turnover' to be increased, indicating that excess synthesis and degradation had been initiated. The demonstration that transformed cells contain increased microsomal (Bosmann, Hagopian and Eylar, 1968; Bosmann, 1970) and surface membrane (Bosmann, 1972a) glycosyltransferase (synthetic) enzymes, as well as increased glycosidase and protease (Bosmann, 1969, 1972b) (degradative) enzymes, seems to support Warren's (1969) suggestion.

Increased degradation by glycosidases, however, does not appear to be a general property of transformed cells, since it has not been observed for hamster, as opposed to mouse, cells (Sela *et al.*, 1970). Moreover, an increased level of glycosyltransferase does not necessarily indicate a faster rate of *turnover*, but may reflect an increase in the *amount* of product formed (that is, increased synthesis is not matched by increased degradation). This is the interpretation put by Warren, Fuhrer and Buck (1972) on their finding of an elevated sialyltransferase level in transformed BHK cells; such cells contain more of a particular sialopeptide than do normal cells (Warren, Critchley and Macpherson, 1972). In the same way, Cumar *et al.* (1970) and Kijimoto and Hakomori (1971) correlated *decreased* levels of specific glycosyltransferases with *decreased* levels of certain higher gangliosides and neutral glycolipids respectively (Section 4.4.3). Interpretation of an altered enzyme activity as altered turnover is therefore unjustified unless the component in question can be shown not to change in amount.

Another suggested alteration is in the *type* of glycosyltransferase involved (Roth and White, 1972); transformed cells appear to have lost the ability to *trans*-glycosylate adjacent cells [postulated as a basis for cellular adhesion (Roseman, 1970)] and to *cis*-glycosylate the same cell. However, cellular adhesion is not increased by transformation in every case (*see* J. G. Edwards, Vol. 4, Chapter 3 of this series).

With regard to protein turnover, there are few data other than those of Warren (1969). These imply (a) that in the case of normal cells, turnover occurs in static but not in growing cells, and (b) that in the case of transformed cells, turnover occurs during growth as well. According to Turner and Burger (1973), the first point is borne out by an experiment (Baker and Humphreys, 1972) designed to explore the finding that proteolytic agents cause normal cells to become as agglutinable as transformed cells (Section 4.3.2). It was found that the agglutinability of static, but not of growing, chick fibroblasts is sensitive to inhibition of protein synthesis by cycloheximide. The result was interpreted (Baker and Humphreys, 1972) as indicating a lack of turnover of an agglutination-preventing protein (i.e. the protein that is degraded by proteolysis) in growing, but not in static, cells. However, this is not really the same as point (a) above, since it implies lack of *synthesis* (and not just lack of *degradation*) during growth. In fact the results suggest that synthesis is initiated only at confluency. This may be true of the agglutination-inhibiting protein. It is not true of the bulk of the surface coat (Sections 4.3.2.1 and 4.4.3) of normal or transformed cells, the synthesis of which *is* sensitive to inhibition by cycloheximide during growth (Mallucci, Poste and Wells, 1972). It is obviously important to distinguish between the behaviour of specific proteins revealed by biological assay, and 'bulk' proteins revealed by isotopic labelling. Furthermore, the possibility that turnover of some proteins and other components actually represents secretion or other form of release into the culture medium (Thomas, 1968; Kraemer and Tobey, 1972; Uhr and Vitetta, 1973) has to be borne in mind.

In summary, it is possible that certain surface components turn over faster in transformed than in normal cells, and this may contribute to a greater fluidity at the cell surface, but generalizations have to be made with caution in view of the fact that each component of the surface membrane probably behaves in a unique manner. That the cell surface is modified by altered enzymatic activity in transformed cells is not in dispute. It is the relation between the relevant anabolic and catabolic processes that is unclear, and it is this that determines the extent of turnover.

4.3.2   Receptors

In this discussion, receptors will be taken to include (a) antigens, that is, molecules capable of eliciting an immune response and of combining with the specific antibody so formed, and (b) lectin-binding sites, that is, molecules which happen to possess the correct configuration for combining with certain agglutinins of plant or other origin. In each case, agglutination can result from binding. Cell surface antigens are of particular importance, since certain theories of malignancy have invoked a failure of immunological surveillance (Klein, 1972).

4.3.2.1   ANTIGENS

Transformation results in the appearance of specific antigens on the cell surface (Burger, 1971). In the case of viral transformation, studies with temperature-sensitive mutants have shown the changes to be of viral origin, though not due to intact virions. However, it should be pointed out that of several antigens displayed by tumour cells *in vivo*, some can be detected in embryonic normal cells and some in masked, or cryptic form, even in adult normal tissue (Alexander, 1972). Such revelation of receptor sites has also been observed in cultured cells (Häyry and Defendi, 1970) and has been studied in particular detail in the case of lectin binding (*see* Section 4.3.2.2).

Not only do new antigens appear on transformation, but others are lost. The histocompatibility or transplantation antigens are particularly interesting since their loss might explain the escape of malignant cells from immune surveillance. However, many cultured cells that are tumorigenic retain histocompatibility antigens. Moreover, experiments with hybrid cells have not revealed any correlation between the loss of histocompatibility antigen and malignancy (Grundner *et al.*, 1971).

The loss of some antigens during the acquisition of others may be due less to genetic change than to a restriction on the number of molecules capable of being expressed at any one time. The cell surface is, after all, a limited structure [and some transformed cells are actually smaller than their normal counterparts (Ben-Bassat, Inbar and Sachs, 1971; Noonan and Burger, 1973)], and there may be 'room' for only so many proteins or glycoproteins in positions sufficiently exposed to be reactive. Others may be forced into positions of non-reactivity—the masking phenomenon referred to earlier. Note that masking does not necessarily imply failure of antibody to make contact with antigen. Although agglutination, which is a common test for the expression of surface components, may be reduced in normal cells, binding of antibody may be unimpaired. This has been well documented in the case of lectin receptors (Section 4.3.2.2). Whatever the means by which it is achieved, an altered antigenic make-up provides strong evidence that the cell surface has become modified during transformation.

The relevance of antigenic variation to the behaviour of transformed cells must await the identification of specific antigens as enzymes or other membrane components (*see* Section 4.4.3). Clarification of structure alone is not sufficient, as illustrated in the case of histocompatibility antigens (Hughes, 1973), the physiological significance of which is still a matter for conjecture (Bodmer, 1972). Since transformed cells are characterized by a *loss* of certain functions such as contact inhibition, the significance of antigenic change may lie in the loss or masking of normal antigens, rather than in the gain or revelation of new ones. The point, which is of special regard to malignancy, may be illustrated by simple analogy with the acquisition of virulence in pneumococci and other bacteria resulting from the formation of an extra surface layer (Dubos, 1955). They, too, escape the host's immune system and thus increase their pathogenicity.

Although the literature contains many references to the *unmasking* of antigenic and other (Section 4.3.2.2) sites in transformed and malignant cells (Burger, 1971), only a few reports (Burger, 1973) of the reverse have appeared. Yet many tumour cells have thicker coats as revealed by histological stains (Martinez-Palomo, 1970) or optical measurement (Mallucci, Poste and Wells, 1972). Most of the stains are indicative of carbohydrate-containing components, but proteins and phospholipids may also be present. Thus SV_{40}-transformed rabbit and hamster cells display surface antigen (measured by binding of fluorescent antibody) only if preincubated with phospholipase C (Collins and Black, 1973); an alternative interpretation of this result is that the antigens are 'buried' more deeply than is generally imagined. However, removal of the surface coat with trypsin releases some phospholipids also (P. Knox, K. J. Micklem and C. A. Pasternak, unpublished work).

Elucidation of the composition of the surface coat and of the mode of its attachment to the rest of the plasma membrane should help to clarify many conflicting results concerning the behaviour of surface antigens in normal and transformed cells.

4.3.2.2   LECTIN-BINDING SITES

The demonstration (Aub, Tieslau and Lankester, 1963) that a component of wheat-germ lipase causes lymphoma (cancer) cells to agglutinate at concentrations much lower than those required to agglutinate thymocyte

Table 4.1   AGGLUTINABILITY BY LECTINS

Wheat-germ agglutinin, $\mu g\ ml^{-1}$ *required for half-max agglutination**			*Concanavalin A, arbitrary scale according to density and size of aggregates*[†]		
Cell line	*Before trypsin*	*After trypsin*	*Cell line*	*Before trypsin*	*After trypsin*
Normal mouse	> 500	10‡	Normal mouse	−	+ + + +
'Py' mouse	10	10‡	'SV$_{40}$' mouse	+ + + +	−
Normal hamster	> 200	30§	Normal hamster	−	+ + + +
'Py' hamster	25	25‖	'Py' hamster	+ + + +	+

* Calculated from the data of Burger (1969). The mouse cells were 3T3 and polyoma-transformed 3T3 cells. The hamster cells were BHK and polyoma-transformed BHK cells.
† Taken from Inbar and Sachs (1969a). The mouse cells were 3T3 and SV$_{40}$-transformed 3T3 cells. The hamster cells were secondary cultures of golden-hamster embryo cells and a polyoma-transformed derivative.
‡ 0.02% trypsin for 3 min.
§ 0.05% trypsin for 3 min.
‖ 0.01% trypsin for 15 min.

(normal) cells has led to extensive use of this and other agglutinins, called *lectins* (Sharon and Lis, 1972), to study the malignant cell surface. Following the purification of the agglutinating factor in wheat-germ (which turned out not to be the lipase) it was soon shown that virally transformed cells *in vitro* can be distinguished from untransformed cells by their sensitivity to agglutination (*Table 4.1*). This has proved to be a rather general phenomenon, extending to chemically and spontaneously transformed cells also. Although the acquisition of agglutinability in virally transformed cells depends on the functioning of some virally coded gene product (Benjamin and Burger, 1970), it does not appear to result from the synthesis of new or specific molecules, since untransformed cells can be rendered agglutinable by treatment with trypsin or other (Burger, 1973) proteolytic agent (*Table 4.1*). Moreover, the agglutinability of transformed cells may be decreased relative to non-transformed cells (*Table 4.1*). Infection by non-transforming viruses also increases agglutinability (Zarling and Tevethia, 1971; Poste and Reeve, 1972). Such exposure of cryptic or masked sites is reminiscent of the appearance of tumour antigens (Section 4.3.2.1) and emphasizes the relevance of lectin binding for studies on malignant transformation.

What, then, is the nature of the masking phenomenon? An initial suggestion, based on the use of fluorescent wheat-germ agglutinin, was that masking is related to the position of cells in the cell cycle (Fox, Sheppard and Burger, 1971): normal cells are unmasked during mitosis but become masked in interphase, whereas transformed cells are permanently unmasked. This is probably an oversimplification (Shoham and Sachs, 1972), although the implication that the cell surface is altered during mitosis is undoubtedly correct. In fact, several reports have shown that tumour, i.e. transformed, cells also express specific antigens more strongly in mitosis than at other times (Kuhns and Bramson, 1968; Cikes and Friberg, 1971; Pasternak, Warmsley and Thomas, 1971), and the differences between growing and non-growing cells (Section 3.1) are even more profound (Thomas, 1971; Thomas and Phillips, 1973). However, the reason appears to lie not so much in the accessibility of antigenic sites (Lerner, Oldstone and Cooper, 1971; Sumner, Collin and Pasternak, 1973) as in more general membrane changes (Graham *et al.*, 1973).

That agglutinability and accessibility of lectin binding sites do not alter in concert has become clear from the reports that various virally transformed cells do not bind more [^{3}H]acetylated concanavalin A (Con A) (Cline and Livingston, 1971) or ^{125}I-labelled Con A (Arndt-Jovin and Berg, 1971; Ozanne and Sambrook, 1971) or ^{125}I-labelled wheat-germ agglutinin (Ozanne and Sambrook, 1971) than do the parent untransformed cells. An earlier report on differences in binding of [^{63}Ni]Con A (Inbar and Sachs, 1969b) has subsequently been amended (Ben-Bassat, Inbar and Sachs, 1971). While recent studies with [^{3}H]acetylated Con A support the notion that lack of agglutinability in normal cells cannot be accounted for by lack of accessibility, they do show a small but significant increase in binding by transformed cells (Noonan and Burger, 1973); in this instance, as in the studies of Arndt-Jovin and Berg (1971), the assay was carried out at 0 °C in order to eliminate non-specific absorption of isotopes due to endocytosis and other processes. Whatever the outcome regarding the exact degree of binding by normal and transformed cells, the concept of masked sites clearly has to be revised.

An alternative proposal takes account of the fact that transformed cells are generally smaller (despite their thicker surface coat) than normal ones. Hence an equivalent number of lectin molecules bound per cell implies an increase in the number bound per surface area and a higher density of receptor sites (Ben-Bassat, Inbar and Sachs, 1971; Noonan and Burger, 1973).

A rather similar explanation for the cryptic behaviour of untransformed cells has resulted from studies in which ferritin-labelled (Nicolson, 1971) or peroxidase-labelled (Martinez-Palomo, Wicker and Bernhard, 1972) Con A is used to visualize the topographical distribution of binding sites. The main conclusion is that the sites are distributed randomly over the surface of normal cells, but become clustered in transformed cells. Use of fluorescent-labelled lectin (Nicolson, 1973a) leads to the same conclusion. Clustered sites result in agglutination, unclustered ones do not (Nicolson, 1972); presumably the strength of a localized array of intercellular bonds formed through a multivalent lectin is greater than that of single, diffusely situated bonds. Or perhaps the very process of clustering alters the surface properties in localized regions so as to overcome some of the forces that normally keep cells apart (Curtis, 1972). Note that at *high* lectin concentration, clustering may no longer be necessary for agglutination (Nicolson, 1974) to occur.

In fact the lectin-binding sites are probably *not* clustered on transformed cells prior to addition of lectin, since (a) fixation of cells before exposure to lectin does not show clustering, and (b) clustering does not occur at 0 °C (Comoglio and Guglielmone, 1972; Inbar *et al.*, 1973a; Nicolson, 1973a;

Rosenblith *et al.*, 1973). These observations can be reconciled with those that have been discussed by assuming that a temperature-dependent rearrangement of sites occurs when lectin is added to transformed cells. Experimental support comes from the fact that antigens and other membrane constituents move more freely at 37 °C than at 0 °C, in the plane of the membrane (Singer and Nicolson, 1972). What has yet to be established is: (1) why free movement leads to clustering, (2) that normal cells have a less fluid membrane and (3) why proteolytic treatment of normal cells should increase their fluidity.

1. Since lectins are generally multivalent molecules, their interaction with receptors may be similar to that of bivalent antibodies with cell surface antigens. The observation that patching of lymphocyte receptors occurs with bivalent but not with monovalent antibody fragments (Taylor *et al.*, 1971) is therefore relevant, especially since monovalent lectin fragments do not cause agglutination (Burger and Noonan, 1970). In other words, intracellular (*cis*) as well as intercellular (*trans*) interactions may be involved in agglutination, but it must also be noted that patching of lymphocyte receptors occurs equally well at 4 °C and 37 °C; only below 4 °C is patching prevented (Raff and de Petris, 1973). This is actually the result expected from measurements of lipid phase transitions in animal membranes (Blazyk and Steim, 1972), though individual membrane components may show different transitions (Esfahani *et al.*, 1971).
2. The indications of an altered phospholipid profile in transformed cells (Sections 4.3.1.3 and 4.4.2) are obviously pertinent but the fact that normal lymphocytes, as just stated, appear to be *more* fluid than transformed cells, in the sense that patching occurs at 4 °C as well as at 37 °C, seems at first sight paradoxical. It can be rationalized if it is assumed that membrane fluidity decreases in the order lymphocytes > transformed fibroblasts > untransformed fibroblasts, and there is some evidence for this (Inbar, Shinitzky and Sachs, 1973b). Alternatively, clustering of antigenic receptors on lymphocytes may not reflect the same kind of fluidity as clustering of lectin receptors on fibroblasts. The fact that in the case of lymphocytes, changes in fluidity as assessed by microviscometry (Shinitzky and Inbar, 1974) do not parallel changes in fluidity as assessed by clustering of Con A receptor sites (Inbar, Shinitzky and Sachs, 1973b), lends support to the latter view.
3. This is perhaps the most difficult to explain, since it is extremely unlikely that brief proteolytic treatment is able to alter the composition of the surface membrane. However, the mobility of membrane components may be controlled by factors other than phospholipid content (Nicolson, 1974); different surface receptors, for example, move at distinct rates.

A rather different explanation for the temperature sensitivity of agglutination is to postulate the existence of an agglutination site distinct from, but close to, the binding site (Inbar, Ben-Bassat and Sachs, 1971b); it is this site that is cryptic in normal cells and sensitive to proteolytic activation. The true picture probably lies somewhere between these two hypotheses; *see*, for example, Willingham and Pastan (1975), and Knutton *et al.* (1976).

The events that appear to be necessary before cells that have lectin bound to them can agglutinate may be summarized as follows:

The first condition is that the temperature has to be raised above 0 °C. As a result, receptor sites become clustered, but the necessity for this is not proved; in fact unclustered lectin receptors can lead to agglutination if a sufficient excess of lectin is provided (Nicolson, 1974). The nature of the temperature-sensitive step is therefore not clear. It *may* be related to a change in membrane fluidity between 0 °C and 37 °C, but it should be noted that agglutination of cells by lectins such as wheat-germ agglutinin (Inbar, Ben-Bassat and Sachs, 1971b) or by certain viruses (Okada, 1962) does not require raising the temperature above 4 °C.

In the case of untransformed cells, a much higher concentration of lectin has generally to be provided than for transformed cells, though there are exceptions where the opposite is true (Burger, 1973). Brief treatment of untransformed cells with proteolytic agents obviates this need; neuraminidase has a similar effect, though in this case the number of accessible receptor sites is increased (Nicolson, 1973b). The relationship of the protease-sensitive molecule to the receptor site is not known, but it must be spatially coordinated, since proteolysis allows lectin-bound receptors on untransformed cells to move as freely as those on transformed cells.

The fact that proteolysis removes much of the surface coat (Onodera and Sheinin, 1970) suggests that it may have some kind of restraint on movement. Since the coat is apparently thicker on transformed cells than on normal cells, it is not clear how this is brought about. Perhaps the coat does not span as much of the surface in transformed cells (*Figure 4.3*); this would explain,

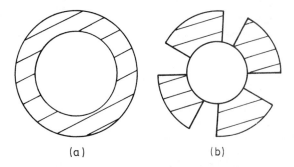

(a)　　　　　　　(b)

Figure 4.3　Possible disposition of surface coat. (a) Normal cell; (b) transformed cell. Highly diagrammatic representation only

for example, why an increased thickness is not accompanied by an increase in total carbohydrate-containing molecules (Section 4.4.3). The exact position of lectin-binding sites has deliberately been omitted from *Figure 4.3*; in addition to equal accessibility in normal and transformed cells, any model has to account for the fact that prolonged proteolysis of transformed cells *decreases* their agglutinability (*Table 4.1*). Another complication is the fact that hyaluronidase treatment increases the agglutinability of some transformed cells without affecting the untransformed parents (Burger and Martin, 1972).

4.4   DIFFERENCES IN COMPOSITION

Analysis of components that are not unique to the surface membrane, such as protein, phospholipids and cholesterol, depends on the prior isolation of the membrane. Since preparation of plasma membrane free of elements of endoplasmic reticulum and Golgi apparatus is technically a rather tedious operation, relatively few studies of the composition of pure surface material have been made. Glycolipids and glycoproteins, on the other hand, are located predominantly on the cell surface, and determination of total cellular carbohydrate has generally been assumed to reflect surface components (but *note* Critchley, Graham and Macpherson, 1973; Keshgegian and Glick, 1973); in that sense, therefore, much more is known about carbohydrate-containing molecules than about the more major constituents of the cell surface.

4.4.1   Proteins

The acquisition or loss of proteins with biological activity, such as antigens and enzymes, has been discussed (Section 4.3). Relatively few reports of changes in chemically defined proteins have appeared. Instead, extracts of plasma membranes have been subjected to polyacrylamide-gel electrophoresis and the resulting patterns scanned for possible differences between normal and transformed cells. While Sakiyama and Burge (1972) were unable to detect any significant changes in transformed cells, Greenberg and Glick (1972) have found that several polypeptides, separated according to molecular weight, decrease in amount. With virally transformed 3T3 cells, on the other hand, Sheinin and Onodera (1972) found that separation according to charge rather than to size reveals the biggest differences; some of the altered poly-peptides are probably glycoproteins. Wickus and Robbins (1973) have refined the technique of size separation, and by using a temperature-sensitive mutant at permitted and restricted temperature, have pinpointed one out of some twenty-six polypeptides as being the only one lost on transformation. The technique of specifically labelling surface proteins from the exterior with ^{125}I is also beginning to yield interesting results (Hynes, 1973; Wickus, Branton and Robbins, 1974).

Protein fractions have been reported to be released from transformed, but not from normal, cells (Rubin, 1970; Bürk, 1973). Whether they are surface components has not been established; what is interesting is that such factors appear to be responsible for many of the characteristic properties of trans-formed cells (Bürk, 1974), i.e. the X of *Figure 4.4* (p. 109).

4.4.2   Phospholipids

No detailed studies of plasma membrane phospholipids of normal and trans-formed cells *in vitro* appear to have been made. Chandrasekhara and Narayan (1970) have compared liver plasma membrane from normal and carcinogen-fed rats and found an increase in phospholipid content in the pre-neoplastic and neoplastic states. This is accompanied by a relative increase in polyunsaturated fatty acids. Emmelot (1973), however, found no consistent differences in phospholipid composition between the plasma membrane of liver and various hepatomas. Another comparison is that of Bergelson *et al.* (1970) on membrane phospholipids of liver and some derived hepatomas. Their main observation was that the phospholipid patterns characteristic of specific organelles disappear in the tumours. Thus cardiolipin, restricted to mitochondria in normal liver, is present also in tumour microsomes and plasma membrane. Sphingomyelin, concentrated in plasma membrane of normal liver, becomes fairly evenly distributed among all cellular membranes in the tumours. Such 'de-differentiation' of tumour membranes is not a consequence of an increased growth rate, since the pattern in regenerating liver is normal. This result, which has not yet been substantiated, does not appear to be true of cells grown *in vitro*: neither the polar headgroups nor the acyl side-chains of phospholipids of isolated plasma membrane were found to be different in transformed, compared with normal, cells (Micklem *et al.*, 1976). So far as *total* phospholipids are concerned, neither Cunningham (1972) nor Diringer, Strobel and Koch (1972) found significant differences between normal and transformed cells either. Nor did Sakiyama and Robbins (1973) observe differences, except in the case of phosphatidylinositol, which appears to decrease in concert with glycolipids (Section 4.4.3). Since phosphatidylinositol resembles glycolipids in having an extremely polar headgroup (five free hydroxyl groups), this may be of significance with respect to the cell surface. However, it must be pointed out that Sakiyama and Robbins (1973) used incorporation of [^{14}C]palmitic acid as a measure of total phospholids, which makes the interpretation subject to some criticism.

4.4.3   Glycolipids and glycoproteins

Sialic acid (*N*-acetylneuraminic acid) is a characteristic residue of the glycoproteins and glycolipids found at the cell periphery, and provides a major contribution to the surface charge (zeta potential) of animal cells (Mehrishi, 1972). Suggestions (Abercrombie and Ambrose, 1962) that the surface charge on tumour cells is higher than that on their normal counterparts have therefore led to many measurements of sialic acid content and accessibility to neuraminidase (Weiss, 1973). The results have been disappointing in that no clear-cut trend has emerged: higher, lower and unchanged values have all been reported (Burger, 1972). Indeed the very correlation between transformation and altered zeta potential seems in doubt (Patinkin, Zaritsky and Doljanski, 1970; Weiss and Sinks, 1970).

Attention has recently shifted to specific molecular entities with the reports that transformed cells contain less complex glycolipids than do normal cells. The results from two laboratories are illustrated in *Table 4.2*. In each case, the altered pattern of glycolipids has been correlated with the loss of a specific glycosyltransferase (Cumar *et al.*, 1970; Kijimoto and Hakomori, 1971). One report has even suggested that lactosylceramide (CDH, *Table 4.2*) is *the* surface antigen characteristic of a large number of human tumour cells (Tal, 1972).

While several reports have confirmed a general decrease in the levels of higher glycolipids in some transformed cells (Robbins and Macpherson,

1971; Yogeeswaran *et al.*, 1972; Brady, Fishman and Mora, 1973), others have not (Warren, Critchley and Macpherson, 1972; Diringer, Strobel and Koch, 1972; Sakiyama, Gross and Robbins, 1972); instead they corroborate (Critchley and Macpherson, 1973; Sakiyama and Robbins, 1973) previous results (Hakomori, 1970; Robbins and Macpherson, 1971; Hakomori *et al.*, 1974) relating glycolipid pattern to the state (growing or confluent) of the cells; that factors other than genetic ones influence glycolipid composition is also indicated by the demonstration that cells growing *in vivo* have a pattern different from that of cells grown *in vitro* (Murray *et al.*, 1973).

Table 4.2 ALTERATIONS IN GLYCOLIPID COMPOSITION

Neutral glycolipids, nmol/mg protein*				*Gangliosides*, nmol/mg protein †				
Cell line	*CTH*	*CDH*	*CMH*	*Cell line*	$G_{D_1}a$	G_{M1}	G_{M3}	*Total*
Normal hamster	1.2–2.9	0.6–0.9	0.6–0.8	Normal mouse	1.8	1.5	0.58	3.9
'I-Py' hamster	<0.1	2.1–2.5	1.1	'SV$_{40}$'-mouse	0.16	0.22	1.9	2.3
'III-Py' hamster	0	1.1	1.0	'Py' mouse	0.14	0.08	2.3	2.5

CTH: Ceramide–Glu–Gal–Gal
CDH: Ceramide–Glu–Gal
CMH: Ceramide–Glu
$G_{D_1}a$: Ceramide–Glu–Gal–GalNAc–Gal
                                    | |
                                NANA   NANA
G_{M1}: Ceramide–Glu–Gal–GalNAc–Gal
                                    |
                                NANA
G_{M3}: Ceramide–Glu–Gal
                                    |
                                NANA

* Calculated from data of Hakomori (1970); the cells were BHK and polyoma-transformed BHK cells. The range of values indicates the difference between growing (low density) and confluent (high density) cells.
† Taken from Brady and Mora (1970); the cells were N-AL/N (epithelial-like) and SV$_{40}$- and polyoma-transformed N-AL/N cells.

Indeed, growth of cells in media containing high concentrations of galactose (Kalckar *et al.*, 1973) can actually modify their properties. Unlike agglutination (Section 4.3.2.2), viral infection, as opposed to transformation, does not affect glycolipid metabolism (Mora, Cumar and Brady, 1971; Mora *et al.*, 1973).

The generality of an altered glycolipid pattern in transformed cells is thus equivocal, especially since some spontaneously transformed, tumorigenic, cell lines do not have an altered glycolipid pattern (Brady and Mora, 1970). Failure to elongate glycolipids cannot therefore be a direct cause of the altered growth characteristics of transformed cells. Rather it is a consequence of their behaviour resulting from some other modification.

In the case of glycoproteins, differences in pattern between normal and transformed cells have also been reported (Wu *et al.*, 1969; Glick, Rabinowitz and Sachs, 1973). A rather specific change, that seems to be typical of many transformed cells, is the appearance of one or more extra sialic acid residues bound to a fucose-containing glycoprotein (Warren, Critchley and Macpherson, 1972); the change is observed only at the permissive temperature in temperature-sensitive mutants and can be correlated with an increased activity of a sialyltransferase enzyme (Section 4.3.1.3).

It will by now have become apparent that several alterations in surface carbohydrates accompany transformation of cultured cells. None of them alone, however, is likely to affect the crucial difference between normal and transformed cells. Rather, the changes are part of an overall alteration in surface structure involving many discrete components; an increase in one may be offset by a decrease in another; the total number of Con A binding sites for example, which presumably represents a mixture (Kraemer, Tobey and van Dilla, 1973) of glycolipids, glycoproteins and other sugar-containing molecules of particular configuration, remains the same (Section 4.3.2.2). In order to construct an accurate picture of the transformed cell surface, the relation of chemical entities to morphological features such as the cell coat or microvilli is of obvious importance.

4.5   RELEVANCE TO GROWTH CONTROL

Lack of growth control has been defined (Section 4.1) as the basic lesion in transformed cells. It appears to be due to a general insensitivity to environmental factors such as the nature of the growth medium or the presence of neighbouring cells. A manifestation of this is a reduction in the density-dependent inhibition of growth (Stoker and Rubin, 1967), referred to in certain instances as 'topoinhibition' (Dulbecco, 1970). A greater degree of cell movement, perhaps associated with invasiveness *in vivo*, is often apparent (but *note* Guelstein *et al.*, 1973). How are the surface changes that have been discussed relevant to these alterations in the behaviour of transformed cells?

4.5.1   Differences in morphology

Loss of close contacts between cells is clearly an important factor in reducing topoinhibition. In addition to faulty assembly of the microtubular–micro-filamentous system at the cell surface, decreased intercellular adhesion may be involved. Adhesion may, of course, itself depend on correctly exposed microfilaments. The fact that (a) the morphological changes brought about by demecolcine or cytochalasin B are accompanied by alterations in growth properties and that (b) growth changes triggered by diBcAMP influence cellular morphology as well, is a measure of the extent to which morphology and growth are coupled.

4.5.2   Differences in biochemical functions

The significance of altered transport mechanisms has been mentioned (Section 4.3.1.1); possible changes in the turnover of phospholipids and carbohydrate-containing membrane components with respect to membrane fluidity or 'stickiness' resulting from *trans*-glycosidation have also been discussed (Section 4.3.1.2). An increased rate of protein turnover due to endogenous proteases is referred to below.

4.5.3   Receptor sites

While the relevance of an altered antigenic make-up to the operation of immune surveillance *in vivo* is clear, there seems no obvious reason why the disposition of particular receptor sites should affect growth in culture. Yet Burger and Noonan (1970) and Shoham, Inbar and Sachs (1970) have each shown that addition of Con A to transformed cells (under conditions of non-agglutinability) reduces the extent of their growth *in vitro*. This is therefore a rather clear-cut example of topoinhibition and suggests that free saccharide groups at the cell periphery are somehow required for continued growth. It may be noted that in lymphocytes lectin binding has the opposite effect, in that it stimulates blast formation and eventual cell division; this may be related to differences in membrane fluidity (Section 4.3.2.2). Whatever the reason, it is a good illustration of the importance of pre-programming of membrane receptors during development (Pasternak, 1970).

Just as the binding of lectin to transformed cells apparently causes them to revert to normal behaviour, so mild proteolysis brings about the opposite change. Not only is the agglutinability of untransformed cells increased (Section 4.3.2.2), but growth of confluent cells is resumed and a new saturation density reached. The suggestion has therefore been made (Burger, 1973) that transformed cells contain endogenous proteases lacking, or decreased, in normal cells. Although the concept of growth stimulation via proteolytically altered agglutinability has been questioned (Glynn, Thrash and Cunningham, 1973), increased protease activity in transformed cells *has* been reported (Ossowski *et al.*, 1973; Unkeless *et al.*, 1973). Since proteolysis removes a substantial amount of material from the cell surface, components other than lectin-binding receptors may be involved. Indeed, following proteolysis, cells bind large amounts of serum protein and this may be related to the decreased serum requirement of transformed cells (P. Knox and C. A. Pasternak, Vol. 5, Chapter 7 of this series); a comparison of binding capacity by normal and transformed cells is under way.

4.5.4   Differences in chemical composition

The possibility that transformed cells have an increased membrane fluidity owing to a different phospholipid composition should not be difficult to test (*see* Addendum, p. 113). Just why this might lead to a lack of growth control (assuming that agglutination itself is an irrelevant factor) is less clear. Perhaps a more fluid membrane promotes a more effective uptake of nutrients (Section 4.3.1.1), such that some intracellular control is lost. Whatever changes in membrane components occur—and it is by no means clear that *major* differences necessarily accompany transformation— correlation with an altered adhesiveness or other contact property is not difficult to imagine.

4.5.5 Conclusion

It is apparent that no single property has yet emerged to account satisfactorily for the behaviour of transformed cells *in vitro*. The more closely one examines differences at the molecular level, the more exceptions to any one hypothesis there appear to be. This is perhaps not surprising, since there is no reason why two cell lines should behave identically on transformation, or why two viruses should transform a cell in the same manner. The control of cellular growth is, after all, a complex phenomenon; defective control, for which transformed cells are selected, is therefore likely to result from an alteration in any one of a large number of contributing processes.

The outcome in the transformed cell is often a simultaneous change in several surface properties. Yet the initial modification may affect no more than one or two genes, for many transforming viruses have a rather small genome and in temperature-sensitive variants (presumed to result from a single mutation), most of the surface properties are affected in concert. In other words, surface properties are controlled in a 'pleiotypic' (Hershko

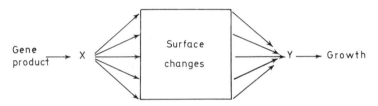

Figure 4.4   Possible relation between surface changes and growth

et al., 1971) manner (*Figure 4.4*). Genetic modification may be at X or at the structural genes corresponding to the various surface modifications. Likewise, modulation of growth control by nutritional and environmental changes, or by effectors such as demecolcine, may be by stimulation or inhibition of X, as well as directly on the cell surface. While this review has attempted to delineate some of the surface changes that accompany transformation, the nature of X and Y remains to be elucidated.

A better definition of the interdependence of events at the cell periphery is clearly desirable. What is emerging is a lack of knowledge concerning the surface coat. The exact disposition of its components and their relation to the microtubular and microfilamentous system and to surface microvilli has now to be determined. For it is here, rather than in the bilayer or plasma membrane proper, that many key differences between normal and transformed cells are likely to reside.

Acknowledgements

The author is grateful to his colleagues for helpful discussions and to the Medical Research Council and the Cancer Research Campaign for financial support.

REFERENCES

ABERCROMBIE, M. and AMBROSE, E. J. (1962). *Cancer Res.*, **22**:525.
ALEXANDER, P. (1972). *Nature, Lond.*, **235**:137.
ARNDT-JOVIN, P. J. and BERG, P. (1971). *J. Virol.*, **8**:716.
AUB, J. C., TIESLAU, C. and LANKESTER, A. (1963). *Proc. natn. Acad. Sci. U.S.A.*, **50**:613.
AZARNIA, R. and LOEWENSTEIN, W. R. (1971). *J. Membrane Biol.*, **6**:368.
BAJER, A. S. and MOLE-BAJER, J. (1972). *Int. Rev. Cytol.*, Suppl. 3.
BAKER, J. B. and HUMPHREYS, T. (1972). *Science, N.Y.*, **175**:905.
BARNETT, R. E., FURCHT, L. T. and SCOTT, R. E. (1974). *Proc. natn. Acad. Sci. U.S.A.*, **71**:1992.
BEN-BASSAT, H., INBAR, M. and SACHS, L. (1971). *J. Membrane Biol.*, **6**:183.
BENJAMIN, T. L. and BURGER, M. M. (1970). *Proc. natn. Acad. Sci. U.S.A.*, **67**:929.
BENNETT, M. V. L. (1973). *Fedn Proc. Fedn Am. Socs exp. Biol.*, **32**:65.
BERGELSON, L. D., DYATLOVITSKAYA, E. V., TORKHOVSKAYA, T. I., SOROKINA, I. B. and GORKOVA, N. P. (1970). *Biochim. biophys. Acta*, **210**:287.
BERWALD, Y. and SACHS, L. (1963). *Nature, Lond.*, **200**:1182.
BLAZYK, J. F. and STEIM, J. M. (1972). *Biochim. biophys. Acta*, **266**:737.
BODMER, W. F. (1972). *Nature, Lond.*, **237**:139.
BOSMANN, H. B. (1969). *Expl Cell Res.*, **54**:217.
BOSMANN, H. B. (1970). *FEBS Lett.*, **8**:29.
BOSMANN, H. B. (1972a). *Biochem. biophys. Res. Commun.*, **48**:523.
BOSMANN, H. B. (1972b). *Biochim. biophys. Acta*, **264**:339.
BOSMANN, H. B., HAGOPIAN, A. and EYLAR, E. H. (1968). *J. cell. Physiol.*, **72**:81.
BRADY, R. O., FISHMAN, P. H. and MORA, P. T. (1973). *Fedn Proc. Fedn Am. Socs exp. Biol.*, **32**:102.
BRADY, R. O. and MORA, P. T. (1970). *Biochim. biophys. Acta*, **218**:308.
BURGER, M. M. (1969). *Proc. natn. Acad. Sci. U.S.A.*, **62**:994.
BURGER, M. M. (1971). *Curr. Topics cell. Regulation*, **3**:135.
BURGER, M. M. (1973). *Fedn Proc. Fedn Am. Socs exp. Biol.*, **32**:91.
BURGER, M. M. and MARTIN, G. S. (1972). *Nature, New Biol.*, **237**:9.
BURGER, M. M. and NOONAN, K. D. (1970). *Nature, New Biol.*, **228**:512.
BÜRK, R. R. (1968). *Nature, Lond.*, **219**:1272.
BÜRK, R. R. (1973). *Proc. natn. Acad. Sci. U.S.A.*, **70**:369.
BÜRK, R. R. (1974). *Control of Proliferation in Animal Cells*, pp. 27–35. Ed. B. CLARKSON and R. BASERGA. New York; Cold Spring Harbor Laboratory.
CHANDRASEKHARA, N. and NARAYAN, K. A. (1970). *Cancer Res.*, **30**:2876.
CIKES, M. and FRIBERG, S. (1971). *Proc. natn. Acad. Sci. U.S.A.*, **68**:566.
CLINE, M. J. and LIVINGSTON, D. C. (1971). *Nature, New Biol.*, **232**:155.
COLLINS, J. J. and BLACK, P. H. (1973). *J. natn. Cancer Inst.*, **51**:115.
COMOGLIO, P. M. and GUGLIELMONE, R. (1972). *FEBS Lett.*, **27**:256.
CRITCHLEY, D. R., GRAHAM, J. M. and MACPHERSON, I. (1973). *FEBS Lett.*, **32**:37.
CRITCHLEY, D. R. and MACPHERSON, I. (1973). *Biochim. biophys. Acta*, **296**:145.
CUMAR, F. A., BRADY, R. O., KOLODNY, E. H., MCFARLAND, V. W. and MORA, P. T. (1970). *Proc. natn. Acad. Sci. U.S.A.*, **67**:757.
CUNNINGHAM, D. D. (1972). *J. biol. Chem.*, **247**:2464.
CURTIS, A. S. G. (1972). *Sub-Cell. Biochem.*, **1**:179.
DANIEL, V., LITWACK, G. and TOMKINS, G. M. (1973). *Proc. natn. Acad. Sci. U.S.A.*, **70**:76.
DE PIERRE, J. W. and KARNOVSKY, M. L. (1973). *J. Cell Biol.*, **56**:275.
DIRINGER, H., STROBEL, G. and KOCH, M. A. (1972). *Hoppe-Seyler's Z. physiol. Chem.*, **353**:1769.

DUBOS, R. J. (1955). *The Bacterial Cell*, pp. 188–228. Cambridge, Massachusetts; Harvard University Press.
DULBECCO, R. (1969). *Science, N.Y.*, **166**:962.
DULBECCO, R. (1970). *Nature, Lond.*, **227**:802.
EDELMAN, G. M. (1974). *Chemical and Immunological Approaches to the Cell Surface*, pp. 245–270. Ed. B. D. KAHAN and R. A. REISFELD. New York; Plenum Press.
EAGLE, H., PIEZ, K. A. and LEVY, M. (1961). *J. biol. Chem.*, **236**:2039.
EDWARDS, J. G., CAMPBELL, J. A. and WILLIAMS, J. F. (1971). *Nature, New Biol.*, **231**:147.
EMMELOT, P. (1973). *Eur. J. Cancer*, **9**:319.
EMMELOT, P. and BOS, C. J. (1969a). *Int. J. Cancer*, **4**:705.
EMMELOT, P. and BOS, C. J. (1969b). *Int. J. Cancer*, **4**:723.
ENDERS, J. F. (1965). *Harvey Lect.*, **59**:113.

ESFAHANI, M., LIMBRICK, A. R., KNUTTON, S., OKA, T. and WAKIL, S. J. (1971). *Proc. natn. Acad. Sci. U.S.A.*, **68**:3180.

ESTENSEN, R. D., ROSENBERG, M. and SHERIDAN, J. D. (1971). *Science, N.Y.*, **173**:356.

FENNER, F. (1968). *The Biology of Animal Viruses*, pp. 641–700. New York; Academic Press.

FERBER, E. (1971). *The Dynamic Structure of Cell Membranes* (22nd Mosbach Colloquium), pp. 129–147. Berlin; Springer-Verlag.

FOSTER, D. O. and PARDEE, A. B. (1969). *J. biol. Chem.*, **244**:2675.

FOX, T. O., SHEPPARD, J. R. and BURGER, M. M. (1971). *Proc. natn. Acad. Sci. U.S.A.*, **68**:244.

FREEMAN, A. E. and HUEBNER, R. J. (1973). *J. natn. Cancer Inst.*, **50**:303.

GAFFNEY, B. J. (1975). *Proc. natn. Acad. Sci. U.S.A.*, **72**:664.

GLICK, M. C., RABINOWITZ, Z. and SACHS, L. (1973). *Biochemistry*, **12**:4864.

GLYNN, R. D., THRASH, C. R. and CUNNINGHAM, D. D. (1973). *Proc. natn. Acad. Sci. U.S.A.*, **70**:2676.

GRAHAM, J. M. (1976). *Surfaces of Normal and Malignant Cells*, in press. Ed. R. O. HYNES. Chichester; John Wiley.

GRAHAM, J. M., SUMNER, M. C. B., CURTIS, D. H. and PASTERNAK, C. A. (1973). *Nature, Lond.*, **246**:291.

GREENBERG, C. S. and GLICK, M. C. (1972). *Biochemistry*, **11**:3680.

GRUNDNER, A., FENYO, E. M., KLEIN, G., KLEIN, E., BREGULA, U. and HARRIS, H. (1971). *Expl Cell Res.*, **68**:315.

GUELSTEIN, V. I., IVANOVA, O. Y., MARGOLIS, L. B., VASILIEV, J. M. and GELFAND, I. M. (1973). *Proc. natn. Acad. Sci. U.S.A.*, **70**:2011.

HADDEN, J. W., HADDEN, E. M., HADDOX, M. K. and GOLDBERG, N. D. (1972). *Proc. natn. Acad. Sci. U.S.A.*, **69**:3024.

HAKOMORI, S. (1970). *Proc. natn. Acad. Sci. U.S.A.*, **67**:1741.

HAKOMORI, S., GAHMBERG, C. G., LAINE, R. and KIJIMOTO, S. (1974). *Control of Proliferation in Animal Cells*, pp. 461–471. Ed. B. CLARKSON and R. BASERGA. New York; Cold Spring Harbor Laboratory.

HATANAKA, M., AUGL, C. and GILDEN, R. V. (1970). *J. biol. Chem.*, **245**:714.

HATANAKA, M. and HANAFUSA, H. (1970). *Virology*, **41**:647.

HATANAKA, M., HUEBNER, R. J. and GILDEN, R. V. (1969). *J. natn. Cancer Inst.*, **43**:1091.

HÄYRY, P. and DEFENDI, V. (1970). *Virology*, **41**:22.

HEIDRICK, M. L. and RYAN, L. (1971). *Cancer Res.*, **31**:1313.

HERSHKO, A., MAMOUT, P., SHIELDS, R. and TOMKINS, G. M. (1971). *Nature, New Biol.*, **232**:206.

HILL, E. E. and LANDS, W. E. M. (1970). *Lipid Metabolism*, pp. 185–279. Ed. S. J. WAKIL. New York; Academic Press.

HOLLEY, R. W. (1972). *Proc. natn. Acad. Sci. U.S.A.*, **69**:2840.

HSIE, A. W. and PUCK, T. T. (1971). *Proc. natn. Acad. Sci. U.S.A.*, **68**:358.

HUGHES, R. C. (1973). *Prog. Biophys. molec. Biol.*, **26**:189.

HYNES, R. O. (1973). *Proc. natn. Acad. Sci. U.S.A.*, **70**:3170.

HYNES, R. O. (1974). *Cell*, **1**:147.

INBAR, M., BEN-BASSAT, H. and SACHS, L. (1971a). *J. Membrane Biol.*, **6**:195.

INBAR, M., BEN-BASSAT, H. and SACHS, L. (1971b). *Proc. natn. Acad. Sci. U.S.A.*, **68**:2748.

INBAR, M. and SACHS, L. (1969a). *Proc. natn. Acad. Sci. U.S.A.*, **63**:1418.

INBAR, M. and SACHS, L. (1969b). *Nature, Lond.*, **223**:710.

INBAR, M., BEN-BASSAT, H., SACHS, L., HUET, C. and OSEROFF, A. R. (1973a). *Biochim. biophys. Acta*, **311**:594.

INBAR, M., SHINITZKY, M. and SACHS, L. (1973b). *J. molec. Biol.*, **81**:245.

ISSELBACHER, K. J. (1972). *Proc. natn. Acad. Sci. U.S.A.*, **69**:585.

JOHNSON, G. S., FRIEDMAN, R. M. and PASTAN, I. (1971). *Proc. natn. Acad. Sci. U.S.A.*, **68**:425.

JOHNSTONE, R. M. and SCHOLEFIELD, P. G. (1965). *Adv. Cancer Res.*, **9**:144.

KALCKAR, H. M., ULLREY, D., KIJIMOTO, S. and HAKOMORI, S. (1973). *Proc. natn. Acad. Sci. U.S.A.*, **70**:839.

KESHGEGIAN, A. A. and GLICK, M. C. (1973). *Biochemistry*, **12**:1221.

KIJIMOTO, S. and HAKOMORI, S. (1971). *Biochim. biophys. Res. Commun.*, **44**:557.

KLEIN, G. (1972). *Clin. Immunobiol.*, **1**:219.

KNUTTON, S., JACKSON, D., GRAHAM, J. M., MICKLEM, K. J. and PASTERNAK, C. A. (1976). *Nature, Lond.*, **262**:52.

KOPROWSKI, H., JENSEN, F., GIRARDI, A. and KOPROWSKA, I. (1966). *Cancer Res.*, **26**:1980.

KRAEMER, P. M. and TOBEY, R. A. (1972). *J. Cell Biol.*, **55**:713.

KRAEMER, P. M., TOBEY, R. A. and VAN DILLA, M. A. (1973). *J. cell. Physiol.*, **81**:305.

KUHNS, W. J. and BRAMSON, S. (1968). *Nature, Lond.*, **219**:938.

KULAS, H. P., MARGGRAF, W. D., KOCH, M. A. and DIRINGER, H. (1972). *Hoppe-Seyler's Z. physiol. Chem.*, **353**:1755.

LELIEVRE, L., PRIGENT, B. and PARAF, A. (1971). *Biochem. biophys. Res. Commun.*, **45**:637.

LERNER, R. A., OLDSTONE, M. B. A. and COOPER, N. R. (1971). *Proc. natn. Acad. Sci. U.S.A.*, **68**:2584.

LEVINE, A. J. and BURGER, M. M. (1972). *J. theor. Biol.*, **37**:435.

LOEWENSTEIN, W. R. (1973). *Fedn. Proc. Fedn Am. Socs exp. Biol.*, **32**:60.

MAKMAN, M. H. (1971). *Proc. natn. Acad. Sci. U.S.A.*, **68**:2127.

MALLUCCI, L., POSTE, G. H. and WELLS, V. (1972). *Nature, New Biol.*, **235**:222.

MARGULIS, L. (1973). *Int. Rev. Cytol.*, **34**:333.

MARTIN, G. S., VENUTA, S., WEBER, M. and RUBIN, H. (1971). *Proc. natn. Acad. Sci. U.S.A.*, **68**:2739.

MARTINEZ-PALOMO, A. (1970). *Int. Rev. Cytol.*, **29**:29.

MARTINEZ-PALOMO, A., WICKER, R. and BERNHARD, W. (1972). *Int. J. Cancer*, **9**:676.

MEHRISHI, J. N. (1972). *Prog. Biophys. molec. Biol.*, **25**:1.

MICKLEM, K. J., ABRA, R. M., KNUTTON, S., GRAHAM, J. M. and PASTERNAK, C. A. (1976). *Biochem. J.*, **154**:561.

MORA, P. T. (Ed.) (1975). *Cell Surfaces and Malignancy.* Fogarty International Center Proceedings No. 28. Washington, DC; U.S. Government Printing Office.

MORA, P. T., CUMAR, F. A. and BRADY, R. O. (1971). *Virology*, **46**:60.

MORA, P. T., FISHMAN, P. H., BASSIN, R. H., BRADY, R. O. and MCFARLAND, V. W. (1973). *Nature, New Biol.*, **245**:226.

MURRAY, R. K., YOGEESWARAN, G., SCHIMMER, B. P. and SHEININ, R. (1973). *Tumor Lipids: Biochemistry and Metabolism*, pp. 285–302. Ed. R. WOOD. Champaign, Illinois; American Oil Chemists' Society.

NICOLSON, G. L. (1971). *Nature, New Biol.*, **233**:244.

NICOLSON, G. L. (1972). *Nature, New Biol.*, **239**:193.

NICOLSON, G. L. (1973a). *Nature, New Biol.*, **243**:218.

NICOLSON, G. L. (1973b). *J. natn. Cancer Inst.*, **50**:1443.

NICOLSON, G. L. (1974). *Control of Proliferation in Animal Cells*, pp. 251–270. Ed. B. CLARKSON and R. BASERGA. New York; Cold Spring Harbor Laboratory.

NOONAN, K. D. and BURGER, M. M. (1973). *J. biol. Chem.*, **248**:4286.

NUMA, S. and YAMASHITA, S. (1974). *Curr. Topics cell. Regulation*, **8**:197.

OKADA, Y. (1962). *Expl Cell Res.*, **26**:98.

O'NEILL, C. H. and FOLLETT, E. A. C. (1970). *J. Cell Sci.*, **7**:695.

ONODERA, K. and SHEININ, R. (1970). *J. Cell Sci.*, **7**:337.

OSSOWSKI, L., UNKELESS, J. C., TOBIA, A., QUIGLEY, J. P., RIFKIN, D. B. and REICH, E. (1973). *J. exp. Med.*, **137**:112.

OTTEN, J., JOHNSON, G. S. and PASTAN, I. (1971). *Biochem. biophys. Res. Commun.*, **44**:1192.

OZANNE, B. and SAMBROOK, J. (1971). *Nature, New Biol.*, **232**:156.

PARDEE, A. B. (1964). *Natn. Cancer Inst. Monogr.*, **14**:7.

PARDEE, A. B. (1971). *In vitro*, **7**:95.

PASTAN, I., ANDERSON, W. B., CARCHMAN, R. A., WILLINGHAM, M. C., RUSSELL, T. R. and JOHNSON, G. S. (1974). *Control of Proliferation in Animal Cells*, pp. 563–570. Ed. B. CLARKSON and R. BASERGA. New York; Cold Spring Harbor Laboratory.

PASTERNAK, C. A. (1970). *Biochemistry of Differentiation*, pp. 94–96. London; Wiley–Interscience.

PASTERNAK, C. A. (1972). *J. Cell Biol.*, **53**:231.

PASTERNAK, C. A. (1973a). *Biochem. Soc. Trans.*, **1**:333.

PASTERNAK, C. A. (1973b). *Tumor Lipids: Biochemistry and Metabolism*, p. 66. Ed. R. WOOD. Champaign, Illinois; American Oil Chemists' Society.

PASTERNAK, C. A. (1974). *Companion to Biochemistry*, pp. 399–414. Ed. A. T. BULL, J. R. LAGNADO, J. O. THOMAS and K. F. TIPTON. London; Longmans.

PASTERNAK, C. A. and GRAHAM, J. M. (1973). *The Biology of Fibroblasts*, pp. 81–82. Ed. E. KULONEN and J. PIKKARAINEN. London; Academic Press.

PASTERNAK, C. A. and MICKLEM, K. J. (1973). *J. Membrane Biol.*, **14**:293.

PASTERNAK, C. A. and MICKLEM, K. J. (1974a). *Biochem. J.*, **140**:405.

PASTERNAK, C. A. and MICKLEM, K. J. (1974b). *Biochem. J.*, **144**:593.

PASTERNAK, C. A., WARMSLEY, A. M. H. and THOMAS, D. B. (1971). *J. Cell Biol.*, **50**:562.

PATINKIN, D., ZARITSKY, A. and DOLJANSKI, F. (1970). *Cancer Res.*, **30**:498.

PEERY, C. V., JOHNSON, G. S. and PASTAN, I. (1971). *J. biol. Chem.*, **246**:5785.

PITOT, H. C. (1963). *Cancer Res.*, **23**:1474.

POLLARD, T. D. and KORN, E. D. (1973). *J. biol. Chem.*, **248**:448.

PONTEN, J. (1971). *Virol. Monogr.*, **8**:1.

POSTE, G. and REEVE, P. (1972). *Nature, New Biol.*, **237**:113.

PUCK, T. T., WALDREN, C. A. and HSIE, A. W. (1972). *Proc. natn. Acad. Sci. U.S.A.*, **69**:1943.

RAFF, M. C. and DE PETRIS, S. (1973). *Fedn Proc. Fedn Am. Socs exp. Biol.*, **32**:48.

ROBBINS, P. W. and MACPHERSON, I. (1971). *Proc. R. Soc., Ser. B*, **177**:49.

ROSEMAN, S. (1970). *Chem. Phys. Lipids*, **5**:270.

ROSENBLITH, J. Z., UKENA, T. E., YIN, H. H., BERLIN, R. D. and KARNOVSKY, M. J. (1973). *Proc. natn. Acad. Sci. U.S.A.*, **70**:1625.

ROTH, S. and WHITE, D. (1972). *Proc. natn. Acad. Sci. U.S.A.*, **69**:485.

RUBIN, H. (1970). *Science, N.Y.*, **167**:1271.

RUDLAND, P. S., GOSPODAROWICZ, D. and SEIFERT, W. (1974). *Nature, Lond.*, **250**:741.

SAKIYAMA, H. and BURGE, B. W. (1972). *Biochemistry*, **11**:1366.

SAKIYAMA, H., GROSS, S. K. and ROBBINS, P. W. (1972). *Proc. natn. Acad. Sci. U.S.A.*, **69**:872.

SAKIYAMA, H. and ROBBINS, P. W. (1973). *Fedn Proc. Fedn Am. Socs exp. Biol.*, **32**:86.

SANFORD, K. K., BARKER, B. E., WOODS, M. W., PARSHAD, R. and LAW, L. W. (1967). *J. natn. Cancer Inst.*, **39**:705.

SCHORR, I., RATHNAM, P., SAXENA, B. B. and NEY, R. L. (1971). *J. biol. Chem.*, **246**:5806.

SEIFERT, W. and RUDLAND, P. S. (1974). *Nature, Lond.*, **248**:138.

SELA, B.-A., LIS, H., SHARON, N. and SACHS, L. (1970). *J. Membrane Biol.*, **3**:267.

SHARON, N. and LIS, H. (1972). *Science, N.Y.*, **177**:949.

SHEININ, R. and ONODERA, K. (1972). *Biochim. biophys. Acta*, **274**:49.

SHEPPARD, J. P. (1972). *Nature, New Biol.*, **2**:14.

SHINITZKY, M. and INBAR, M. (1974). *J. molec. Biol.*, **85**:603.

SHOHAM, J., INBAR, M. and SACHS, L. (1970). *Nature, Lond.*, **227**:1244.

SHOHAM, J. and SACHS, L. (1972). *Proc. natn. Acad. Sci. U.S.A.*, **69**:2472.

SINGER, S. J. and NICOLSON, G. L. (1972). *Science, N.Y.*, **175**:720.

STOKER, M. G. P. (1972). *Proc. R. Soc., Ser. B*, **181**:1.

STOKER, M. G. P. and RUBIN, H. (1967). *Nature, Lond.*, **215**:171.

SUMNER, M. C. B., COLLIN, R. C. L. S. and PASTERNAK, C. A. (1973). *Tissue Antigens*, **3**:477.

TAL, C. (1972). *Embryonic and Fetal Antigens in Cancer*, Vol. 2, pp. 53–65. Ed. N. G. ANDERSON, J. H. COGGIN, E. COLE and J. W. HOLLERNAN. Oak Ridge, Tennessee; Oak Ridge National Laboratory.

TAYLOR, R. B., DUFFUS, W. P. H., RAFF, M. C. and DE PETRIS, S. (1971). *Nature, New Biol.*, **233**:225.

THOMAS, D. B. (1968). *Biochem. J.*, **109**:79.

THOMAS, D. B. (1971). *Nature, Lond.*, **233**:317.

THOMAS, D. B. and PHILLIPS, B. (1973). *J. exp. Med.*, **138**:64.

TURNER, R. S. and BURGER, M. M. (1973). *Ergebn. Physiol.*, **68**:121.

UHR, J. W. and VITETTA, E. S. (1973). *Fedn. Proc. Fedn Am. Socs exp. Biol.*, **32**:35.

UNKELESS, J. C., TOBIA, A., OSSOWSKI, L., QUIGLEY, J. P., RIFKIN, D. B. and REICH, E. (1973). *J. exp. Med.*, **137**:85.

WALLACH, D. F. H. (1973). *Biological Membranes*, Vol. 2, pp. 253–294. Ed. D. CHAPMAN and D. F. H. WALLACH. New York; Academic Press.

WARREN, L. (1969). *Curr. Topics dev. Biol.*, **4**:197.

WARREN, L., CRITCHLEY, D. and MACPHERSON, I. (1972). *Nature, Lond.*, **235**:275.

WARREN, L., FUHRER, J. P. and BUCK, C. A. (1972). *Proc. natn. Acad. Sci. U.S.A.*, **69**:1838.

WEISS, L. (1973). *J. natn. Cancer Inst.*, **50**:3.

WEISS, L. and SINKS, L. F. (1970). *Cancer Res.*, **30**:90.

WESSELLS, N. K., SPOONER, B. S., ASH, J. F., BRADLEY, M. O., LUDUENA, M. A., TAYLOR, E. L., WRENN, J. T. and YAMADA, K. M. (1971). *Science, N.Y.*, **171**:135.

WICKUS, G. G., BRANTON, P. E. and ROBBINS, P. W. (1974). *Control of Proliferation in Animal Cells*, pp. 541–546. Ed. B. CLARKSON and R. BASERGA. New York; Cold Spring Harbor Laboratory.

WICKUS, G. G. and ROBBINS, P. W. (1973). *Nature, New Biol.*, **245**:65.

WILLINGHAM, M. C. and PASTAN, I. (1975). *Proc. natn. Acad. Sci. U.S.A.*, **72**:1263.

WU, H. C., MEEZAN, E., BLACK, P. H. and ROBBINS, P. W. (1969). *Biochemistry*, **8**:2509.

YOGEESWARAN, G., SHEININ, R., WHERRETT, J. R. and MURRAY, R. K. (1972). *J. biol. Chem.*, **247**:5146.

ZARLING, J. M. and TEVETHIA, S. S. (1971). *Virology*, **45**:313.

ADDENDUM

Some major advances that have occurred since the completion of this review may be summarized as follows. (a) The participation of microtubules and microfilaments in surface phenomena (Section 4.2.1) has been further documented; a mechanism by which this may be achieved has been postulated to involve the following interactions: microtubules (in the cytoplasm)–microfilaments (at the inner surface of the plasma membrane)–membrane protein (spanning the lipid bilayer)–carbohydrate or other residue (at the outer surface coat) (Edelman, 1974). Ways in which this and other biochemical parameters may be altered by transformation have been reviewed (Mora, 1975). (b) The suggestion (Hadden *et al.*, 1972) that it is an increase in intracellular cGMP, rather than a decrease in intracellular cAMP, that controls growth and cell division has been confirmed (Rudland, Gospodarowicz and Seifert, 1974; Seifert and Rudland, 1974). As indicated in Section 4.3.1.2, therefore, altered levels of cAMP in transformed cells may be of secondary importance. (c) Membrane lipid fluidity (Section 4.3.2.2) may not, despite reports to the contrary (Barnett, Furcht and Scott, 1974; Shinitzky and Inbar, 1974), necessarily be altered by transformation (Gaffney, 1975; Micklem *et al.*, 1976); a more exact definition of 'fluidity' should help to resolve present discrepancies. (d) The loss of specific plasma membrane proteins from transformed cells (Section 4.4.1) and its significance have been reviewed (Hynes, 1974).

5

The tumor cell periphery: carbohydrate components*

David F. Smith†
The American National Red Cross Blood Research Laboratory, Bethesda, Maryland

Earl F. Walborg, Jr.
Department of Biochemistry,
The University of Texas System Cancer Center,
M. D. Anderson Hospital and Tumor Institute, Houston, Texas

5.1 CANCER: A SOCIAL DISORDER AT THE CELLULAR LEVEL

In spite of centuries of observations and research by physicians and scientists, neoplastic growth still defies definition. A search for the invariant characteristics of cancer cells has led instead to the conclusion that the neoplastic state is characterized by its pleiomorphism and temporal variation. Even those characteristics most frequently associated with the neoplastic state— such as the loss of growth control and the propensity of cancer cells to invade and spread to sites remote from the primary focus—are subject to temporal and quantitative variation. Moreover, one or more of these characteristics is constantly or temporarily displayed by certain embryonic or differentiated normal cells. The following statement by Foulds (1969) offers considerable insight into the problem of neoplasia: 'The central problem of neoplasia is not merely one of growth, but also one of morphological pattern. The central problem is one of biological organization in space and time. Its essential character can be briefly, if imperfectly, expressed by saying it constitutes a breach in the integral "wholeness" of the organism.'

* Contribution No. 324 from The American National Red Cross Blood Research Laboratory.
† Supported, in part, by PHS Fellowship No. 5 FO2 CA55287.

115

Morphological pattern results from the specific association of cells into multicellular units. The process of cellular association, which is spatially and temporally programmed, presumably operates through recognitive macromolecules present at the cell periphery (Moscona, 1968). Thus it is not surprising that neoplasia has been considered a disorder of the social interactions among cells which evolves from chemical alterations at the cell periphery (Kalckar, 1965; Wallach, 1968; Pardee, 1971; Burger, 1971a; Emmelot, 1973).

A number of observations lend support to such a view of cancer as a social disease at the cellular level; for example, the altered mutual adhesiveness of tumor cells, the decreased incidence of junctional specializations and defective intercellular communication exhibited by neoplastic tissue, and the loss of postconfluence inhibition of cell division displayed by cancer cells grown *in vitro*. Studies on a limited number of normal and malignant tissues have revealed that the force required to separate cancerous cells is significantly less than that required for separation of their normal counterparts (Coman, 1944; McCutcheon, Coman and Moore, 1948). Coman (1953) proposed that the decreased mutual adhesiveness of cancer cells partially explains their propensity to passive dissemination. More recently Modjanova and Malenkov (1973) demonstrated that the decrease in mutual adhesiveness of hepatoma cells is directly related to the progression of the hepatomas to a more malignant state. Coman (1960) also demonstrated that epithelial cells subjected to short exposure to a chemical carcinogen exhibit decreased mutual adhesiveness, suggesting that this property is not limited to tumors which have undergone progression to a malignant form, but may appear early in carcinogenesis. Because of the fundamental importance of intercellular adhesion to metastasis and conceivably to neoplastic transformation, these observations deserve further investigation.

The loss of mutual cellular adhesiveness in tumor tissue should be reflected in ultrastructural alterations at cellular junctions; and indeed such alterations have been observed. Chang (1967) investigated the ultrastructure of normal rat liver and azo-dye-induced hepatomas (both primary tumors and transplantable tumors in early passages). The loss of adhesion between cells was evidenced by wider separation of adjacent cells and a decreased incidence of desmosomes. Emmelot and Benedetti investigated the ultrastructure of plasma membranes of normal liver from rat and mouse, a transplantable azo-dye-induced rat hepatoma, and transplantable mouse hepatomas derived from spontaneous tumors. The only common morphological features which distinguished the hepatomas from liver cells were the loss of junctional specializations *in situ* and the decreased incidence of tight junctions observed in purified plasma membrane preparations (Emmelot and Benedetti, 1967; Benedetti and Emmelot, 1967). These observations have been confirmed in a number of epithelial tumors, including mouse mammary carcinoma, mouse skin carcinoma and human cervical carcinoma (Shingleton *et al.*, 1968; Martinez-Palomo, 1970). Martinez-Palomo, Braislovsky and Bernhard (1969) extended these investigations to cultures of normal and transformed cells. Whereas tight junctions were frequently observed in the nontransformed cell cultures, these structures were only infrequently observed in the cultures of virally and spontaneously transformed cells. Although substantial morphological evidence indicates that diminished cell–cell contacts are associated with many neoplastic tissues, it is still premature to consider this alteration a general feature of cancer cells.

The decreased incidence of junctional complexes in cancer cells may be indicative of an alteration in the cell surface which affects the interchange of information between cells, information which may be involved in the control of cellular growth. The pioneering research of W. R. Loewenstein focused on the role of membrane junctions in mediating ionic and molecular interchange between cells (*see* Burton, 1971). Loewenstein (1966) observed that the junctional membranes of epithelial cells exhibit high permeability to ions and low-molecular-weight molecules. This observation indicated that 'the entire connected cell system, rather than the single cell, constitutes the functional unit', thus providing 'a means for tissue homeostasis and functional control'. Such a view of a connected cell system raises the possibility that cell growth may be modulated by the intercellular flow of substances regulating gene activity (Loewenstein, 1968).

Neoplastic transformation of epithelial cells is accompanied by a loss of intercellular communication. For example, Loewenstein and Kanno (1967) found that a variety of rat hepatomas, differing in growth rates, invasiveness and degree of cellular differentiation, exhibit high junctional membrane resistances to ion flow, whereas adult and regenerating rat liver cells exhibited low junctional membrane resistances (Loewenstein and Penn, 1967). These initial studies were subsequently confirmed for other animal and human epithelial tumors (Kanno and Matsui, 1968; Jamakosmanović and Loewenstein, 1968). Epithelial cells of adult rat liver and hamster embryo in tissue culture also exhibit ionic communication through membrane junctions; however, cancerous counterparts of these cells do not communicate under comparable conditions of growth *in vitro* (Borek, Higashino and Loewenstein, 1969)*. These observations suggest that an essential feature of the cancer cell may be its ability to escape the modulating control of its neighbors. One mechanism of escape might involve blockage of the switchboard (junctional membrane) which controls the flow of regulatory messages. Burton and Canham (1973) have recently proposed a general theory of growth control based on the above observations concerning intercellular communication in normal and neoplastic cells. To ascertain whether the phenotypic characteristics of malignancy and loss of intercellular communication are genetically linked, hybrid cells derived by fusing a non-malignant communicating cell with a noncommunicating malignant cell are being investigated (Loewenstein, 1973; Azarnia and Loewenstein, 1973). The introduction of a normal genome into malignant cells resulted in hybrids which were nontumorigenic and which displayed intercellular communication characteristic of the normal cell. The segregants of the hybrids are now being analyzed.

Although the association between neoplastic transformation and loss of ion-permeable junctional membranes appears well established for cells of epithelial origin, such a correlation has not been found in fibroblastic cells (Potter, Furshpan and Lennox, 1966; Borek, Higashino and Loewenstein, 1969; Rubin, 1970; Sheridan, 1970). These findings raise the possibility that transformation of fibroblasts involves growth-regulating processes

* The contradictory results obtained using cultured Novikoff hepatoma cells have not been resolved (Borek, Higashino and Loewenstein, 1969; Sheridan, 1970).

distinct and independent from junctional communication or that the techniques for measurement of junctional communication are not sufficiently refined to permit detection of specific alterations of the membrane junction. The junctional membrane may serve as a switchboard for many conversations, only some of which are growth-regulatory messages involved in neoplastic transformation.

The altered growth behavior of malignant cells grown *in vitro* has suggested that the cell surface is involved in the control of cell locomotion and growth (Abercrombie, 1970). When normal cells are grown in culture they reach a stationary density shortly after attainment of monolayer confluency. Cells in this stationary phase exhibit a greatly reduced mitotic rate. This pheno-menon has been denoted 'contact inhibition of mitosis' (Abercrombie, 1970), 'density-dependent inhibition of growth' (Stoker and Rubin, 1967) and 'postconfluence inhibition of cell division' (Martz and Steinberg, 1972). The last of these designations has been suggested on the basis of recent data which indicate that the inhibition of cell division is not dependent on simple cell–cell contact or local cell density. The observation was made that 40 percent of the cells which experienced cell contact around their entire perimeter or a superconfluent local cell density throughout the G_1 stage of the cell cycle proceeded to divide after a relatively short generation time. These findings do not support hypothetical growth-control mechanisms which operate through the establishment of local diffusion gradients, such as a local concentration gradient of an inhibitor of cell division or of a substance essential for cell division (Rein and Rubin, 1968). Although simple cell–cell contact is not an effective inhibitor of cell division, the observations of Martz and Steinberg (1972) do not exclude the possibility that cell contacts become effective only after a period of maturation.

Cancer cells, in contrast to their normal counterparts, exhibit loss of postconfluence inhibition of cell division; that is to say, a high mitotic rate continues well past monolayer confluence with the resultant formation of a multilayered cell mass. In some cases a more or less stationary density may be attained, although such densities are subject to high variability dependent upon growth conditions. The relationship of this growth pheno-menon *in vitro* to malignancy *in vivo* has been substantiated by the strong, although not invariable, correlation which exists between loss of post-confluence inhibition of cell division and tumorigenicity (Aaronson and Todaro, 1968; Pollack, Green and Todaro, 1968; Eagle *et al.*, 1970).

The manner in which tumor cells interact with normal cells in tissue culture indicates an alteration in the adhesive properties of the periphery of the tumor cell. When moving fibroblasts of the same or different species collide, their continued locomotion in the direction of their contact is inhibited (Abercrombie and Heaysman, 1953). This phenomenon has been designated 'contact inhibition of overlapping' (Martz and Steinberg, 1973), and can be adequately explained by Steinberg's differential adhesion hypothesis, proposed to explain the spatial organization of mixed cell populations (Steinberg, 1970). This hypothesis states that a mixed population of cells seeks an equilibrium configuration which maximizes adhesive strengths. On the other hand, the locomotion of sarcoma cells was found not to be inhibited by contact with normal fibroblasts; thus malignant cells exhibited a loss of contact inhibition of overlapping on normal cells (Abercrombie, Heaysman and Karthauser, 1957; Abercrombie, 1960; Barski and Belehradek, 1965). These observations indicated that the adhesive properties of malignant cells had been altered such that they adhered less strongly to the substratum and/or more strongly to normal cells. Presumably the altered adhesive properties reflect chemical alterations at the periphery of the cancer cell, alterations which may contribute to the propensity of cancer cells to invade and metastasize.

Chemical alterations at the cell periphery may explain many of the biological manifestations of cancerous cells, particularly their ability to escape the spatial and temporal order normally imposed upon them by neighboring cells. The determination of the molecular basis of the derangement(s) will depend on the elucidation of the chemical nature of the plasma membrane and its associated structures.

Any review dealing with the cell surface or plasma membrane must initially dispose of the question of what morphological structures are included in such a definition. A major difficulty lies in distinguishing between the plasma membrane and intercellular material, as well as the arbitrary distinction between bound structural elements and free secretory products or adsorbed components. While the terms 'glycocalyx', suggested by Bennett (1963), and 'cell periphery', proposed by Weiss (1967), do not make such distinctions, the term 'greater membrane' (Revel and Ito, 1967) has been used to describe the carbohydrate-rich 'fuzz' located exterior to the lipid bilayer, thus defining two regions of the plasma membrane, namely the hydrophobic interior region and the exterior hydrophilic region (*see also* Curtis, 1967). The 'fluid mosaic' model of membrane structure proposed by Singer and Nicolson (1972) has been an important development in the resolution of the many different concepts of membrane structure. In light of this development and recent studies of the dynamic nature of surface topography, Oseroff, Robbins and Burger (1973) have extended the concept of the 'greater membrane' to include submembrane material, but indicated the necessity to distinguish between the carbohydrate-rich molecules strongly bound to or extending through the lipid bilayer and extrinsic factors such as secreted products and adsorbed components. In agreement with these workers, we will utilize in this review the terms 'cell periphery' or 'plasma membrane' in their broader functional connotation.

The object of the following discussion will be the nature of some of the cellular components, particularly carbohydrate-rich molecules located at the cell periphery, which may be involved in the normal processes of cell–cell and cell–environment interactions and their alteration after malignant transformation.

5.2 THE ROLE OF CARBOHYDRATES IN INTERCELLULAR RECOGNITIVE FUNCTIONS

Although it has been demonstrated that the plasma membrane is composed of lipids, proteins, carbohydrates, and even nucleic acids, carbohydrates have become perhaps the most popular membrane components in investigations of membrane alterations associated with the neoplastic process. This popularity is well founded in light of the evidence which implicates cell surface carbohydrates in several normal cell-surface-related functions, as well as altered cell surface carbohydrate associated with some of the manifestations of transformed cells. Evidence for the involvement of carbohydrate in many of the normal cell-surface-related functions of various cell types has been reviewed by the late Richard J. Winzler (1970, 1972). It is interesting that in nearly every instance carbohydrates are involved as recognition sites on cell surfaces. Examples of such phenomena involving cell surface carbohydrates include the following: the immunological properties of cell surfaces and binding of viruses and lectins, tissue-specific reaggregation of dissociated embryonic cells, 'homing' of small lymphocytes to lymphatic tissue, and species-specific attachment of egg cells and sperm in several animal species. In some cases, on the other hand, cells have demonstrated the ability to recognize specific carbohydrates. This has been observed in the species-specific aggregation of sponge cells, where aggregation occurs presumably as a result of cellular recognition of carbohydrate moieties on extracellular aggregating factors (Margolish et al., 1965; Turner and Burger, 1973; Weinbaum and Burger, 1973). The ability of certain monosaccharides to modify the morphology and growth behavior of mammalian cells in culture (Cox and Gesner, 1968a, b) also suggests that cells are capable of recognizing carbohydrate structures. The recognition of nonreducing terminal galactose residues, present on the oligosaccharide moieties of neuraminidase-treated serum glycoproteins, is involved in the binding and removal of these glycoproteins from the circulation by liver cells (Morell et al., 1971; Pricer and Ashwell, 1971). This recognition process, which involves the liver cell periphery and extracellular glycoproteins, suggests that similar mechanisms may be operative in cell–cell recognition.

While it is true that no single complex carbohydrate structure at the cell surface has yet been clearly associated with a specific physiological cellular function (Kramer, 1971), it is clear that at least one role of surface carbohydrate is to serve as cellular recognition sites. As discussed by Winzler (1970), carbohydrates are well suited for such a role owing to their ability to form stereospecific structures capable of having high information content. Oligosaccharides, especially branched structures, may form relatively rigid molecules with many groups available for hydrogen bonding. Alterations of the sequence of the monosaccharides, the position of the glycosidic linkages, and their α or β nature allow the biosynthesis of many different stereospecific structures using only a few monosaccharide units. This property contributes to the specificity observed in many of the interactions of cell surface carbohydrates with environmental agents such as antibodies, viruses and lectins.

Much of the current speculation on the involvement of cell surface carbohydrates in cell–cell adhesion stems from the hypothesis presented by Roseman (1970), which envisions cell–cell adhesion as being mediated by enzyme–substrate interactions between cell surface glycosyltransferases on one cell with oligosaccharide acceptors on a corresponding cell. This hypothesis has been supported by the demonstration of glycosyltransferases on the cell surface of embryonic neural retina cells (Roth, McGuire and Roseman, 1971), cultured mouse fibroblasts (Roth and White, 1972) and human platelets (Jamieson, Urban and Barber, 1971). Roth and White (1972) have demonstrated an apparent contact dependence of the incorporation of galactose into surface oligosaccharide acceptors when 3T3 cells are incubated with UDP–galactose ('*trans*-glycosylation'), while incorporation of galactose by malignant 3T12 cells shows no dependence on contact. These authors suggested that on the malignant cell, the transferase and acceptor may be close enough to permit glycosylation of acceptors on a single cell surface ('*cis*-glycosylation'). More recently, Chipowsky, Lee and Roseman (1973) have shown that 3T3 mouse fibroblasts specifically adhere to galactose-derivatized Sephadex beads; however, SV_{40}-transformed 3T3 cells adhere to the derivatized beads better than do the nontransformed 3T3 cells.

Pat and Grimes (1974) have confirmed many of the findings of Roth, McGuire and Roseman (1971) and Roth and White (1972) and have shown that exogenously supplied sugar nucleotides participate in the synthesis of endogenous glycolipid and glycoprotein without entering the intracellular pool of sugar nucleotides. It would, therefore, appear that investigations using intact cells and exogenously supplied sugar nucleotides permit observation of synthesis of glycolipid and glycoprotein *in vivo*. On the other hand, Pat and Grimes (1974) were not able to confirm that nontransformed 3T3 cells possess only a *trans*-glycosylation mechanism for synthesis of endogenous glycolipid and glycoprotein using exogenous nucleotide sugar substrate, as reported by Roth and White (1972).

Since the involvement of cell surface glycosyltransferase in intercellular adhesion presumably requires the formation of enzyme–substrate complexes between the enzyme and cell surface complex carbohydrates, these interactions would be subject to the wide variety of enzymatic control mechanisms. Smith, Kosow and Jamieson (1975) have recently investigated the mechanism of the collagen:glucosyltransferase thought to be involved in the adhesion of blood platelets to collagen, and provided evidence that the kinetic mechanism of this enzyme may be ordered with UDP–glucose as the first substrate. These authors suggested that since there is no evidence for the existence of UDP–glucose in extracellular fluids, platelet or plasma cofactors may be required to maintain this cell surface enzyme in a state which could form a ternary complex with collagen. These studies point out the necessity of investigating the control mechanisms involved in the synthesis of glycoprotein and glycolipid and clearly indicate that more definitive experimentation in this area will be required to establish glucosyltransferase–substrate interaction as a mechanism for cellular adhesion and recognition.

5.3   CARBOHYDRATE COMPONENTS OF NEOPLASTIC CELLS

5.3.1   Sialic acid

The suggestion by Abercrombie and Ambrose (1962) that the adhesiveness of surfaces depends upon the relative magnitude of attractive and repulsive forces between them, and that the reduction in adhesiveness of tumor cells may be explained in terms of interactions of charged groups at the cell surface, placed increased importance on investigations concerned with the chemical nature of the periphery of the tumor cell. Early investigations of the surface charge of tumor cells by estimations of their electrophoretic mobility resulted in the conclusion that tumor cells have a higher negative charge than their normal counterparts (Ambrose, James and Lowick, 1956). In one instance a progressive increase in the negative surface charge was associated with an increase in the invasive properties of a mouse sarcoma (Purdom, Ambrose and Klein, 1958). Since the electrophoretic mobility of many cell types, including erythrocytes and normal and transformed cells, was reduced by treatment with neuraminidase (see Winzler, 1970), it was concluded that sialic acid is important in the negative surface charge of all cell types and elevated in tumor cells, resulting in their increased electrophoretic mobility.

This notion was supported by histochemical data (Gasic and Gasic, 1962a; Gasic and Berwick, 1963) which demonstrated the peripheral distribution of sialic acid in ascites tumor cells by colloidal iron staining sensitive to neuraminidase treatment. Furthermore, several histochemical investigations indicated increases in the amounts of carbohydrate staining material at the surface of tumor cells relative to normal cells (Defendi and Gasic, 1963; Martinez-Palomo and Braislovsky, 1968; Martinez-Palomo, Braislovsky and Bernhard, 1969). Other studies, however, either failed to show significant differences in the electrophoretic mobilities of normal and tumor cells (Vassar, 1963) or correlated increased electrophoretic mobilities with rapidly dividing normal cells (Eisenberg, Ben-Or and Doljanski, 1962; Heard, Seaman and Simon-Rouss, 1961; Ruhenstroth-Bauer and Fuhrman, 1961) or normal cells during mitosis (Mayhew, 1966). Mayhew's observation (1966) that the elevation of electrophoretic mobility which occurred just prior to and during cell division was interpreted by Kramer (1967) to be a change in the spatial arrangement of sialic acid in the electrokinetic shear layer. This conclusion was based on the observation that no abrupt change occurred in cell surface sialic acid during the division of synchronized cell cultures. In light of more recent observations which suggest that cell surface topographical alterations associated with tumor cells may be similar to those of the mitotic cell surface of normal cells (see Section 5.4), Kramer's interpretation may not only explain many of the discrepancies which prevented the generalization that tumor cells have a higher negative surface charge resulting from increased cell surface sialic acid levels, but also suggests that surface topography may play a significant role in surface charge.

Chemical investigations of virally transformed hamster and mouse cells have demonstrated a lower sialic acid content in transformed cells (Ohta et al., 1968). Other investigations (Wu et al., 1969; Grimes, 1970, 1973) have demonstrated that the amounts of neutral sugars and hexosamine, as well as sialic acid, are lower in transformed cells. These data raise the as yet unresolved question of whether tumor cells have more or less carbohydrate at their cell surface than normal cells.

Hartman et al. (1972) have investigated one aspect of this problem and shown that if the loss of glycoprotein by repeated cell washings is taken into consideration, the only consistent difference in the carbohydrate content of cells before and after transformation by DNA- or RNA-containing oncogenic viruses was that the level of fucose was significantly lower in the transformed cells. These authors suggested that the relatively rough handling of cells for biochemical analysis might remove carbohydrate-containing components, giving false low values for the carbohydrate content, while histochemical data (obtained by fixing cells *in situ*) suggest an increase in carbohydrate-containing surface material of tumor cells. It was concluded from these data that transformed cells contain as much or more surface carbohydrate but that this material is more loosely bound in malignant cells. These observations, however, do not preclude the possibility of increased carbohydrate in some surface components with concomitant decreases in others or the possibility of structural alterations with no significant change in total carbohydrate.

The fact that sialic acid is located on the exterior surface of the plasma membrane suggests its involvement in interactions with the immediate environment of the cell. This has been supported by reports of the involvement of sialic acid in various surface-related phenomena, such as aggregation of trypsin-dissociated fibroblasts (Kemp, 1968), 'homing' of rat lymphocytes (Gesner and Woodruff, 1969; Woodruff and Gesner, 1969), adhesive properties of tumor cells (Gasic and Gasic, 1962b), binding of various viruses to cell surfaces (*see* Winzler, 1970) and the masking of certain surface antigens in murine lymphoid cells (Schlesinger and Amos, 1971), glial cells (Herschman, Breeding and Nedrud, 1972) and human lymphoid cells (Rosenberg, Polcinik and Rogentine, 1972).

Treatment with *Vibrio cholera* neuraminidase has been observed by several investigators to decrease the transplantability of tumors, and this is discussed in Section 5.6.

5.3.2   Glycolipids

The chemistry of the glycosphingolipids has been reviewed by several authors (Svennerholm, 1964; Leeden, 1965; McKibbin, 1970; Christie, 1973). This class of lipids contains mono-, di- or oligosaccharides glycosidically linked to the primary hydroxyl group of ceramide. The conjugation of the hydrophobic, hydrocarbon chains of ceramide with the hydrophilic saccharide moiety confers unique solubility properties upon these molecules, thereby dictating their orientation in the plasma membrane. The hydrophobic portion is buried in the hydrophobic interior of the plasma membrane while the saccharide moiety extends into the aqueous environment immediately surrounding the cell.

Although glycosphingolipids were first isolated from mammalian nervous tissue, it has since been demonstrated that these substances are present in virtually all mammalian tissues and demonstrate a higher degree of species and tissue specificity than perhaps any other lipid class (Svennerholm, 1964; Carter, Johnson and Weeber, 1965; Leeden, 1965). Marked differences have been observed in the carbohydrate portions of glycosphingolipids isolated from various ascites tumors (Gray, 1965), normal kidney of various mouse strains (Adams and Gray, 1968) and erythrocytes of different animal species (Sweeley and Dawson, 1969). Weinstein *et al.* (1970) have suggested that the immunological properties and tissue specificity of glycosphingolipids may be an important part of the surface pattern of cells which is involved in such biological properties as the recognition of cell types and the interaction of the cell surface with its immediate environment.

The localization of glycolipids at the cell periphery was inferred from early experiments which indicated that cellular glycolipids were responsible for the blood-group specificities borne on the erythrocyte cell surface. It is now known that A, B, H(O) and Lewis blood-group antigens are specific carbohydrate structures of glycosphingolipids in the erythrocyte membrane and that the blood-group glycolipids may be isolated from erythrocyte stroma (see Hakomori, 1970a).

The immunologic behavior of glycosphingolipids as haptens has been known for many years. Glycolipid haptens have been isolated from human cancer tissue (Rapport et al., 1959; Hakomori and Jeanloz, 1964), human brain (Jaffe, Rapport and Graf, 1963; Rapport, Graf and Leeden, 1968), human kidney (Rapport, Graf and Schneider, 1964) and rat lymphosarcoma (Rapport, Schneider and Graf, 1967).

Dodd and Gray (1968) provided the first direct evidence that glyco-sphingolipids are associated with the plasma membrane of rat liver cells. Klenk and Choppin (1970) demonstrated a three- to eightfold purification of neutral glycolipid and gangliosides in the isolated plasma membrane of cultured bovine kidney cells and a fourfold purification of the gangliosides in the plasma membrane of cultured hamster kidney cells. While these investigations pointed to a plasma membrane localization of glycosphingo-lipid, more detailed investigations (Renkonen et al., 1972; Weinstein et al., 1970) revealed that glycolipid may be located on intracellular membranes as well. Weinstein et al. (1970) showed that while hematoside, disialoganglio-side, monosialoganglioside and lactosylceramide were the major glycolipid components of cultured L cells, only hematoside and disialoganglioside were found associated with the plasma membrane of these cells. In some cases, therefore, glycosphingolipids may demonstrate membrane specificity as well as tissue and species specificity. While much of the work on glyco-sphingolipids in complex mammalian cells has been done on material isolated from whole-cell extracts, these investigations indicate the necessity of rigorously defining the subcellular location prior to the interpretation of chemical comparison of cell surface glycolipids isolated from various cell types.

Specific antisera prepared against purified glycolipids may cause aggluti-nation or immunolysis of cells, presumably by interacting with the glycolipid determinant at the cell surface. This technique has been used to study the presence of Forssman antigen on normal and transformed BHK cells (Makita and Seyama, 1971; Burger, 1971b). In these studies, however, the Forssman glycolipid was not isolated, and Kijimoto and Hakomori (1972) have suggested this reactivity may be due to a carbohydrate determinant on glycoprotein rather than glycolipid since they were unable chemically to identify Forssman glycolipid as a component of BHK cells.

Gahmberg and Hakomori (1973a) have recently developed a method for ascertaining the cell surface location of glycolipid without the necessity of isolating the purified plasma membrane. These authors have treated intact erythrocytes with galactose oxidase followed by reduction of the resulting 6-aldogalactosyl residue with sodium [^{3}H]borohydride, thus introducing a radioactive label into surface heteroglycans that are accessible to galactose oxidase and that possess galactosyl residues unsubstituted at C-6. By utilizing this technique it may be possible to follow plasma membrane purification by a more sensitive method than the conventional use of marker enzymes, while labeled, cell-surface, carbohydrate-containing components may be isolated without the necessity for first purifying the plasma membrane. This technique has recently been utilized to demonstrate changes in cell surface glycolipids and glycoproteins associated with altered growth behavior of transformed 3T3 and NIL cells (Gahmberg and Hakomori, 1973b).

The isolation from human epidermoid carcinoma of cytolipin H, later shown to be identical to lactosylceramide, and the demonstration that this glycolipid would react with rabbit antibodies produced against different types of human tumors (Rapport *et al.*, 1958, 1959) provided early evidence suggesting that glycosphingolipids were important components of tumor cell surfaces. The first demonstration of chemical differences in glycosphingolipids of normal and transformed cultured cells was provided by Hakomori and Murakami (1968). These authors demonstrated that when BHK 21 cells are transformed by polyoma virus, hematoside is decreased while lactosylceramide shows a concomitant increase. Spontaneously transformed BHK 21 cells, which were more malignant than the control clone and less malignant than the virally transformed clone, showed a glycolipid pattern similar to that of the virally transformed clone, but to an intermediate degree. Since lactosylceramide is a precursor of hematoside, Hakomori and Murakami (1968) advanced the hypothesis that one of the molecular events which accompanies spontaneous or viral transformation of cultured hamster fibroblasts is lack of completion of the oligosaccharide chain in the glycosphingolipids. This hypothesis was supported by similar observations for virally transformed 3T3 cells and human heteroploid fibroblasts (Hakomori, Teather and Andrews, 1968).

In studying the ganglioside patterns of cultured normal and transformed mouse cell lines, Mora *et al.* (1969) observed a decrease in the disialo- and monosialogangliosides of virally transformed cells, but not of spontaneously transformed cells. This decrease in higher or more complex gangliosides was accompanied by the appearance of hematoside as the principal ganglioside of the transformed cell type (Brady and Mora, 1970). This alteration was postulated to be the result of a block in the synthesis of the higher gangliosides resulting from decreased activity of a UDP–*N*-acetylgalactosamine:hematoside *N*-acetylgalactosaminyltransferase (Cumar *et al.*, 1970). While the observation of decreased complexity of the glycosphingolipids of the virally transformed cells was in basic agreement with the work of Hakomori and coworkers, Cumar *et al.* suggested that such changes occurred only in those cells which contained the viral genome.

More recently, Den *et al.* (1974) investigated the ganglioside pattern and glycosyltransferase activity of transformed hamster cells and their revertants, which showed normal growth characteristics *in vitro*. These investigators demonstrated the presence of similar blocks in UDP–*N*-acetyl-D-galactosamine:hematoside *N*-acetylgalactosaminyltransferase activity and ganglioside synthesis in both cell types. They interpreted these data as indicating that such blocks were therefore not associated with growth regulation, but were perhaps due to a common genetic change in transformed and revertant cells not found in their normal counterparts.

The loss of higher gangliosides has also been observed in several slow-growing and nonglycolytic Morris hepatomas when compared with normal rat hepatocytes (Brady, Borek and Bradley, 1969; Cheema *et al.*, 1970; Siddiqui and Hakomori, 1970). These results indicate that the altered patterns of glycosphingolipids are not the result of rapidly dividing cells or localized pH changes associated with the increased lactate production of cells with a high glycolytic rate, but are probably due to inherent changes in glycosphingolipid metabolism associated with the malignant state.

Observations of density dependent changes in carbohydrate include the presence of greater amounts of heteroglycan associated with cells grown in monolayer relative to cells grown in suspension in the absence of cell–cell contact (Shen and Ginsburg, 1968) and the increase in a certain membrane glycopeptide fraction of mouse fibroblasts at high cell density (Meezan *et al.*, 1969). Hakomori (1970b) investigated the alterations of glycospingolipids of normal cultured BHK fibroblasts as compared with virally transformed cells at different cell densities. The results of this investigation indicated that the amounts of certain glycolipids increase when normal cells reach confluency, presumably owing to extension of the carbohydrate by addition of monosaccharides to precursor glycolipid. Interestingly, the contact-sensitive glycolipids were found to be significantly reduced or totally absent in transformed cells, and their concentrations were independent of cell density.

The most sensitive extension response was described by Hakomori (1970b) as the addition of the terminal α-galactosyl residue to ceramide trihexoside in hamster kidney cells and the terminal sialyl residue of disialohematoside of human diploid cells. Kijimoto and Hakomori (1971) subsequently demonstrated that the enzyme responsible for the addition of the terminal α-galactosyl residue to ceramide trihexoside (UDP–galactose: lactosylceramide α-galactosyltransferase) was increased two- to threefold in contact-inhibited cells when compared with cell groups at lower density. In polyoma-transformed cells the enzyme activity was less than 50 percent that of normal growing cells and was not influenced by population density. Thus, the incomplete synthesis of oligosaccharide chains observed in transformed cells may be a result of the lack of glycosyl extension of density-dependent glycolipids.

The presence of density-dependent glycolipids has also been described in another hamster fibroblast line (NIL) by Robbins and Macpherson (1971) and Sakiyama, Gross and Robbins (1972). These workers have concluded that, in general, the density-dependent response in nontransformed cells results in the increased synthesis of more complex neutral glycolipids, while virally transformed cells show no density-dependent glycolipid synthesis. Sakiyama, Gross and Robbins (1972) have studied several such lines of NIL cells which have differing glycolipid patterns and different saturation densities. While they could not demonstrate a correlation between the complexity of the glycolipid pattern and saturation density, they were able to demonstrate that in cell lines containing Forssman glycolipid, globoside and ceramide trihexoside, the density-dependent response was limited to the neutral glycolipids. One very interesting observation was that in one nontransformed clone (NIL 1c1), none of these neutral glycolipids was present. In this instance the density-dependent response was observed in hematoside. Virally transformed cells which share the same glycolipid pattern as the nontransformed NIL 1c1 line did not demonstrate the density-dependent response in hematoside.

While the mechanism involved in the glycosyl extension of glycolipid and possibly glycoprotein is far from being understood, this interesting phenomenon has resulted in considerable speculation concerning the involvement of cell surface carbohydrate in the control of cell division and cell–cell recognition. If this phenomenon were interpreted in light of the hypothesis that has been presented by Roseman and coworkers (Roseman, 1970; Roth and White, 1972), it may be that the carbohydrate moiety of cell surface glycolipid and/or glycoprotein serves as substrate for cell surface glycosyltransferases when cells make contact. As suggested by Critchley and Macpherson (1973), this type of contact-dependent extension could result in the accumulation of specific carbohydrate moieties as 'byproducts' of reactions involved in cellular adhesion and/or recognition (*trans*-glycosylation). The altered cellular adhesion and recognition observed in tumor cells may result from insufficient glycosyltransferase activity at the cell surface rather than altered cell surface carbohydrate (substrate) as suggested by Roseman (1970). However, Critchley, Graham and Macpherson (1973) have shown that the density-dependent glycolipids of NIL 2 hamster cells are present at the cell surface but are not confined to it. These authors suggested that *trans*-glycosylation is unlikely to be the only mechanism for the synthesis of density-dependent glycolipids since they are present in intracellular membranes. The *trans*-glycosylation hypothesis also suffers from the lack of evidence for available glycosyl donors at the cell surface since sugar nucleotides are not found in extracellular fluids.

An alternative explanation, suggested by Critchley and Macpherson (1973), was that the accumulation of density-dependent oligosaccharide chains triggers cellular events which inhibit cell division. The decreased levels or absence of density-dependent glycolipid in tumor cells might prevent initiation of these cellular events. Hakomori (1971) suggested earlier that the synthesis of contact-sensitive oligosaccharide chains may change the conformation of a 'repeating unit' of cell membrane, resulting in a message being transferred to the nucleus by means of a 'cooperative' change in the plasma membrane. In support of this hypothesis, Laine and Hakomori (1973) demonstrated that the artificial enhancement of the glycolipid content in plasma membrane of NIL cells (achieved by growing the cells in media enriched with exogenous globoside) results in a greatly reduced growth rate of the cultured cells. While these observations certainly suggest an association between cell surface oligosaccharide moieties and the density-dependent inhibition of cell division, a more detailed understanding of the control mechanisms involved in the biosynthesis of cell surface oligosaccharide moieties is necessary if their role in cellular growth control is to be elucidated.

More recently, Yogeeswaran, Laine and Hakomori (1974) investigated the mechanism of contact-dependent enhancement of glycolipid synthesis using glycolipid covalently attached to glass surfaces as substrates. These authors demonstrated a contact-dependent transfer of galactose residues to these substrates by intact cells in the absence of exogenously supplied sugar nucleotide or detergents. This represented the first observation of the transfer of monosaccharide residues from intracellular metabolites to exogenous acceptors through the plasma membrane. The mechanism for this transfer phenomenon has been suggested to involve a lipid-soluble, glyco-phospholipid intermediate in the form of retinol-phosphate–galactose

(Yogeeswaran, Laine and Hakomori, 1974). The involvement of glyco-phospholipid intermediates in the synthesis of glycolipid and glycoprotein has been investigated by other workers (Baynes, Hsu and Heath, 1973; Behrens *et al.*, 1973; Waechter, Lucas and Lennarz, 1973; Wolf, Brecken-ridge and Shelton, 1974; Hsu, Baynes and Heath, 1974; Lucas, Waechter and Lennarz, 1975; Spiro, 1975). These intermediates may be involved in '*trans*-glycosylation' reactions as proposed by Roseman (1970) and Roth and White (1972).

5.3.3   Glycoproteins

Although glycoproteins have been successfully isolated from the plasma membrane of erythrocytes, similar approaches applied to tumor cells have been somewhat less successful. The isolation of purified plasma membrane from tumor cells suffers from several disadvantages: (a) the separation of the plasma membrane from the more abundant intracellular membranes; (b) the enzymatic degradation of the plasma membrane by intracellular glycosidases and proteases released during rupture of the cells, and (c) the redistribution and/or loss of membrane components during isolation (Kramer, 1971).

Enzymatic removal of surface material from viable, intact cells, on the other hand, represents a procedure which avoids cell rupture and is therefore thought to be specific for cell surface carbohydrates accessible to enzymes. This method is especially useful for studying surface carbohydrates of isolated cells, including erythrocytes, cultured cells and ascites tumor cells. Investigations involving the proteolytic release of surface glycopeptides rely on the relative resistance of intact, viable cells to proteolytic digestion. This resistance to proteolytic lysis has been referred to as the 'trypsin barrier' (Northrup, 1926; Kramer, 1971). The use of proteolytic enzymes for nondisruptive studies of cell surface carbohydrates has been recently reviewed by Kramer (1971) and Wallach (1972). While this approach has been valuable in the isolation of glycopeptide fragments from intact cells, the absolute cell surface specificity of such structures has been questioned by Jett and Jamieson (1973) from studies of the surface glycopeptides of cultured RAJI lymphoid cells (*see also* Chapter 3). More accurate interpre-tations of nondisruptive studies will require a combination of the non-disruptive enzymatic approach and studies of protease digestion of isolated plasma membranes.

Langley and Ambrose (1964) demonstrated that glycopeptides could be isolated from the surface of tumor cells by treatment of intact Ehrlich ascites cells with trypsin. Shen and Ginsburg (1968) subsequently analyzed the carbohydrate released from HeLa cells by the action of trypsin. While these investigations demonstrated the presence of glycoprotein at the surface of tumor cells, they were not amenable to comparative studies on normal homologous cell types.

Wu *et al.* (1969) performed a comparative study on the carbohydrate-containing components of the subcellular membrane fractions of normal and SV_{40}-virus-transformed 3T3 mouse fibroblasts. These workers observed a decrease in the content of most neutral and amino sugars, as well as qualitative differences in the carbohydrates in the particulate fractions of SV_{40}-transformed cells relative to nontransformed cells. Since the changes observed were reflected in all particulate fractions (nuclei, mitochondria, surface membrane and endoplasmic reticulum), these authors suggested that one defect of transformed cells manifested itself in lower carbohydrate content in not only surface membrane, but all the membranes of the cell. Meezan et al. (1969) subsequently examined the change in the composition of membrane glycoproteins by fractionating the labeled membrane fractions derived from 3T3 and transformed 3T3 cells on Sephadex G-150. This investigation revealed that significant quantitative differences existed in the membrane glycoproteins and that these differences were again reflected in all subcellular fractions studied. One glycoprotein fraction corresponding to a molecular weight of 20 000 was found in much higher quantities in normal than in transformed 3T3 cells. This was of particular interest since this glycoprotein fraction was present at elevated levels in confluent relative to growing cultures of 3T3 cells, suggesting a possible physiological variation of the amount of this fraction and a loss of this glycopeptide in transformed cells. When the labeled subcellular fractions were digested with pronase and submitted to chromatography on Sephadex G-50, both quantitative and qualitative differences were again observed in the glycopeptides of normal and transformed cell membranes. Similarly, Onodera and Sheinin (1970) have demonstrated that qualitative differences exist in the glyco-protein-containing surface components from SV_{40}-transformed and nontransformed 3T3 cells. DEAE-cellulose chromatography of the trypsin-treated surface component revealed the presence of two gluco-samine-containing fractions that were not present in the SV_{40}-transformed cells.

Buck, Glick and Warren (1970) demonstrated that major differences exist between the cell surface glycoproteins and glycopeptides from pronase-digested membranes of nontransformed hamster cells and those from such cells transformed by Rous sarcoma virus. In all cases, regardless of the state of growth during labeling with L-fucose, the transformed cells, relative to nontransformed controls, were enriched in a glycopeptide fraction which eluted more quickly from Sephadex G-50 gels. These results were extended and confirmed by investigations comparing normal and polyoma-virus-transformed hamster cells, normal mouse fibroblasts and the same cells transformed by murine sarcoma virus, and normal chick embryo fibroblasts and the same cells transformed by Rous sarcoma virus (Buck, Glick and Warren, 1971).

Using the same technique with chick embryo fibroblasts transformed by a temperature-sensitive mutant of Rous sarcoma virus, Warren, Critchley and Macpherson (1972) demonstrated a similar, early-eluting glycopeptide fraction. This fraction was obtained from cells grown at their permissive temperature (36 °C), at which they appeared to be morphologically trans-formed. When studied at the nonpermissive temperature (41 °C), the cells behaved normally in culture and the elution profiles of the cell surface glycopeptides were the same as control cells. These data, combined with the fact that cells infected with a Rous-associated virus (a non-oncogenic virus) produced no change in the glycopeptide (Warren, Critchley and Macpherson, 1972), suggested that the alterations observed were a manifestation of the transformed cell surface and not simply a consequence of virus infection.

More recently, Warren, Fuhrer and Buck (1973) have shown that neuraminidase digestion of the glycopeptide fraction from control and transformed cells results in the disappearance of the more rapidly eluting glycopeptides associated with the transformed cell line. They also demonstrated the presence of a sialyltransferase in both normal and transformed cells which transferred sialic acid from CMP–sialic acid to the desialylated, early-eluting, glycopeptide fraction. This activity, which appeared to be associated with the surface membrane, was increased 2.5- to 11-fold in the transformed cells relative to nontransformed cells. In chick embryo fibroblasts transformed by the temperature-sensitive virus, the sialyltransferase activity was higher in cells grown at the permissive temperature than in those grown at the nonpermissive temperature.

While these investigations have demonstrated alterations in cell surface glycoprotein associated with viral transformation, Glick, Rabinowitz and Sachs (1973) have demonstrated that similar alterations are associated with tumorigenesis. The amount of this specific group of fucose-containing cell surface glycopeptides shows a direct correlation with the tumorigenicity of various cell lines derived from tumors produced *in vivo* by dimethyl-nitrosamine-transformed hamster cells. This group of fucose-containing glycopeptides was found to be present only in very low quantities in the chemically transformed cells prior to passage *in vivo* and was absent in variants of tumors which were only weakly tumorigenic.

One explanation for the differences observed in the carbohydrate content of normal and transformed cells might be changes in levels of glycosyl-transferase activities as a result of malignant transformation. Grimes (1970, 1973) has demonstrated an inverse correlation in the levels of sialyltransferase activities and the saturation densities of normal, SV_{40}-transformed and spontaneously transformed 3T3 cells in culture. Furthermore, quantitative measurements of sialic acid indicated that a decrease in sialic acid associated with transformed cells correlated with the reduced sialyltransferase activity. This led to the suggestion that glycosyltransferases might determine the amounts of sugar on glycoproteins and glycolipids (Grimes, 1970). Similar observations were made by Den *et al.* (1971), who reported decreases in glucosaminyl-, galactosyl- and sialyltransferases in polyoma-transformed BHK relative to nontransformed BHK cells.

It may be concluded, therefore, that alterations of cell surface glyco-proteins are truly a manifestation of the transformed state, possibly as a result of altered glycosyltransferase activities. However, unlike the investigations of glycolipid alterations in tumor cells, work with cell surface glyco-peptides has failed to reveal specific structural changes; rather, these investigations indicate quantitative, and/or qualitative differences. Indeed, there has not yet been reported a comparative structural analysis of a cell surface glycopeptide from normal and transformed cell types, as has been the case for the glycosphingolipids. There are several technical difficulties that explain this lack of information. First, unlike glycolipids, which represent a series of similar structures synthesized from a common precursor, glycopeptide components represent a more complex group which, in general, contain a larger variety and number of monosaccharide residues.

The heterosaccharide chains of glycoproteins are covalently linked to peptide and are not as easily extracted and purified as are glycosphingolipids. Secondly, while the heterosaccharide structures of glycolipid are generally assigned by their chromatographic behavior on thin-layer chromatography relative to known standards and confirmed by chemical analysis, the more complex heterosaccharide structures of glycopeptides require more rigorous examination owing to the lack of available glycopeptide standards and chromatographic techniques which provide adequate separation. Thirdly, perhaps the greatest obstacle is the fact that at least milligram quantities of purified glycopeptide are required for structural analysis. This is a large order since cell surface glycoprotein represents only a minor portion of the total cell weight, and the glycoprotein must be further purified, protease-digested, and fractionated into component glycopeptides. This obstacle may, however, be overcome by large-scale cell cultures or the use of ascites tumor cells in comparative studies.

Milligram quantities of cell surface glycopeptides have been isolated from Novikoff (Walborg, Lantz and Wray, 1969) and AS-30D rat ascites hepatoma cells (Smith and Walborg, 1972). These systems are amenable to comparative studies with readily available cells such as normal, regenerating or neonatal rat liver, and may prove useful in detecting structural alterations in cell surface glycopeptides associated with neoplastic transformation.

5.4   DIFFERENTIAL LECTIN-INDUCED AGGLUTINATION OF NORMAL AND TRANSFORMED CELLS

The increased agglutinability of tumor cells by certain saccharide-binding proteins, termed *lectins*, suggests that neoplastic transformation is accompanied by an altered expression of cell surface oligosaccharide moieties. Transformed cells, in contrast to their normal counterparts, express lectin-binding sites at their periphery which are responsible for agglutination by several plant lectins, such as concanavalin A (Con A) (Inbar and Sachs, 1969a), wheat-germ agglutinin (WGA) (Aub, Sanford and Cote, 1965; Burger and Goldberg, 1967), soybean agglutinin (Sela *et al.*, 1970) and *Ricinus communis* or castor-bean agglutinin (Nicolson and Blaustein, 1972). Because of their defined saccharide-binding specificities, these proteins have become powerful biochemical probes for the detection of alterations in the expression of cell surface oligosaccharide moieties in transformed cells (*see* Sharon and Lis, 1973).

Since Con A and WGA have been most intensively studied and purportedly possess the highest degree of specificity for the agglutination of tumor cells (Burger, 1973), they will be the primary object of the following discussion. Increased agglutinability by Con A and WGA has been shown to occur upon transformation of cells by viruses, by chemical carcinogens, or by X-irradiation, indicating that a similar cell surface alteration is common to transformed cells of etiologically different origin (Biddle, Cronin and Sanders, 1970; Inbar and Sachs, 1969a). Since lectins interact with specific saccharide moieties—i.e. Con A with α-D-manno- or α-D-glucopyranosyl residues (Poretz and Goldstein, 1970) and WGA with 2-acetamino-2-deoxy-β-D-glucopyranosyl residues (Allen, Neuberger and Sharon, 1973)—

the altered agglutinability is thought to involve changes in the expression of oligosaccharide moieties at the cell surface.

A large body of indirect evidence indicates that the expression of Con A and WGA binding sites is related to cellular growth control. The degree of agglutinability of virally transformed cell lines by WGA is directly related to their saturation densities (Pollock and Burger, 1969). Revertants of transformed cells, i.e. cells which exhibit recovery of contact inhibition, lose their agglutinability by Con A (Inbar, Rabinowitz and Sachs, 1969) while revertants selected on the basis of resistance to Con A toxicity demonstrate the recovery of contact inhibition of growth (Culp and Black, 1972; Wollman and Sachs, 1972). Similarly, investigations utilizing cells transformed by a temperature-sensitive mutant of polyoma virus (Eckhart, Dulbecco and Burger, 1971) and temperature-sensitive SV_{40}-transformed cells (Noonan et al., 1973) have demonstrated that the enhanced expression of WGA and Con A binding sites is correlated with the loss of cellular growth control at the permissive temperatures.

Another line of investigation which suggests the involvement of lectin binding sites in growth control is the demonstration that nontransformed cells are released from their contact-inhibited state by brief exposure to protease, a treatment which also results in the agglutinability of the nontransformed cells by WGA and Con A (Burger and Noonan, 1970). It has also been demonstrated that growth control of transformed cells is apparently reestablished if exposed Con A binding sites are covered with monovalent Con A (Burger and Noonan, 1970; Burger, 1973). The mechanism of this phenomenon is not clear, but Burger and Noonan suggested that very subtle and specific surface changes may result in altered cellular growth control. One explanation for the loss of growth control of transformed cells is that the surface of the tumor cell appears to be locked in a mitotic configuration (Burger, 1973). Evidence for this hypothesis stems from the work of Fox, Sheppard and Burger (1971), who demonstrated that WGA binding sites were permanently expressed on transformed cells and detected in normal cells only during mitosis.

The relationship between tumorigenicity and cytoagglutinability by Con A and WGA is unclear. While positive correlations between agglutinability of transformed cells by lectins and their tumorigenicity have been demonstrated (Inbar, Ben-Bassat and Sachs, 1972; de Micco and Berebbi, 1972), a number of authors have reported the lack of such a correlation. The highly malignant, non-strain-specific TA3-Ha murine mammary adenocarcinoma was not agglutinable by Con A in contrast to the strain-specific subline (TA3-st) and other murine ascites tumor cells (Friberg, 1972). Gantt, Martin and Evans (1969) were unable to show any correlation between malignancy and agglutinability of several mouse tumors by WGA, while Dent and Hillcoat (1972) observed a similar lack of correlation when relating Con A agglutinability to the malignancy of three mouse lymphoma lines. It is also interesting that revertants of transformed cells which regained their ability to be contact-inhibited in vitro and lost their agglutinability by Con A, retained a high degree of tumorigenicity in animals (Inbar, Rabinowitz and Sachs, 1969).

However, the implied role of Con A and WGA binding sites in the regulation of cellular growth and the usefulness of lectins in studies of cell surface topography have provided the impetus for investigation of the molecular mechanism operative in the altered agglutinability of transformed cells. Early hypotheses explained the increased agglutinability of tumor cells in terms of 'cryptic' lectin binding sites which became expressed as a result of topological rearrangements of the plasma membrane associated with neoplastic transformation (Burger, 1969; Inbar and Sachs, 1969a). This conclusion was based primarily on the observation that nontransformed cells became agglutinable following treatment with proteases and that proteases temporarily induced the same surface alterations in nontransformed cells as seen with oncogenic agents (Burger, 1970a). Burger (1973) and Schnebli and Burger (1972) have discussed the role of proteases as possible mediators in the loss of growth control of cells in culture.

Subsequent investigations of this mechanism have been repeatedly frustrated by the lack of a quantitative assay for cytoagglutination. Immunological assays utilizing microscopic analysis of cytoagglutination are difficult and semiquantitative at best (Burger, 1973; Sharon and Lis, 1973). Radioassays which attempt to determine the number of isotopically labeled lectin molecules bound to cell surfaces have been developed utilizing labeled WGA (Ozanne and Sambrook, 1971), Con A (Inbar and Sachs, 1969b; Ozanne and Sambrook, 1971; Cline and Livingston, 1971; Arndt-Jovin and Berg, 1971; Noonan and Burger, 1973a), *Ricinus communis* or castor-bean agglutinin (Nicolson, 1973a) and soybean agglutinin (Sela, Lis and Sharon, 1971). While these appeared to be excellent assays for the quantitation of lectin binding sites on the cell surface, early results utilizing this technique, with one exception (Inbar and Sachs, 1969b), not only failed to demonstrate a positive correlation between amount of lectin bound and the cytoagglutinability of cells by the corresponding lectin, but concluded that the number of binding sites on normal and transformed cells was similar (Cline and Livingston, 1971; Ozanne and Sambrook, 1971; Arndt-Jovin and Berg, 1971; Ben Bassat, Inbar and Sachs, 1971). Noonan and Burger (1973a), however, have recently questioned this conclusion with experimentation utilizing a [^{3}H]Con A binding assay which eliminated endocytosis of Con A and nonspecific binding. With this assay it was demonstrated that virally transformed 3T3 cells bind 2.5–5 times as much labeled lectin as do normal 3T3 cells. The 2.5 to 5-fold increase in binding, however, may still not account for the 12 to 20-fold increase in agglutinability of these transformed lines. This may be concluded from the fact that even when Con A sites on transformed cells were saturated, as determined by the binding assay, only half-maximal agglutination of these cells was observed (Noonan and Burger, 1973a). This suggested that even when very controlled conditions are used for binding assays, these techniques may not be able to measure the most important binding sites, namely those involved in cytoagglutination. This would be expected if cytoagglutination is a result of weak binding forces which are lost or broken by the repeated washings necessary for the assay of labeled-lectin binding. In order accurately to measure the amount of lectin necessary for maximal agglutination, therefore, it may be necessary to perform these investigations under equilibrium conditions. Thus, until methods which measure 'agglutinating or productive' binding sites are developed, the question of whether normal and transformed cells possess equal or unequal numbers of binding sites may remain unresolved.

Nonetheless, the suggestion that normal and transformed cells have similar numbers of binding sites available for interaction with labeled lectin prompted the development of models other than the 'cryptic site' hypothesis. Ben-Bassat, Inbar and Sachs (1971) proposed that the altered cytoagglutin-ability of transformed cells could be a result of changes in the topological arrangement of the lectin receptors at the cell surface, for example an increased density (sites per μm^2) of lectin receptors resulting from a decrease in cell volume or a localized clustering of receptors without a change in cell volume. The occurrence of the latter rearrangement was observed by Nicolson (1971) using a method which permitted electron-microscopic investigation of the two-dimensional distribution of ferritin-conjugated lectins or antibodies on cell surfaces (Nicolson and Singer, 1971). Ferritin-conjugated Con A appeared randomly dispersed on nontransformed cells while transformed cells, or trypsin-treated nontransformed cells, showed a clustered arrangement of the electron-dense lectin. Nicolson (1971) suggested that increased agglutinability could be explained by the observed clustering of Con A binding sites which appeared after transformation or protease treatment of normal cells. This hypothesis was supported by the observation that thin-sectioned samples of cells agglutinated by ferritin-conjugated Con A demonstrated a concentration of lectin at contact points between cells (Nicolson, 1972). Similar conclusions were drawn from electron microscopic investigations of the surface distribution of Con A on normal and polyoma-transformed hamster embryo cells treated with peroxidase-coupled Con A. The peroxidase product was less uniformly distributed in transformed as compared with nontransformed cells (Martinez-Palomo, Wicker and Bernhard, 1972; Bretten, Wicker and Bernhard, 1972).

More recently, Nicolson (1973b) has shown by direct and indirect fluorescent labeling techniques that the clustering of Con A binding sites on the surface of malignant 3T12 or virally transformed 3T3 cells was a function of the temperature at which cells were exposed to Con A and fluorescent label prior to fixation of the membrane with formaldehyde. These results indicated that at low temperatures (0 °C) the Con A binding sites of transformed cells were immobilized and after fixation the fluorescent label appeared uniformly distributed. At higher temperatures (22–37 °C) the binding sites appeared clustered, presumably by Con-A-induced cross-linking of receptors in the same cell. In nontransformed 3T3 cells, on the other hand, the distribution of binding sites was uniform regardless of the temperature at which the cells were labeled with Con A. The ability of Con A to induce clustering of surface binding sites was subsequently shown by Rosenblith et al. (1973), who demonstrated that the Con A binding sites on normal, trypsinized-normal or transformed 3T3 cells were uniformly distributed at 4 °C while a rapid rearrangement of Con A binding sites to a clustered configuration occurred at 37 °C in the transformed and trypsinized cells but not in normal intact cells. This rearrangement was inhibited if cells were fixed with formaldehyde prior to the temperature shift. It was concluded from these observations (Nicolson, 1973b) that increased membrane fluidity permitted clustering of lectin binding sites and thus explained the enhanced agglutinability of transformed cells.

This hypothesis is, at present, dependent on two important assumptions: (a) that the increased agglutinability of transformed cells is directly and primarily related to the clustering of lectin receptors, and (b) that normal and transformed cells possess structurally identical lectin receptors. With respect to the first assumption it appears that clustered binding sites and increased agglutinability are associated phenomena. However, neither Nicolson (1973b) nor Rosenblith *et al.* (1973) investigated the agglutinability of transformed cells which carried clustered sites (cells incubated with Con A at 37 °C followed by fixation) as compared with transformed cells which carried uniformly distributed sites (cells incubated with Con A at 0 °C followed by fixation). In fact, little direct evidence has been reported to prove that clustering of sites is an obligatory requirement for agglutination, nor has there been a report, based on experimentation, of a mechanism whereby clustered sites would result in increased agglutinability. The best direct evidence, however, comes from Inbar, Rabinowitz and Sachs (1973), who demonstrated the inhibition of Con A agglutination of a mouse lymphoma by prolonged fixation of the cells, a treatment which was shown to inhibit the Con-A-induced clustering of Con A binding sites (Inbar and Sachs, 1973). The mouse lymphoma, however, may be an exception in this case since Inbar and Sachs (1973) have suggested that transformation of cells (e.g. lymphocytes and lymphoma cells) which grow in suspension rather than from solid tissues *in vivo*, may demonstrate different responses in membrane fluidity; that is, decreased fluidity on transformation as opposed to increased fluidity. Earlier investigations by Inbar, Rabinowitz and Sachs (1971) demonstrated that the agglutinability of transformed cells was inhibited at low temperatures (4 °C). This could perhaps be explained by immobilization of the binding sites at low temperature as suggested by more recent investigations (Nicolson, 1973b; Noonan and Burger, 1973a; Inbar and Sachs, 1973). However, trypsin-treated normal and transformed cells were agglutinable even at low temperatures, which suggested two interesting possibilities. First, as suggested by Inbar, Rabinowitz and Sachs (1971), there may be more than one type of binding site. Temperature-sensitive sites may be present only on intact, transformed cells while temperature-insensitive, trypsin-exposed (activated) sites may be present on both normal and transformed cells. Secondly, since agglutination of trypsin-treated cells occurred at 4 °C, where the binding sites are thought to be immobilized, agglutination may occur in some cases without clustered sites.

However, Gordon and Marquardt (1974) have demonstrated that trypsinized human erythrocytes are agglutinated by Con A at both 24 and 4 °C. The rate of agglutination in a cold environment is significantly slower than at 24 °C. These authors suggest that the inability of cells to agglutinate in the presence of Con A at low temperatures may reflect changes in the agglutinating ability of Con A rather than changes in the cell surface. This hypothesis is supported by the known temperature dependence of the Con A tetramer–dimer dissociation (Kalb and Lustig, 1968; McKenzie, Sawyer and Nichol, 1972). Gordon and Marquardt (1974) also demonstrated that soybean agglutinin-mediated hemagglutination is faster at lower temperatures and that no temperature-dependent change in its molecular form is observed. These authors further suggested that since chemical derivatization of Con A can convert the protein to a dimer with decreased ability to agglutinate erythrocytes and no alteration of its binding ability (Gunther *et al.*, 1973), the temperature-dependent tetramer–dimer dissociation may explain observations that Con A binds to normal and transformed cells equally well at high and low temperatures while the agglutinability of the cells differs (Inbar, Ben-Bassat and Sachs, 1971; Noonan and Burger, 1973b). In light of these observations it may not be unreasonable to suggest that the clustering of Con A sites induced by temperature shifts in the presence of Con A may in some cases be a result of dimer–tetramer association.

At the present time we conclude that no mechanism for the increased agglutinability by lectins of transformed cells which accurately and completely embodies all of the existing data is yet available. In the midst of a voluminous literature which implicates surface architectural rearrangements as a basis for this mechanism, it should be emphasized that the available evidence does not exclude the possibility of qualitatively or structurally different binding sites on normal and transformed cells.

An alternative to the hypothesis that increased membrane fluidity results in increased agglutinability postulates structural differences between the lectin receptors of normal and neoplastic cells. Transformed cells may possess receptors which allow binding of lectin in a manner such that it can bind with another receptor on the same cell or a receptor on another cell, thus promoting both clustering of lectin binding sites and cytoagglutination. Nontransformed cells may possess receptors which allow binding of lectin in such a manner that it cannot bind to other receptors on the same cell or receptors on another cell, resulting in cells that would not exhibit clustering of receptor sites in the presence of lectin (nor would they agglutinate). Another possibility, not excluded by this hypothesis, is the existence of receptors which allow binding of lectin to two adjacent receptors on the same cell but not with a receptor on other cells, resulting in clustering of binding sites without cytoagglutination. As discussed previously, evidence for qualitative differences in cell surface carbohydrates of normal and malignant cells is well documented. It seems, therefore, only logical to conclude that investigations of the chemical structure and organization of lectin binding sites of normal and transformed cells may lead to further insights into the mechanism of their increased lectin-induced agglutination.

5.5   CELL SURFACE LECTIN RECEPTORS

In an effort to characterize the cell surface aberration responsible for the altered agglutinability of transformed cells by Con A and WGA, cellular components which bind these lectins have been isolated. Hakomori et al. (1967) described a glycolipid isolated from human adenocarcinomas which inhibited cytoagglutination by WGA. However, its location in the plasma membrane has not been established so its relationship to cytoagglutination by WGA is uncertain. Burger (1968) demonstrated that L1210 mouse leukemia cells and polyoma-virus-transformed BHK cells lose their agglutinability by WGA when subjected to brief hypotonic shock. Concomitant with the loss of agglutinability, a particulate fraction was released into the supernatant buffer. This particulate component, solubilized by sonication, inhibited cytoagglutination by WGA. Although no evidence has been presented, Burger (1970b, 1973) has stated that similar quantities of WGA receptor were also isolated from nontransformed cells. Janson and Burger (1973) further purified the WGA receptor from L1210 cells and demonstrated that it contains glycoproteins. L1210 cells, which are agglutinable by WGA but poorly agglutinable by Con A, yielded receptor material which possessed WGA receptor activity but no detectable Con A receptor activity (Janson and Burger, 1973; Janson, Sakamoto and Burger, 1973).

Another approach to the isolation of cell surface lectin receptors utilizes limited proteolysis of intact cells (Wallach, 1972). This approach has proved particularly successful for the isolation of human erythrocyte cell surface glycopeptides which bear receptors for a variety of lectins (Leseney, Bourrillon and Kornfeld, 1972; Kornfeld and Kornfeld, 1969; Kornfeld, Roger and Gregory, 1971). This experimental approach has also been applied to the investigation of lectin receptors present on the surface of neoplastic rat liver cells (Neri, Smith and Walborg, 1974a; Neri et al., 1974b, c; Walborg et al., 1975). These investigations employed two transplantable rat ascites hepatomas: Novikoff (Novikoff, 1957) and AS-30D (Smith, Walborg and Chang, 1970). These two hepatomas possess different cytoagglutination properties: Novikoff cells are readily agglutinated by Con A and WGA (Wray and Walborg, 1971; Neri et al., 1974c; Walborg et al., 1975) whereas AS-30D cells are agglutinated by WGA, but only weakly agglutinated by high concentrations of Con A (Smith, Neri and Walborg, 1973; Neri et al., 1974b; Walborg et al., 1975).

Incubation of hepatoma cells with papain releases sialoglycopeptides from the cell surface (Walborg, Lantz and Wray, 1969; Smith and Walborg, 1972; Neri et al., 1974b; Smith, Neri and Walborg, 1973). These glycopeptides contain 30–50 percent of the cell surface sialic acid and can be obtained in high yield (ca. 3.5 mg per 10 g of tumor cells). These cell surface glycopeptides possess receptor activity for Con A and WGA (Smith, Neri and Walborg, 1973; Neri, Smith and Walborg, 1974a), as evidenced by their ability to inhibit lectin-induced hemagglutination. Following papain treatment, rat hepatoma cells become or remain highly agglutinable by Con A or WGA (Smith, Neri and Walborg, 1973; Neri et al., 1974b). This fact, coupled with the observations that the cell surface glycopeptides possess lectin receptor activity, indicates the existence of two qualitatively distinct classes of lectin receptors: glycopeptide receptors cleaved from the cell surface by protease and protease-stable receptors which function in lectin-induced agglutination of protease-digested cells (Smith, Neri and Walborg, 1973; Neri et al., 1974b, c). Such a situation is not unique to tumor cells, since protease-labile and protease-stable lectin receptors exist on the surface of human erythrocytes (Kornfeld and Kornfeld, 1969; Kornfeld, Roger and Gregory, 1971). The protease-stable receptors may be present on plasma membrane glycoproteins or glycolipids.

If cell-surface glycoproteins function in lectin-induced cytoagglutination, glycopeptides solubilized by protease digestion, or a portion thereof, should possess lectin receptor activities reflecting the cytoagglutination properties of the cells from which they were derived. A glycopeptide fraction possessing such properties has been isolated. The cell surface glycopeptides were digested with pronase and the limit sialoglycopeptides resolved by gel filtration on Sephadex G-50. Two sialoglycopeptide fractions were resolved, one having a molecular weight of >4200, which was excluded from the gel, and the other of lower molecular weight (2000–3500) partially excluded from the gel matrix. The fraction of higher molecular weight possessed all the WGA receptor activity and >90 percent of the Con A receptor activity. Furthermore, this sialoglycopeptide fraction exhibited lectin receptor activities closely reflecting the agglutination properties of the cells from which it was derived (Smith, Neri and Walborg, 1973; Neri *et al.*, 1974b, c). This class of cell surface glycopeptides may represent 'productive binding sites' (Sharon and Lis, 1972) which function in the agglutination of rat hepatoma cells by Con A and WGA. The lower-molecular-weight glycopeptides possessed only Con A receptor activity (Smith, Neri and Walborg, 1973; Neri *et al.*, 1974b).

Since the sialoglycopeptide fraction of molecular weight >4200 possessed the major portion of the Con A and WGA receptor activity, this fraction was further resolved by gel filtration and ion-exchange chromatography. A macrosialoglycopeptide fraction was isolated and partially characterized (Smith, Neri and Walborg, 1973; Neri *et al.*, 1974b). The macrosialoglyco-peptide fraction from Novikoff cells possessed potent and specific receptor activity for Con A (Neri *et al.*, 1974b), while a similar fraction from AS-30D cells exhibited potent and specific WGA receptor activity (Smith, Neri and Walborg, 1973). These lectin receptors were designated *macrosialoglyco-peptides* on the basis of the following properties: (a) their high apparent molecular weight, as judged by their behavior on Sephadex G-200; (b) their compositional analysis, particularly the enrichment of those amino acids (aspartic acid, threonine and serine) known to be involved in the linkage of carbohydrate to the peptide chains, and (c) their resistance to proteolysis.

Comparative investigations of the glycopeptides cleaved from the surface of rat liver cells by papain suggested that neoplastic transformation is accompanied by structural alterations of the plasma membrane glycoproteins (Neri, Smith and Walborg, 1974a; Walborg *et al.*, 1975). Rat liver cells, prepared by perfusion with buffer containing Ca^{2+} chelators followed by mechanical dispersal, were digested with papain. Sialoglycopeptides were released from the cell surface in yields comparable to those obtained for hepatoma cells. The gel filtration profile of liver cell surface glycopeptides on Sephadex G-50 was found to differ significantly from that of the hepatoma glycopeptides. Unlike the hepatoma sialoglycopeptides, most of the liver cell surface sialoglycopeptides had partial access to the gel matrix. Further-more, the liver glycopeptides possessed no detectable WGA receptor activity. Although Con A receptor activity was detected in a low-molecular-weight glycopeptide fraction, no Con A receptor activity was detected in the glycopeptide fraction excluded from the gel. Intact or papain-digested liver cells were not agglutinated by high concentrations of Con A or WGA (Neri, Smith and Walborg, 1974a; Becker, 1974). As discussed previously, nontransformed cells (primarily fibroblasts) grown *in vitro* are only weakly agglutinated by high concentrations of Con A or WGA. The non-agglutinability of protease-treated liver cells by Con A or WGA is clearly at variance with this pattern, and suggests the existence of significant differ-ences between the surfaces of cells growing *in vivo* and those growing *in vitro*.

These investigations demonstrate the utility of lectins for the detection and localization of structural alterations of plasma membrane glycoproteins

—alterations temporally related to neoplastic transformation. It is imperative that the structure of the membrane glycoproteins be elucidated to localize the molecular determinants for binding of the lectins and to explain the molecular basis for the altered lectin-induced agglutination of malignantly transformed cells.

5.6 ANTIGENIC ALTERATIONS OF THE TUMOR CELL PERIPHERY

Structural or topological alterations at the tumor cell periphery are now being utilized in the search for new approaches to the detection and therapy of cancer. Recent experimentation concerning the chemistry and immunology of the tumor cell periphery gives reason for optimism, even though practical application to the diagnosis and treatment of human cancer may be slow in developing. Immunological techniques have demonstrated the existence of tumor-associated, cell surface antigens on a variety of malignantly transformed cells. These cell surface antigens include the tumor-associated rejection antigens commonly present on chemically induced tumors (Baldwin, 1972a), surface antigens associated with viral transformation (Pasternak, 1969; Deichman, 1969) and the oncofetal antigens (Alexander, 1972). As in the case of lectin receptors, these antigenic alterations appear to involve the expression of altered glycoproteins at the surface of the malignant cell.

The tumor-associated rejection antigens, which are characterized by their unique specificities for individual tumors and their quantitative variability of expression, have been isolated from aminoazo-dye-induced rat hepatomas (Baldwin, 1972a) and from diethylnitrosamine-induced hepatomas of guinea-pigs (Leonard *et al.*, 1972). Chemical evidence indicates that these tumor-associated rejection antigens are glycoproteins chemically similar to the histocompatibility antigens. This conclusion is based on their localization in the plasma membrane, their solubilization by methods usually employed for the isolation of mouse H-2 or human H-LA antigens, and their amino acid and carbohydrate composition (Baldwin, 1972b). Baldwin and Glaves (1972) demonstrated that the deletion of liver-specific, cell-surface-associated antigens was accompanied by the expression of new tumor-associated rejection antigens. Similar results were obtained in mouse sarcomas induced with 3-methylcholanthrene (Haywood and McKhann, 1971). These observations suggest that the tumor-associated rejection antigens result from the replacement or modification of normal cell surface components.

The chemical nature of the virus-specific antigens present on the surface of cells infected with oncogenic viruses has only recently been investigated. Kennel *et al.* (1973) have isolated an oncornavirus-specific antigen from mouse lymphoblasts which produce the Scripps leukemia virus. This antigen was identified as a plasma membrane glycoprotein which has a molecular weight of approximately 70 000 and which accounts for 10 percent of the cellular glucosamine. This glycoprotein reacts with sera which neutralize Moloney, Kirsten, Rauscher, AKR and Scripps viruses. The methodology developed by Kennel *et al.* (1973) will undoubtedly be applied to the isolation of virus-specific surface antigens resulting from the infection of cells with other oncogenic viruses.

During neoplastic transformation some tumors acquire cell surface antigens which are normally expressed only early in the ontogeny of a particular organ or cell (Alexander, 1972). These antigens, designated *oncofetal antigens*, may offer a means of detecting early cancer in humans. Such a substance (or group of substances with a common antigenic determinant), detected with heterologous antisera, was found to be present in adenocarcinomas of the human digestive tract (Gold, 1971). Because this substance was also detected in the digestive tract of the human fetus (aged 2 to 6 months), the substance was designated *carcinoembryonic antigen* (CEA). Its physicochemical properties and its ease of extraction suggested that it was a glycoprotein similar to those found in the gastrointestinal mucous secretions. Although immunofluorescence techniques have localized the antigen at or near the plasma membrane, this may be a transitory localization associated with its secretion from the cell. However, it should be emphasized that the distinction between metabolic turnover and secretion of membrane glycoproteins is subtle and may reflect only a difference in the relative rates of turnover, dependent in part upon their degree of interaction with the hydrophobic portion of the plasma membrane (Warren, 1969; Kapeller *et al.*, 1973; Doljanski, 1973). A CEA-like substance has been detected in the serum of patients with colonic and pancreatic cancer, thereby suggesting a clinical application in diagnosis (Gold, 1971; Moore *et al.*, 1971). The present method for the detection of CEA has proved useful in the diagnosis of metastasis, in the assessment of effectiveness of surgical resection, and in monitoring the effects of chemotherapy (Zamcheck *et al.*, 1972).

Although there is ample evidence for the occurrence of tumor-associated antigens which can and do elicit an immune response by the host, the appearance and growth of tumors in their autochthonous hosts indicates that, at least some of the time, the host is unsuccessful in rejecting a tumor. This failure may be caused either by an inadequate immune response by the host or resistance of the tumor cells to the host's immune response (Klein, 1972). The resistance of tumor cells to humoral or cell-mediated cytotoxicity has been attributed to adsorption of blocking factors on the surface of target cells (Hellström and Hellström, 1971), antigen loss (Möller, 1964) or masking of antigens by glycosaminoglycans or sialoglycoproteins (Currie and Bagshawe, 1967). This last possibility has recently received support by the isolation of a mucin-like sialoglycoprotein present on the surface of non-strain-specific TA3-Ha mouse mammary adenocarcinoma cells (Codington, Sanford and Jeanloz, 1972a; Slayter and Codington, 1973). This glycoprotein was not detected on the surface of the strain-specific subline (TA3-St) of the same tumor (Codington, Sanford and Jeanloz, 1972b). It has been proposed that this glycoprotein masks the expression of the H-2 histocompatibility antigens and thereby allows transplantation in allogeneic hosts (Sanford *et al.*, 1973). Similar mechanisms may be operative in masking tumor-associated rejection antigens, thereby contributing to the decreased immunogenicity of tumor cells in their syngeneic or autochthonous host.

One approach to increasing the effectiveness of active immunization against tumor cells has been enhancement of their antigenicity, particularly the unmasking of the tumor-associated rejection antigens (Prager and Baechtel, 1973). Neuraminidase digestion of TA3 mouse mammary carcinoma cells (Sanford, 1967), Landschütz ascites tumor cells (Currie and Bagshawe, 1968) and L1210 mouse leukemia cells (Bekesi, St Arneault and Holland, 1971) increased their immunogenicity, indicating that such an approach was feasible. However, since these studies were performed using tumors transplanted into allogeneic hosts, it was not possible to distinguish whether rejection or inhibition of tumor growth resulted from the exposure of histocompatibility or tumor-associated rejection antigens. The effect of neuraminidase on the immunogenicity of syngeneic methylcholanthrene-induced murine sarcomas has been investigated by Currie and Bagshawe (1969) and Simmons et al. (1971). These studies indicated that neuraminidase increases the immunogenicity of tumor cells, presumably by exposing cryptic tumor-associated rejection antigens.

These latter investigations of Simmons et al. (1971) were extended to the immunotherapy of firmly established methylcholanthrene-induced murine fibrosarcomas (Simmons and Rios, 1971; Rios and Simmons, 1972). Treatment of tumor-bearing hosts with living tumor cells which had been exposed to neuraminidase in vitro produced partial or total regression of the existing tumors. The therapeutic effect was magnified by the simultaneous administration of the nonspecific immunostimulant, Mycobacterium bovis BCG. Tumor rejection was immunospecific since rejection could be induced only with neuraminidase-treated cells identical in type to those of the growing tumor. More recently this therapeutic approach proved successful in inducing regression of other syngeneic murine tumors, for example two mammary adenocarcinomas and a melanoma (Rios and Simmons, 1973).

Structural or topological alterations at the cell periphery are responsible for the altered antigenicity of the tumor cell. The demonstration of these antigenic alterations has provided the experimental basis for the development of immunological approaches to the diagnosis and therapy of cancer. Future progress will require an understanding of the molecular basis of this altered antigenicity, as well as the humoral and cell-mediated mechanisms relevant to the host's immune response to these altered antigens. The recent development of methods for the isolation of histocompatibility and tumor-associated rejection antigens provides an experimental system to investigate the molecular nature of the antigenic alterations which accompany neoplastic transformation. Finally, the physiological role which these antigenic alterations play in the aberrant social behavior of the cancer cell requires intensive investigation. This latter underlying problem will require the marshalling of a variety of scientific disciplines to investigate the importance of cell–cell interactions in the control of growth and differentiation.

REFERENCES

AARONSON, S. A. and TODARO, G. J. (1968). Science, N.Y., 162:1024.
ABERCROMBIE, M. (1960). Can. Cancer Congr., 4:101.
ABERCROMBIE, M. (1970). Eur. J. Cancer, 6:7.
ABERCROMBIE, M. and AMBROSE, E. J. (1962). Cancer Res., 22:525.
ABERCROMBIE, M. and HEAYSMAN, J. E. M. (1953). Expl Cell Res., 5:11.

ABERCROMBIE, M., HEAYSMAN, J. E. M. and KARTHAUSER, H. M. (1957). *Expl Cell Res.*, **13**:276.
ADAMS, E. P. and GRAY, G. M. (1968). *Chem. Phys. Lipids*, **2**:147.
ALEXANDER, P. (1972). *Nature, Lond.*, **235**:137.
ALLEN, A. K., NEUBERGER, A. and SHARON, N. (1973). *Biochem. J.*, **131**:155.
AMBROSE, E. J., JAMES, A. M. and LOWICK, J. H. B. (1956). *Nature, Lond.*, **177**:576.
ARNDT-JOVIN, D. J. and BERG, P. (1971). *J. Virol.*, **8**:716.
AUB, J. C., SANFORD, B. H. and COTE, M. N. (1965). *Proc. natn. Acad. Sci. U.S.A.*, **54**:396.
AZARNIA, R. and LOEWENSTEIN, W. R. (1973). *Nature, Lond.*, **241**:455.
BALDWIN, R. W. (1972a). *Ser. Haemat.*, **4**:67.
BALDWIN, R. W. (1972b). *Natn. Cancer Inst. Monogr.*, **35**:135.
BALDWIN, R. W. and GLAVES, D. (1972). *Int. J. Cancer*, **9**:76.
BARSKI, G. and BELEHRADEK, J. (1965). *Expl Cell Res.*, **37**:464.
BAYNES, J. W., HSU, A.-F. and HEATH, E. C. (1973). *J. biol. Chem.*, **248**:5693.
BECKER, F. F. (1974). *Proc. natn. Acad. Sci. U.S.A.*, **71**:4307.
BEHRENS, N. H., CARMINATTI, H., STANELONI, R. J., LELOIR, L. F. and CANTARELLA, A. S. (1973). *Proc. natn. Acad. Sci. U.S.A.*, **70**:3390.
BEKESI, J. G., ST ARNEAULT, G. and HOLLAND, J. K. (1971). *Cancer Res.*, **31**:2130.
BEN BASSAT, H., INBAR, M. and SACHS, L. (1971). *J. Membrane Biol.*, **6**:183.
BENEDETTI, E. L. and EMMELOT, P. (1967). *J. Cell Sci.*, **2**:499.
BENNETT, H. S. (1963). *J. Histochem. Cytochem.*, **11**:2.
BIDDLE, F., CRONIN, A. P. and SANDERS, F. K. (1970). *Cytobios*, **5**:9.
BOREK, C., HIGASHINO, S. and LOEWENSTEIN, W. R. (1969). *J. Membrane Biol.*, **1**:274.
BRADY, R. O., BOREK, C. and BRADLEY, R. M. (1969). *J. biol. Chem.*, **244**:6552.
BRADY, R. O. and MORA, P. T. (1970). *Biochim. biophys. Acta*, **218**:308.
BRETTON, R., WICKER, R. and BERNHARD, W. (1972). *Int. J. Cancer*, **10**:397.
BUCK, C. A., GLICK, M. C. and WARREN, L. (1970). *Biochemistry*, **9**:4567.
BUCK, C. A., GLICK, M. C. and WARREN, L. (1971). *Science, N.Y.*, **172**:169.
BURGER, M. M. (1968). *Nature, Lond.*, **219**:499.
BURGER, M. M. (1969). *Proc. natn. Acad. Sci. U.S.A.*, **62**:994.
BURGER, M. M. (1970a). *Nature, Lond.*, **227**:170.
BURGER, M. M. (1970b). *Permeability and Function of Biological Membranes*, p. 107. Ed. L. BOLIS. Amsterdam; North-Holland.
BURGER, M. M. (1971a). *Curr. Topics cell. Regul.*, **3**:135.
BURGER, M. M. (1971b). *Nature, New Biol.*, **231**:125.
BURGER, M. M. (1973). *Fedn Proc. Fedn Am. Socs exp. Biol.*, **32**:91.
BURGER, M. M. and GOLDBERG, A. R. (1967). *Proc. natn. Acad. Sci. U.S.A.*, **57**:359.
BURGER, M. M. and NOONAN, K. D. (1970). *Nature, Lond.*, **288**:512.
BURTON, A. C. (1971). *Perspect. Biol. Med.*, **14**:301.
BURTON, A. C. and CANHAM, P. B. (1973). *J. theor. Biol.*, **39**:555.
CARTER, H. E., JOHNSON, P. and WEEBER, E. J. (1965). *A. Rev. Biochem.*, **34**:109.
CHANG, J. P. (1967). *Carcinogenesis: A Broad Critique*, pp. 535–586. Baltimore; Williams & Wilkins.
CHEEMA, P. YOGECSWARAN, G., MORRIS, P. M. and MURRAY, R. K. (1970). *FEBS Lett.*, **11**:181.
CHIPOWSKY, S., LEE, Y. C. and ROSEMAN, S. (1973). *Proc. natn. Acad. Sci. U.S.A.*, **70**:2309.
CHRISTIE, W. W. (1973). *Lipid Analysis: Isolation, Separation, Identification, and Structural Analysis of Lipids*. New York; Pergamon Press.
CLINE, M. J. and LIVINGSTON, D. C. (1971). *Nature, New Biol.*, **232**:156.
CODINGTON, J. F., SANFORD, B. H. and JEANLOZ, R. W. (1972a). *Biochemistry*, **11**:2559.
CODINGTON, J. F., SANFORD, B. H. and JEANLOZ, R. W. (1972b). *Fedn Proc. Fedn Am. Socs exp. Biol.*, **31**:465 Abstr.
COMAN, D. R. (1944). *Cancer Res.*, **4**:625.
COMAN, D. R. (1953). *Cancer Res.*, **13**:397.
COMAN, D. R. (1960). *Cancer Res.*, **20**:1202.
COX, R. P. and GESNER, B. M. (1968a). *Expl Cell Res.*, **49**:682.
COX, R. P. and GESNER, B. M. (1968b). *Cancer Res.*, **28**:1162.
CRITCHLEY, D. R., GRAHAM, J. M. and MACPHERSON, I. (1973). *FEBS Lett.*, **32**:37.
CRITCHLEY, D. R. and MACPHERSON, I. (1973). *Biochim. biophys. Acta*, **296**:145.
CULP, L. A. and BLACK, P. H. (1972). *J. Virol.*, **9**:611.
CUMAR, F. A., BRADY, R. O., KOLODNY, E. H., MCFARLAND, V. W. and MORA, P. T. (1970). *Proc. natn. Acad. Sci. U.S.A.*, **67**:757.
CURRIE, G. A. and BAGSHAWE, K. D. (1967). *Lancet*, **1**:708.

CURRIE, G. A. and BAGSHAWE, K. D. (1968). *Br. J. Cancer*, **22**:588.
CURRIE, G. A. and BAGSHAWE, K. D. (1969). *Br. J. Cancer*, **23**:141.
CURTIS, A. S. G. (1967). *The Cell Surface: Its Molecular Role in Morphogenesis.* London; Logos Press, Academic Press.
DEFENDI, V. and GASIC, G. (1963). *J. cell. comp. Physiol.*, **62**:23.
DEICHMAN, G. I. (1969). *Adv. Cancer Res.*, **12**:101.
DE MICCO, P. and BEREBBI, M. (1972). *Int. J. Cancer*, **10**:249.
DEN, H., SCHULTZ, A. M., BASU, M. and ROSEMAN, S. (1971). *J. biol. Chem.*, **246**:2721.
DEN, H., SELA, B., ROSEMAN, S. and SACHS, L. (1974). *J. biol. Chem.*, **249**:659.
DENT, P. B. and HILLCOAT, B. L. (1972). *J. natn. Cancer Inst.*, **49**:373.
DODD, B. J. and GRAY, G. M. (1968). *Biochim. biophys. Acta*, **150**:397.
DOLJANSKI, F. (1973). *Israel J. med. Sci.*, **9**:251.
EAGLE, H., FOLEY, G. E., KOPROWSKI, H., LAZARUS, H., LEVINE, E. M. and ADAMS, R. A. (1970). *J. exp. Med.*, **131**:863.
ECKHART, W., DULBECCO, R. and BURGER, M. M. (1971). *Proc. natn. Acad. Sci. U.S.A.*, **68**:283.
EISENBERG, S., BEN-OR, S. and DOLJANSKI, F. (1962). *Expl Cell Res.*, **26**:451.
EMMELOT, P. (1973). *Eur. J. Cancer*, **9**:319.
EMMELOT, P. and BENEDETTI, E. L. (1967). *Carcinogenesis: A Broad Critique*, pp. 471–533. Baltimore; Williams & Wilkins.
FOULDS, L. (1969). *Neoplastic Development*, p. 224. New York; Academic Press.
FOX, T. O., SHEPPARD, J. R. and BURGER, M. M. (1971). *Proc. natn. Acad. Sci. U.S.A.*, **68**:244.
FRIBERG, S., JR. (1972). *J. natn. Cancer Inst.*, **48**:1463.
GAHMBERG, C. G. and HAKOMORI, S. (1973a). *J. biol. Chem.*, **248**:4311.
GAHMBERG, C. G. and HAKOMORI, S. (1973b). *Proc. natn. Acad. Sci. U.S.A.*, **70**:3329.
GANTT, R. R., MARTIN, J. R. and EVANS, V. J. (1969). *J. natn. Cancer Inst.*, **42**:369.
GASIC, G. and BERWICK, L. (1963). *J. Cell Biol.*, **19**:223.
GASIC, G. and GASIC, T. (1962a). *Nature, Lond.*, **196**:170.
GASIC, G. and GASIC, T. (1962b). *Proc. natn. Acad. Sci. U.S.A.*, **48**:1172.
GESNER, B. M. and WOODRUFF, J. J. (1969). *Cellular Recognition*, pp. 79–90. Ed. R. T. SMITH and R. A. GOOD. New York; Appleton–Century–Crofts.
GLICK, M. C., RABINOWITZ, Z. and SACHS, L. (1973). *Biochemistry*, **12**:4864.
GOLD, P. (1971). *Prog. exp. Tumor Res.*, **14**:43.
GORDON, J. A. and MARQUARDT, M. D. (1974). *Biochim. biophys. Acta*, **332**:136.
GRAY, G. M. (1965). *Nature, Lond.*, **207**:505.
GRIMES, W. J. (1970). *Biochemistry*, **9**:5083.
GRIMES, W. J. (1973). *Biochemistry*, **12**:990.
GUNTHER, G. R., WANG, J. L., YAHARA, I., CUNNINGHAM, B. A. and EDELMAN, G. M. (1973). *Proc. natn. Acad. Sci. U.S.A.*, **70**:1012.
HAKOMORI, S.-I. (1970a). *Chem. Phys. Lipids*, **5**:96.
HAKOMORI, S.-I. (1970b). *Proc. natn. Acad. Sci. U.S.A.*, **67**:1741.
HAKOMORI, S.-I. (1971). *The Dynamic Structure of Cell Membranes*, pp. 65–96. Ed. D. F. H. WALLACH and H. FISCHER. New York; Springer-Verlag.
HAKOMORI, S.-I. and JEANLOZ, R. W. (1964). *J. biol. Chem.*, **239**:Pc. 3606.
HAKOMORI, S.-I. and MURAKAMI, W. T. (1968). *Proc. natn. Acad. Sci. U.S.A.*, **59**:254.
HAKOMORI, S.-I., TEATHER, C. and ANDREWS, H. (1968). *Biochem. biophys. Res. Commun.*, **33**:563.
HAKOMORI, S.-I., KOSCIELAK, J., BLOCK, K. J. and JEANLOZ, R. W. (1967). *J. Immun.*, **98**:31.
HARTMAN, J. F., BUCK, C. A., DEFENDI, V., GLICK, M. C. and WARREN, L. (1972). *J. cell. Physiol.*, **80**:159.
HAYWOOD, G. R. and MCKHANN, C. F. (1971). *J. exp. Med.*, **133**:1171.
HEARD, D. H., SEAMAN, G. V. F. and SIMON-ROUSS, I. (1961). *Nature, Lond.*, **190**:1009.
HELLSTRÖM, I. and HELLSTRÖM, K. E. (1971). *J. reticuloendothelial Soc.*, **10**:131.
HERSCHMAN, H. R., BREEDING, J. and NEDRUD, J. (1972). *J. cell. Physiol.*, **79**:249.
HSU, A.-F., BAYNES, J. W. and HEATH, E. C. (1974). *Proc. natn. Acad. Sci. U.S.A.*, **71**:2391.
INBAR, M., BEN-BASSAT, H. and SACHS, L. (1971). *Proc. natn. Acad. Sci. U.S.A.*, **68**:2748.
INBAR, M., BEN-BASSAT, H. and SACHS, L. (1972). *Nature, New Biol.*, **236**:3.
INBAR, M., RABINOWITZ, Z. and SACHS, L. (1969). *Int. J. Cancer*, **4**:690.
INBAR, M. and SACHS, L. (1969a). *Proc natn. Acad. Sci. U.S.A.*, **63**:1418.
INBAR, M. and SACHS, L. (1969b). *Nature, Lond.*, **223**:710.
INBAR, M. and SACHS, L. (1973). *FEBS Lett.*, **32**:124.
INBAR, M., BEN-BASSAT, H., SACHS, L., HUET, C. and OSEROFF, A. R. (1973). *Biochim. biophys. Acta*, **311**:594.

JAFFE, S., RAPPORT, M. M. and GRAF, L. (1963). *Nature, Lond.*, **197**:60.
JAMAKOSMANOVIĆ, A. and LOEWENSTEIN, W. R. (1968). *J. Cell Biol.*, **38**:556.
JAMIESON, G. A., URBAN, C. L. and BARBER, A. J. (1971). *Nature. New Biol.*, **234**:5.
JANSON, V. K. and BURGER, M. M. (1973). *Biochim. biophys. Acta*, **291**: 127.
JANSON, V. K., SAKAMOTO, C. K. and BURGER, M. M. (1973). *Biochim. biophys. Acta*, **291**:136.
JETT, M. and JAMIESON, G. A. (1973). *Biochem. biophys. Res. Commun.*, **55**:1225.
KALB, A. J. and LUSTIG, A. (1968). *Biochim. biophys. Acta*, **168**:366.
KALCKAR, H. M. (1965). *Science, N.Y.*, **150**:305.
KANNO, Y. and MATSUI, Y. (1968). *Nature, Lond.*, **218**:775.
KAPELLER, M., GAL-OZ, R., GROVER, N. B. and DOLJANSKI, F. (1973). *Expl Cell Res.*, **79**:152.
KEMP, R. B. (1968). *Nature, Lond.*, **218**:1255.
KENNEL, S. J., DEL VILLANO, B. C., LEVY, R. L. and LERNER, R. A. (1973). *Virology*, **55**:464.
KIJIMOTO, S. and HAKOMORI, S. (1971). *Biochem. biophys. Res. Commun.*, **44**:557.
KIJIMOTO, S. and HAKOMORI, S. (1972). *FEBS Lett.*, **25**:38.
KLEIN, E. (1972). *Annls Inst. Pasteur, Paris*, **122**:593.
KLENK, H. and CHOPPIN, P. W. (1970). *Proc. natn. Acad. Sci. U.S.A.*, **66**:57.
KORNFELD, S. and KORNFELD, R. (1969). *Proc. natn. Acad. Sci. U.S.A.*, **63**:1439.
KORNFELD, S., ROGER, J. and GREGORY, W. (1971). *J. biol. Chem.*, **246**:6581.
KRAMER, P. M. (1967). *J. Cell Biol.*, **33**:197.
KRAMER, P. M. (1971). *Biomembranes*, Vol. 1, pp. 67–190. Ed. L. A. MANSON. New York; Plenum Press.
LAINE, R. A. and HAKOMORI, S. (1973). *Biochem. biophys. Res. Commun.*, **54**:1039.
LANGLEY, O. K. and AMBROSE, E. J. (1964). *Nature, Lond.*, **204**:53.
LEEDEN, R. (1965). *J. Am. Oil Chem. Soc.*, **43**:57.
LEONARD, E. J., MELTZER, M. S., BORSOS, T. and RAPP, H. J. (1972). *Natn. Cancer Inst. Monogr.*, **35**:129.
LESENEY, A. M., BOURRILLON, R. and KORNFELD, S. (1972). *Archs Biochem. Biophys.*, **153**:831.
LOEWENSTEIN, W. R. (1966). *Ann. N.Y. Acad. Sci.*, **137**:441.
LOEWENSTEIN, W. R. (1968). *Devl Biol.*, Suppl., **2**:151.
LOEWENSTEIN, W. R. (1973). *Fedn Proc. Fedn Am. Socs exp. Biol.*, **32**:60.
LOEWENSTEIN, W. R. and KANNO, Y. (1967). *J. Cell Biol.*, **33**:225.
LOEWENSTEIN, W. R. and PENN, R. D. (1967). *J. Cell Biol.*, **33**:235.
LUCAS, J. J., WAECHTER, C. J. and LENNARZ, W. J. (1975). *J. biol. Chem.*, **250**:2887.
MAKITA, A. and SEYAMA, Y. (1971). *Biochim. biophys. Acta*, **241**:403.
MARGOLISH, E., SCHENK, J. R., HARGIE, M. P., BUROKAS, S., RICHTER, W. R., BORLOW, G. H. and MOSCONA, A. A. (1965). *Biochem. biophys. Res. Commun.*, **20**:383.
MARTINEZ-PALOMO, A. (1970). *Lab. Invest.*, **22**:605.
MARTINEZ-PALOMO, A. and BRAISLOVSKY, C. (1968). *Virology*, **34**:379.
MARTINEZ-PALOMO, A., BRAISLOVSKY, C. and BERNHARD, W. (1969). *Cancer Res.*, **29**:925.
MARTINEZ-PALOMO, A., WICKER, R. and BERNHARD, W. (1972). *Int. J. Cancer*, **9**:676.
MARTZ, E. and STEINBERG, M. S. (1972). *J. cell. Physiol.*, **79**:189.
MARTZ, E. and STEINBERG, M. S. (1973). *J. cell. Physiol.*, **81**:25.
MAYHEW, E. (1966). *J. gen. Physiol.*, **49**:717.
MCCUTCHEON, M., COMAN, D. R. and MOORE, F. B. (1948). *Cancer*, **1**:460.
MCKENZIE, G. H., SAWYER, W. H. and NICHOL, L. W. (1972). *Biochim. biophys. Acta*, **263**:283.
MCKIBBIN, J. M. (1970). *The Carbohydrates: Chemistry and Biochemistry*, Vol. 2B, pp. 711–738. Ed. W. PIGMAN and D. HORTON. New York; Academic Press.
MEEZAN, E., WU, H. C., BLACK, P. H. and ROBBINS, P. W. (1969). *Biochemistry*, **8**:2518.
MODJANOVA, E. A. and MALENKOV, A. G. (1973). *Expl Cell Res.*, **76**:305.
MÖLLER, E. (1964). *J. natn. Cancer Inst.*, **33**:979.
MOORE, T. L., KUPCHIK, Z., MARCON, N. and ZAMCHECK, N. (1971). *Am. J. dig. Dis.*, **16**:1.
MORA, P. T., BRADY, R. O., BRADLEY, R. M. and MCFARLAND, V. W. (1969). *Proc. natn. Acad. Sci. U.S.A.*, **63**:1290.
MORELL, A. G., GREGORIADIS, G., SCHEINBERG, H. I., HICKMAN, J. and ASHWELL, G. (1971). *J. biol. Chem.*, **246**:1461.
MOSCONA, A. A. (1968). *Devl Biol.*, **18**:250.
NERI, G., SMITH, D. F. and WALBORG, E. F., JR. (1974a). *Internationaux du Centre National de la Recherche Scientifique. No. 221: Methodologie de la Structure et du Metabolisme des Glycoconjuges*, pp. 869–880. Paris; Editions C.N.R.S.
NERI, G., SMITH, D. F., GILLIAM, E. and WALBORG, E. F., JR. (1974b). *Archs Biochem. Biophys.*, **165**:323.

NERI, G., SMITH, D. F., GILLIAM, E. and WALBORG, E. F., JR. (1974c). *Comparative Biochemistry and Physiology of Transport*, pp. 41–48. Ed. L. BOLIS, K. BLOCH, S. E. LURIA and F. LYNEN. Amsterdam; North-Holland.

NICOLSON, G. L. (1971). *Nature, New Biol.*, **233**:244.

NICOLSON, G. L. (1972). *Nature, New Biol.*, **239**:193.

NICOLSON, G. L. (1973a). *J. natn. Cancer Inst.*, **50**:1443.

NICOLSON, G. L. (1973b). *Nature, New Biol.*, **243**:218.

NICOLSON, G. L. and BLAUSTEIN, J. (1972). *Biochim. biophys. Acta*, **266**:543.

NICOLSON, G. L. and SINGER, S. J. (1971). *Proc. natn. Acad. Sci. U.S.A.*, **68**:942.

NOONAN, K. D. and BURGER, M. M. (1973a). *J. biol. Chem.*, **248**:4286.

NOONAN, K. D. and BURGER, M. M. (1973b). *J. Cell Biol.*, **59**:134.

NOONAN, K. D., RENGER, H. C., BASILICO, C. and BURGER, M. M. (1973). *Proc. natn. Acad. Sci. U.S.A.*, **70**:347.

NORTHRUP, J. H. (1926). *J. gen. Physiol.*, **9**:497.

NOVIKOFF, A. B. (1957). *Cancer Res.*, **17**:1010.

OHTA, N., PARDEE, A. B., MCAUSLAN, B. R. and BURGER, M. M. (1968). *Biochim. biophys. Acta*, **158**:98.

ONODERA, K. and SHEININ, R. (1970). *J. Cell Sci.*, **7**:337.

OSEROFF, A. R., ROBBINS, R. W. and BURGER, M. M. (1973). *A. Rev. Biochem.*, **42**:647.

OZANNE, B. and SAMBROOK, J. (1971). *Nature, New Biol.*, **232**:156.

PARDEE, A. B. (1971). *In vitro*, **7**:95.

PASTERNAK, G. (1969). *Adv. Cancer Res.*, **12**:1.

PAT, L. M. and GRIMES, W. J. (1974). *J. biol. Chem.*, **249**:4157.

POLLACK, R. E. and BURGER, M. M. (1969). *Proc. natn. Acad. Sci. U.S.A.*, **62**:1074.

POLLACK, R. E., GREEN, H. and TODARO, G. J. (1968). *Proc. natn. Acad. Sci. U.S.A.*, **60**:126.

PORETZ, R. D. and GOLDSTEIN, I. J. (1970). *Biochemistry*, **9**:2890.

POTTER, D. D., FURSHPAN, E. J. and LENNOX, E. S. (1966). *Proc. natn. Acad. Sci. U.S.A.*, **55**:328.

PRAGER, M. D. and BAECHTEL, F. S. (1973). *Meth. Cancer Res.*, **9**:339.

PRICER, W. E. and ASHWELL, G. (1971). *J. biol. Chem.*, **246**:4825.

PURDOM, L., AMBROSE, E. J. and KLEIN, G. (1958). *Nature, Lond.*, **181**:1586.

RAPPORT, M. M., GRAF, L. and LEEDEN, R. (1968). *Fedn Proc. Fedn Am. Socs exp. Biol.*, **27**:463.

RAPPORT, M. M., GRAF, L. and SCHNEIDER, H. (1964). *Archs Biochem. Biophys.*, **105**:431.

RAPPORT, M. M., SCHNEIDER, H. and GRAF, L. (1967). *Biochim. biophys. Acta*, **137**:409.

RAPPORT, M. M., GRAF, L., SKIPSKI, V. P. and ALONZO, N. F. (1958). *Nature, Lond.*, **181**:1803.

RAPPORT, M. M., GRAF, L., SKIPSKI, V. P. and ALONZO, N. F. (1959). *Cancer*, **12**:438.

REIN, A. and RUBIN, H. (1968). *Expl Cell Res.*, **49**:666.

RENKONEN, O., GAHMBERG, C. G., SIMONS, K. and KÄÄRIÄINEN, L. (1972). *Biochim. biophys. Acta*, **225**:66.

REVEL, J.-P. and ITO, S. (1967). *The Specificity of Cell Surfaces*, pp. 213–234. Ed. D. DAVIS and L. WARREN. Englewood Cliffs, New Jersey; Prentice-Hall.

RIOS, A. and SIMMONS, R. L. (1972). *Cancer Res.*, **32**:16.

RIOS, A. and SIMMONS, R. L. (1973). *J. natn. Cancer Inst.*, **51**:637.

ROBBINS, P. W. and MACPHERSON, I. A. (1971). *Proc. R. Soc., Ser. B.*, **177**:49.

ROSEMAN, S. (1970). *Chem. Phys. Lipids*, **5**:270.

ROSENBERG, S. A., POLCINIK, B. A. and ROGENTINE, G. N., JR. (1972). *J. natn. Cancer Inst.*, **48**:1271.

ROSENBLITH, J. Z., UKENA, T. E., YIN, H. H., BERLIN, R. D. and KARNOVSKY, M. J. (1973). *Proc. natn. Acad. Sci. U.S.A.*, **70**:1625.

ROTH, S., MCGUIRE, E. J. and ROSEMAN, S. (1971). *J. Cell Biol.*, **51**:536.

ROTH, S. and WHITE, D. (1972). *Proc. natn. Acad. Sci. U.S.A*, **69**:485.

RUBIN, H. (1970). *Permeability and Function of Biological Membranes*, pp. 103–106. Ed. L. BOLIS. LONDON; NORTH-HOLLAND.

RUHENSTROTH-BAUER, G. and FUHRMAN, G. F. (1961). *Z. Naturf.*, **16b**:252.

SAKIYAMA, H., GROSS, S. K. and ROBBINS, P. W. (1972). *Proc. natn. Acad. Sci. U.S.A.*, **69**:872.

SANFORD, B. H. (1967). *Transplantation*, **5**:1273.

SANFORD, B. H., CODINGTON, J. B., JEANLOZ, R. W. and PALMER, P. D. (1973). *J. Immun.*, **110**:1233.

SCHLESINGER, M. and AMOS, D. B. (1971). *Transplantn Proc.*, **3**:895.

SCHNEBLI, H. P. and BURGER, M. D. (1972). *Proc. natn. Acad. Sci. U.S.A.*, **69**:3825.

SELA, B., LIS, H. and SHARON, N. (1971). *Biochim. biophys. Acta*, **249**:564.

SELA, B., LIS, H., SHARON, N. and SACHS, L. (1970). *J. Membrane Biol.*, **3**:267.

SHARON, N. and LIS, H. (1972). *Science, N.Y.*, **177**:949.

SHARON, N. and LIS, H. (1973). *A. Rev. Biochem.*, **42**:541.

SHEN, L. and GINSBURG, V. (1968). *Biological Properties of Mammalian Surface Membranes*, pp. 67–85. Ed. L. A. MANSON. Philadelphia; Wistar Institute.

SHERIDAN, J. D. (1970). *J. Cell Biol.*, **45**:91.

SHINGLETON, H. M., RICHART, R. M., WERNER, J. and SPIRO, D. (1968). *Cancer Res.*, **28**:695.

SIDDIQUI, B. and HAKOMORI, S. (1970). *Cancer Res.*, **30**:2930.

SIMMONS, R. L. and RIOS, A. (1971). *Science, N.Y.*, **174**:591.

SIMMONS, R. L., RIOS, A., RAY, P. K. and LUNDGREN, G. (1971). *J. natn. Cancer Inst.*, **47**:1087.

SINGER, S. J. and NICOLSON, G. L. (1972). *Science, N.Y.*, **175**:720.

SLAYTER, H. S. and CODINGTON, J. F. (1973). *J. biol. Chem.*, **248**:3405.

SMITH, D. F., KOSOW, D. P. and JAMIESON, G. A. (1975). *Fedn Proc. Fedn Am. Socs exp. Biol.*, **34**:241.

SMITH, D. F., NERI, G. and WALBORG, E. F., JR. (1973). *Biochemistry*, **12**:2111.

SMITH, D. F. and WALBORG, E. F., JR. (1972). *Cancer Res.*, **32**:543.

SMITH, D. F., WALBORG, E. F., JR. and CHANG, J. P. (1970). *Cancer Res.*, **30**:2306.

SPIRO, R. G. (1975). *J. biol. Chem.*, **250**:2842.

STEINBERG, M. S. (1970). *J. exp. Zool.*, **173**:395.

STOKER, M. G. P. and RUBIN, H. (1967). *Nature, Lond.*, **215**:171.

SVENNERHOLM, L. (1964). *J. Lipid Res.*, **5**:145.

SWEELEY, C. C. and DAWSON, G. (1969). *Red Cell Membrane—Structure and Function*, pp. 172–227. Ed. G. A. JAMIESON and T. J. GREENWALT. Philadelphia; J. B. Lippincott.

TURNER, R. S. and BURGER, M. M. (1973). *Nature, Lond.*, **244**:509.

VASSAR, P. S. (1963). *Lab. Invest.*, **12**:1072.

WAECHTER, C. J., LUCAS, J. J. and LENNARZ, W. J. (1973). *J. biol. Chem.*, **248**:7570.

WALBORG, E. F., JR., LANTZ, R. S. and WRAY, V. P. (1969). *Cancer Res.*, **29**:2034.

WALBORG, E. F., JR., WRAY, V. P., SMITH, D. F. and NERI, G. (1975). *Immunological Aspects of Neoplasia*, pp. 123–147 (The University of Texas M.D. Anderson Hospital and Tumor Institute Twenty-Sixth Annual Symposium on Fundamental Cancer Research). Baltimore; Williams & Wilkins.

WALLACH, D. F. H. (1968). *Proc. natn. Acad. Sci. U.S.A.*, **61**:868.

WALLACH, D. F. H. (1972). *Biochim. biophys. Acta*, **265**:61.

WARREN, L. (1969). *Curr. Topics dev. Biol.*, **4**:197.

WARREN, L., CRITCHLEY, D. and MACPHERSON, I. (1972). *Nature, Lond.*, **235**:275.

WARREN, L., FUHRER, J. P. and BUCK, C. A. (1973). *Fedn Proc. Fedn Am. Socs exp. Biol.*, **32**:80.

WEINBAUM, G. and BURGER, M. M. (1973). *Nature, Lond.*, **244**:510.

WEINSTEIN, D. B., MARSH, J. B., GLICK, M. C. and WARREN, L. (1970). *J. biol. Chem.*, **245**:3928.

WEISS, L. (1967). *The Cell Periphery: Metastasis and Other Contact Phenomena*. New York; American Elsevier.

WINZLER, R. J. (1970). *Int. Rev. Cytol.*, **29**:77.

WINZLER, R. J. (1972). *Glycoproteins*, pp. 1268–1293. Ed. A. GOTTSCHALK. Amsterdam; Elsevier.

WOLF, L. S., BRECKENRIDGE, W. C. and SHELTON, P. P. C. (1974). *J. Neurochem.*, **23**:175.

WOLLMAN, Y. and SACHS, L. (1972). *J. Membrane Biol.*, **10**:1.

WOODRUFF, J. J. and GESNER, B. M. (1969). *J. exp. Med.*, **129**:551.

WRAY, V. P. and WALBORG, E. F., JR. (1971). *Cancer Res.*, **31**:2072.

WU, H. C., MEEZAN, E., BLACK, P. H. and ROBBINS, P. W. (1969). *Biochemistry*, **8**:2509.

YOGEESWARAN, G., LAINE, R. A. and HAKOMORI, S.-I. (1974). *Biochem. biophys. Res. Commun.*, **59**:591.

ZAMCHECK, N., MOORE, T. L., DHAR, P. and KUPCHIK, H. Z. (1972). *Natn. Cancer Inst. Monogr.*, **35**:433.

ADDENDUM

The literature survey on which this chapter is based was completed in 1973 and, consequently, many recent developments in this research area have been omitted. A discussion of these more recent developments is contained in the Proceedings of the 28th Annual Symposium on Fundamental Cancer Research, sponsored by the University of Texas System Cancer Center, M. D. Anderson Hospital and Tumor Institute in February 1975: *Cellular Membranes and Tumor Cell Behavior* (1975). Baltimore: Williams and Wilkins. Within the last several years increasing attention has focused on the dynamic aspects of membrane structure to cytoskeletal components (Nicolson, G. J. (1976). *Biochim. Biophys. Acta*, **457**, 57; **458**, 1).

E. F. W. was supported, in part, by grants from the National Cancer Institute (CA 11710), The Robert A. Welch Foundation and The Paul and May Haas Foundation.

6

The plasma membrane and sarcoplasmic reticulum of muscle

Winifred G. Nayler
Cardiothoracic Institute, London

6.1   INTRODUCTION

It is now generally agreed that the transformation between the diastolic and systolic states in muscle cells reflects a sudden change in the intracellular availability of ionized calcium. However, there can be little doubt that the accompanying biochemical and biophysical events, shown diagrammatically in *Figures 6.1* and *6.2*, are not yet fully understood. According to this outline, depolarization of the cell membrane results in an inward displacement of Na^+ and Ca^{2+} across the sarcolemmal complex, and a displacement of Ca^{2+} from intracellular binding sites. If, as a result of these changes, the intracellular concentration of Ca^{2+} exceeds 10^{-7} M (Weber, Herz and Reiss, 1967) and provided that the appropriate ATPase is active, that sufficient ATP is available, and that the necessary cofactors are present, contraction should occur (Ebashi and Endo, 1968; Katz, 1970; Nayler, 1973a). The reverse phenomenon—that is, the change from the systolic to the diastolic state—requires only a reduction in the intracellular concentration of Ca^{2+} below the critical level (*Figure 6.2*) needed to facilitate the actin-induced activation of the myosin ATPase.

Obviously the sequence of events outlined in *Figures 6.1* and *6.2* depends for its validity on a number of assumptions, each one of which warrants careful consideration and further investigation before being finally accepted. The data presented here have been selected and integrated so as to illustrate the relative importance of the role which both the plasma membrane fraction of the sarcolemma and the sarcoplasmic reticulum play in the events associated with the transformation of the diastolic to the systolic state in striated muscle cells and, conversely, of the systolic to the diastolic state.

Although this chapter forms part of a series dealing with mammalian membranes some data relating to non-mammalian membranes have been

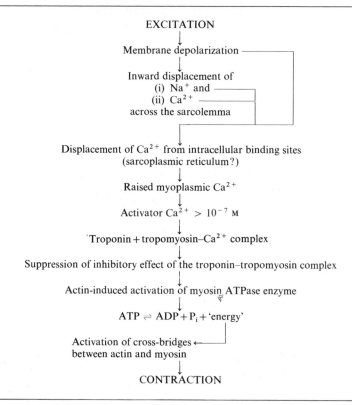

EXCITATION

Membrane depolarization

Inward displacement of
(i) Na$^+$ and
(ii) Ca^{2+}
across the sarcolemma

Displacement of Ca^{2+} from intracellular binding sites
(sarcoplasmic reticulum?)

Raised myoplasmic Ca^{2+}

Activator Ca^{2+} > 10^{-7} M

Troponin + tropomyosin–Ca^{2+} complex

Suppression of inhibitory effect of the troponin–tropomyosin complex

Actin-induced activation of myosin ATPase enzyme

ATP $\rightleftharpoons$ ADP + P$_i$ + 'energy'

Activation of cross-bridges
between actin and myosin

CONTRACTION

Figure 6.1   Schematic representation of the events involved in the transition from diastole to systole in striated muscle cells

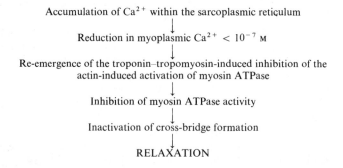

Accumulation of Ca^{2+} within the sarcoplasmic reticulum

Reduction in myoplasmic Ca^{2+} < 10^{-7} M

Re-emergence of the troponin–tropomyosin-induced inhibition of the
actin-induced activation of myosin ATPase

Inhibition of myosin ATPase activity

Inactivation of cross-bridge formation

RELAXATION

Figure 6.2   Schematic representation of the events involved in the transition from systole to diastole in striated muscle cells

included for the purposes of comparison. Data relating to the fine morphology and subcellular localization of these membranous profiles likewise have been included because of the need to correlate the fine structure of muscle cells with their function.

6.2   THE PLASMA MEMBRANE OF MUSCLE

6.2.1   Morphology of the sarcolemmal complex

The plasma membrane must be considered as an integral part of the *sarcolemmal complex*. This complex, which forms the outermost layer of muscle cells, is made up of three parts: (a) a diffuse reticulum of collagen fibrils; (b) an extracellular protein–polysaccharide coating which corresponds to the basement membrane of epithelial cells, and (c) the plasma membrane, which is contiguous with the cytoplasm. These various subcomponents of the cell

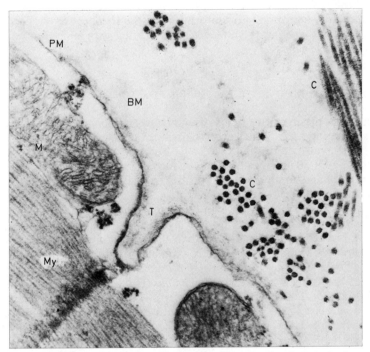

Figure 6.3   Section through the wall of a rabbit heart muscle cell. Note the three components, collagen (C), basement membrane (BM) and plasma membrane (PM) of the sarcolemmal complex. M, mitochondrion; My, myofilaments; T. transverse tubule. × 32 000

wall are readily identifiable in the electron micrograph reproduced in *Figure 6.3*. From a functional viewpoint the plasma membrane proper can be distinguished from its extraneous coats in the following ways (Stoeckenius and Engelman, 1969):

1. The main permeability barrier is located at the innermost layer of the complex, that is, at the plasma membrane.
2. This innermost layer shows typical membrane properties, including the presence of firmly bound multienzyme complexes.
3. The basement coat—that is, the protein–polysaccharide coating and

the associated collagen fibrils—can be removed from the rest of the membrane, which is composed of lipids and proteins, without seriously impairing the viability of the cell.

6.2.1.1   THE COLLAGEN FIBRILS

Although fine collagen fibrils are usually aggregated in clusters or bundles at the *outer* surface of the protein–polysaccharide coating, collagen fibrils are occasionally found deep within the protein–polysaccharide layer (*Figure 6.3*). Irrespective of the exact location of these collagen fibrils, whether within the polysaccharide coating or at its outer margin, their appearance is remarkably uniform. This uniformity persists irrespective of species and irrespective of whether the collagen is associated with smooth, skeletal or cardiac muscle cells.

6.2.1.2   THE PROTEIN–POLYSACCHARIDE COATING

Whereas the plasma membrane is usually only about 9 nm thick, the protein–polysaccharide layer which coats it is much thicker. For example, in cat ventricular and papillary muscle cells this protein–polysaccharide layer is approximately 50 nm thick, and can be further subdivided into an inner zone 20 nm thick and an outer one of width 30 nm (Fawcett and McNutt, 1969). In the atrial cells of the turtle the protein–polysaccharide layer is approximately 40–50 nm wide and, as in the cat ventricular and papillary muscle cells, can be further divided into two layers, the outer layer of width 30 nm being more dense and osmophilic than the 15–20 nm inner layer (Fawcett and Selby, 1958). Some muscle cells, such as those found in the walking-leg muscle fibres of the crayfish *Procambarus* (Brandt *et al.*, 1965) and the giant muscle fibres of the barnacle *Balanus nubilus* (Hoyle, McNeill and Selverston, 1973), have protein–polysaccharide layers which are even wider, and may be between 100 and 300 nm thick. Irrespective of its thickness it is this protein–polysaccharide layer which provides the immediate environment of the outer surface of the plasma membrane and which insulates and, to some extent, protects the plasma membrane from the extracellular phase. Substances which are added to the extracellular phase, whether they be drugs, cations or anions, obviously must penetrate this thick basement layer before making contact with the enzymatically active plasma membrane. The possible significance of the role played by the basement membrane must, therefore, not be overlooked.

6.2.1.3   THE PLASMA MEMBRANE PROPER

Whilst the protein–polysaccharide layer provides a relatively thick coat for the plasma membrane, the plasma membrane itself is relatively thin. For example, cat and sheep ventricular muscle cells have plasma membranes which, respectively, are approximately 9 nm (Fawcett and McNutt, 1969) and 8 nm (Simpson and Oertelis, 1962) wide.

Other muscles may have even thinner plasma membranes since, for example, the plasma membrane of the walking-leg muscle cells of the crayfish is only 6 nm wide (Brandt *et al.*, 1965) whilst the body wall cells of *Ascaris lumbricoides* have plasma membranes which are only 7.5 nm wide (Rosenbluth, 1969). Estimates of plasma membrane thickness are usually based on electron-microscopic observations of fixed, dehydrated and stained preparations. Whether all the components of the membrane are visualized under these conditions will ultimately depend upon the manner in which the heavy metal ions which are used in the staining techniques react with the various subunits of the membrane. These subunits probably vary from species to species and some caution must accordingly be observed in stating absolute values for membrane thickness. This problem has been discussed in detail elsewhere and need not be emphasized here, except to repeat that presently available measurements relating to the thickness of the plasma membranes in muscle cells are based on electron-microscopic examination of fixed, dehydrated and stained tissue and therefore may be inaccurate.

The distribution of the plasma membrane in muscle cells is complex and it need not necessarily be confined to the periphery of the cell. Electron-microscopic studies of a wide variety of muscle cells have consistently revealed that, in many, periodically spaced tubular invaginations extend from the sarcolemma into the cytoplasm. By using electron-dense markers, such as colloidal lanthanum, horseradish peroxidase and ferritin (Huxley, 1965; Forssmann and Girardier, 1970; Nayler and Merrillees, 1971), it has been possible to show unequivocally that the lumina of these tubules are continuous with the extracellular phase (Simpson, 1965; Nelson and Benson, 1963; Simpson, 1965; Rayns, Simpson and Bertaud, 1968; Fawcett and McNutt, 1969). These inwardly directed tubular extensions of the sarcolemma are lined with a membrane which is approximately 9 nm wide, and which is continuous with, and is almost certainly a functional part of, the plasma membrane itself (Peachey, 1965; Brandt *et al.*, 1965; Franzini-Armstrong, 1972, 1973). Typical examples of such 'intracellular' extensions of the cell membrane are shown in electron micrographs of rabbit heart muscle (*Figures 6.3* and *6.4*). Similar invaginations have been described for a wide variety of species, including *Paramecium* (Allen and Eckert, 1969), the basilar wing and tibial extensor muscles of the lepidopteran *Achalarus lyciades* (Reger and Cooper, 1967), and the bat cricothyroid muscle.

It is now well established that these periodically spaced tubular invaginations of the cell membrane usually branch, ramify and anastomose, and that they course throughout the cell. However, their original 'T-tubular' designation, which was based on the assumption that they were oriented perpendicular to the long axis of the muscle (transversely oriented, Anderson-Cedergren, 1959), is still retained.

Even a casual perusal of the relevant literature relating to this particular subject reveals that there are many species-determined variations in the location and morphology of these T-tubules which are lined with plasma membranes. For example, in the giant skeletal muscle fibres of the barnacle (*Balanus nubilus*) and the crab (*Portunus depurator*) (Selverston, 1967; Hoyle, McNeill and Selverston, 1973), and in the walking-leg muscles of the crayfish (Brandt *et al.*, 1965), relatively large clefts intrude into the muscle mass from the sarcolemma and they branch, ramify and anastomose in it.

The limiting membranes of these intrusions are approximately 6 nm thick, and undoubtedly are contiguous with the plasma membrane of the sarcolemma. Their lumina are filled with mucopolysaccharides similar to those which are associated with the sarcolemmal complex. In barnacle muscles these extensions of the sarcolemma occupy as much as 8 percent of the total fibre volume. Tubules which are lined with plasma membranes extend from

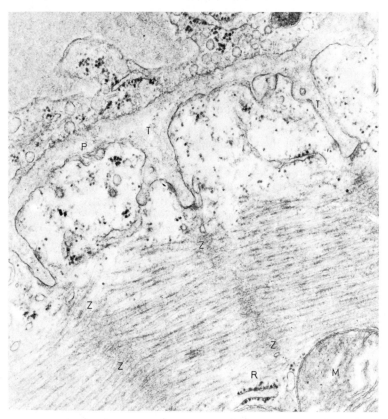

Figure 6.4   Section through part of a cardiac muscle cell, showing the branching T system and the plasma membrane (P); M, mitochondrion; Z, Z band; R, rough endoplasmic reticulum. ×30 000

these invaginations anywhere along their course, and they infiltrate the muscle fibre in all directions. Morphologically, these fine tubular ramifications of the cleft system cannot be differentiated from other plasma-membrane-lined tubules which originate at, or closely adjacent to, the level of the sarcolemma. Although the *cleft lumen* is filled with mucopolysaccharides (Hoyle, McNeill and Selverston, 1973), these are generally absent from the lumina of the T-tubules. Photomicrographs of Hoyle, McNeill and Selverston (1973) indicate that the lumen of the T-tubules in the giant skeletal muscle fibres of the barnacle is approximately 30 nm wide and Brandt *et al.* (1965) have reported a similar dimension (20 nm) for the T-tubule lumen of the crayfish.

In invertebrate skeletal muscle cells the diameter of the T-tubules is greatest close to the sarcolemma and diminishes as the tubules ramify, branch and anastomose within the cell. Comparison of the morphology of these tubules in terms of the diameter of their lumina is, therefore, difficult, irrespective of whether the comparison is made on a cell to cell or species basis. If such a comparison is to be attempted then a particular segment of the T-tubular invagination must be selected for measurement. The data summarized in *Table 6.1* relate to the diameter of the lumina of the T-tubules measured at the periphery of the cell and close, but not contiguous to the sarcolemma; that is, in a peripheral segment of the T-tubule. These data show

Table 6.1 DIAMETER OF THE T-TUBULE LUMEN IN SKELETAL AND CARDIAC MUSCLE CELLS

Species	*Muscle*	*Diameter, nm*	*Reference*
Skeletal muscle			
Rana pipiens	Sartorius	26	Peachey (1965)
Rana pipiens	Sartorius	21–24	Franzini-Armstrong (1970)
Rana temporaria	Twitch	30	Page (1965)
Rana temporaria	Slow	30	Page (1965)
Amblyostoma	Skeletal	26	Porter and Palade (1957)
Toad-fish	Swim bladder (fast muscle)	30	Franzini-Armstrong (1972)
Balanus nubilus	Giant striated	30	Rosenbluth (1969)
Eptesicus fuscus *Myotis lucifugus*	Cricothyroid	30–40	Revel (1962)
Achalarus lyciades	Basilar wing and tibial extensor	40–50	Reger and Cooper (1967)
Procambarus	Walking-leg muscle	15	Brandt *et al.* (1965)
Cardiac muscle			
Ferret	Ventricular	50–80	Simpson and Rayns (1968)
Human	Papillary	150	W. G. Nayler, unpublished work
Cat	Papillary	250	Fawcett and McNutt (1969)
Rat	Papillary	200–250	W. G. Nayler, unpublished work

that, whereas in skeletal muscle cells the lumina of peripheral segments of T-tubules are usually 20–30 nm wide, the lumina of similar located segments of T-tubules in mammalian heart muscle cells are much wider (*ca.* 250 nm). T-tubules are not present in all heart muscle cells; in many of the lower vertebrates careful electron-microscopic examination has failed to reveal their presence. Frog atrium (Barr, Dewey and Berger, 1965) and ventricle (Staley and Benson, 1966; Page and Niedergerke, 1972), toad ventricle (Nayler and Merrillees, 1964), turtle atrium (Fawcett and Selby, 1958), snake (Leak, 1967) and lizard ventricular muscle (Forbes and Sperelakis, 1971) all lack an identifiable T system. Chicken and humming bird heart muscle cells (Sommer *et al.*, 1972) and some atrial cells in the cat (Fawcett and McNutt, 1969) likewise are without a T system. Those heart muscle cells which do not have a T system characteristically display marked pinocytotic activity but the significance of this is obscure. Some authors have attempted to explain the failure of certain cardiac muscle cells to develop a T-tubular system in terms of cell size. Thus Fawcett and McNutt (1969) noted that in cat atrial muscle T-tubules are present only in unusually large cells. It seems doubtful, however, if the presence or absence of these sarcolemmal invaginations depends simply upon cell size. The cells of turtle atrial muscle are relatively small (3 μm diam.) as are snake and toad ventricular cells (5–6 μm diam.) (Leak, 1967; Nayler and Merrillees, 1964) and these cells certainly do not have a T system but other cells which are also relatively small, for example cat ventricular cells (10–12 μm diam.), contain a well-developed T system. Certainly, the presence of T-tubules results in a relatively large increase in the sarcolemmal, and hence plasma membrane, surface area. For instance, Peachey (1965) has calculated that the surface area of the T system in frog sartorius muscle cells is seven times as great as the outer surface area provided by the cell membrane, based on an assumed cell diameter of 100 μm. Irrespective of the precise diameter of the cell it seems probable that the additional surface area which is provided by the T-tubular system results in a highly significant increase in the area which is available for the transmembrane exchange of ions and metabolic substrates. The fact that many of these T-tubular invaginations are in immediate proximity to the mitochondria (*Figure 6.5*) may be significant in terms of reducing the distance across which substrates must diffuse before becoming available for metabolic utilization.

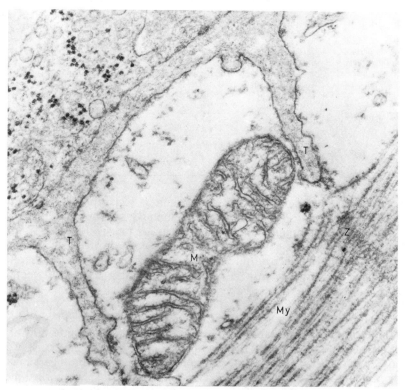

Figure 6.5   Section of a rabbit heart muscle cell showing the close proximity of the T-tubules to the mitochondria. M, T, My and Z as in Figures 6.3 and 6.4

6.2.2   Isolation and characterization of plasma membranes from muscle

During the past few years numerous attempts have been made to separate plasma membranes from other subcellular components of skeletal and cardiac muscle cells. The most commonly used techniques are those of homogenization and differential centrifugation followed by a variety of purification steps (Portius and Repke, 1967; Stam et al., 1970; Tada et al., 1972; Severson, Drummond and Sulakhe, 1972; Kidwai et al., 1973; Sulakhe, Drummond and Ng, 1973a, b). Although it is difficult to establish meaningful criteria for assessing the purity of such preparations, the three most commonly used are: (a) morphological appearance, using either phase-contrast or electron microscopy, (b) chemical composition and (c) enzyme markers, in particular Na^+,K^+-ATPase and adenyl cyclase.

The use of enzyme markers as a meaningful criterion of purity can be hazardous, because the enhanced activity of a specific enzyme may simply reflect the activation of a latent enzyme rather than the removal of a contaminant or inert protein. Moreover, as discussed in detail by Tada et al. (1972), the assignment of a specific marker enzyme to a particular subcellular organelle and the subsequent use of the activity of that enzyme as an index of purity, introduces a certain amount of circular reasoning. For example, adenyl cyclase is commonly used as a plasma membrane marker because plasma membranes yield a relatively high level of adenyl cyclase activity.

The use of morphological data as an index of purity may be equally hazardous, because *isolated* membranes take on a remarkably uniform appearance. If the isolated plasma membranes retain their protein–polysaccharide coating then, of course, this will provide a basis for their identification on morphological grounds. It is conceivable that the chemical composition of these membranes may well provide the most reliable assessment of their purity; for example, either the cholesterol:phospholipid ratio (Kidwai et al., 1973) or the sialic acid content, which is high in plasma membranes, can be used. For example, plasma membranes which have been isolated from cultured cells of chicken breast wing muscle contain as much as 64 nmol of sialic acid per milligram of protein (Streter, 1969).

Phospholipid content may provide an alternative method for assessing the purity of individual plasma membrane preparations. This criterion of purity has already been successfully applied in the separation of plasma membrane from endoplasmic reticulum in rat brain cells (Hemminki, 1973) and obviously can be applied to striated muscle cells. Alternatively, the ratio between cholesterol and total phospholipid content can be used (Kidwai et al., 1973), because this ratio is specific for plasma membrane, reticulum and mitochondrial fractions. Hemminki's studies showed that the plasma membranes isolated from brain cells of newly born rats were enriched in phosphatidylcholine and phosphatidylethanolamine, and that endoplasmic reticulum isolated from the same cells contained concentrations of phosphatidylserine higher than that found in the plasma membrane fractions. He found that plasma membranes from these cells contained 53.5 ± 3.4 percent phosphatidylcholine and 33.6 ± 0.6 percent phosphatidylethanolamine, but only 8.4 ± 0.6 percent phosphatidylserine and 4.5 ± 1.4 percent phosphatidylinositol. Since the apparent reactivities of the phospholipids to Ca^{2+} vary in the order phosphatidylserine > phosphatidylethanolamine $\gg$ phosphatidylcholine (Seimiya and Ohki, 1973) the phospholipid profile exhibited by the various membranous preparations may be significant, not only in terms of the identification of these membranes but also with respect to their ability to bind certain ions, particularly the divalent cations.

6.2.3　Enzymatic activities of muscle plasma membrane

(a)　*5'-Nucleotidase.* This enzyme has been used as a plasma membrane marker for a wide variety of cells. However, because of the recent demonstration of its extreme solubility in muscle homogenates (Severson, Drummond and Sulakhe, 1972), its continued use as a plasma membrane marker is of doubtful value.

(b)　Na$^+$,K$^+$-*ATPase.* Although it is widely assumed that this enzyme is localized within the plasma membrane, unequivocal evidence to substantiate such a conclusion is completely lacking. Because of the comparatively small amount of stimulation which results from the combined use of Na$^+$ and K$^+$ relative to the activity which is detected when both these ions are omitted, the accurate assay of the activity which is directly attributable to the Na$^+$,K$^+$-activated enzyme is difficult. It is generally assumed that this particular enzyme is vectorial, that is, that it spreads across the width of the plasma membrane in such a way that activation by Na$^+$ involves the activation of sites at the intercellular surface of the membrane whilst activation by K$^+$ involves the activation of receptor sites which are associated with the extracellular surface. In general, acceptance of the hypothesis that this particular Na$^+$,K$^+$-ATPase is vectorial in nature must be tempered by an appreciation of the fact that the pattern of enzymatic activity which is exhibited by isolated membranes probably differs significantly from that which occurs *in situ.*

(c)　*Adenyl cyclase.* It is now generally agreed that the plasma membranes of most cells contain adenyl cyclase (Sutherland, Robinson and Butcher, 1968). Its use as a plasma membrane marker (Tada *et al.,* 1972) may therefore be justified, particularly when it is coupled with a hormone-induced stimulation of the enzyme.

(d)　*Ca^{2+}-ATPase.* Isolated sarcolemmal preparations are capable of hydrolysing ATP in the presence of Mg^{2+} and Ca^{2+}, and of binding Ca^{2+} (Dietze and Hepp, 1971; Tada *et al.,* 1972; Kidwai *et al.,* 1973; Stam *et al.,* 1970, 1973; Sulakhe, Drummond and Ng, 1973a, b). This is not surprising, because energy-dependent Ca^{2+} transport mechanisms are known to be present in the plasma membranes of a wide variety of cells, including red cells, fibroblasts, liver and intestinal smooth muscle cells. The functional significance of this superficially located Ca^{2+} transport system in muscle cells will be discussed in a later section of this chapter, both with respect to the events involved in excitation–contraction coupling and the mode of action of certain drugs. The enzyme is mentioned here merely because it provides another possible marker enzyme for the plasma membrane fraction. Although some differences do exist between the Ca^{2+}-binding capacities of sarcolemmal and sarcoplasmic reticulum preparations (*Table 6.2*), and although these fractions differ with respect to their stability and their specificity for phosphate donors (Sulakhe, Drummond and Ng, 1973a, b), their similarities must be emphasized, including their response to oxalate.

Table 6.2 CALCIUM BINDING AND ACCUMULATING ACTIVITY OF RABBIT SKELETAL SARCOLEMMAL AND SARCOPLASMIC RETICULUM FRAGMENTS
(From data of Sulakhe, Drummond and Ng, 1973a)

	Ca^{2+}, nanomoles per mg protein per 30 sec	
Assay conditions	*Sarcolemma*	*Sarcoplasmic reticulum*
ATP-dependent Ca^{2+} *binding* (in the absence of oxalate)	20	180
ATP-dependent Ca^{2+} *uptake* (in the presence of oxalate)	200	2000

their pH dependency, pattern of Ca^{2+} release, affinity for Ca^{2+} and ATP, and the dependency on Ca^{2+} and Mg^{2+} of their respective ATPases (Sulakhe, Drummond and Ng, 1973a, b). Quantitatively, the Ca^{2+} pump of the sarcoplasmic reticulum is much more active than is that of the sarcolemma (*Table 6.2*). It follows, therefore, that measurements of the Ca^{2+}-accumulating activity of the plasma membrane fraction are of doubtful value when used as a criterion of its purity, because of possible contamination with fragments from the more active sarcoplasmic reticulum.

6.3   THE SARCOPLASMIC RETICULUM

6.3.1   Morphology

The sarcoplasmic reticulum (sR) consists of an intricate, lace-like network of branching and anastomosing tubules which encircle the myofilaments (Franzini-Armstrong and Porter, 1964; Peachey, 1965; Martonosi, 1972). In most skeletal muscle cells the lumen of this tubular network accounts for about 11–13 percent of the total cell volume (Martonosi, 1972), but in cardiac muscle cells the volume of the cell occupied by it is considerably less and usually ranges between 6 and 9 percent. However, in some 'fast' muscle cells, such as the striated muscle cells of the toad-fish swim bladder, nearly one-third of the total fibre volume is occupied by the lumen of the sR.

Substances of low molecular weight, including sucrose and chloride ions, penetrate into the lumen of the reticulum while larger molecules, such as ferritin (Huxley, 1964), neutral thorium dioxide (Birks, 1965), lanthanum (Revel and Karnovsky, 1967; Nayler and Merrillees, 1971) and horseradish peroxidase (Forssmann and Girardier, 1970) are excluded and are confined to the T-tubular system. Hence, in marked contrast to the situation described above for the transverse tubular system, the lumen of the sR does not communicate directly with the extracellular space.

Because the distribution of the sR with respect to the other subcellular organelles of muscle cells, particularly the sarcolemma and the T-tubular system, is of considerable importance in relation to its physiological function, it is appropriate to spend some time discussing these morphological details before considering the biochemical properties exhibited by the reticulum.

6.3.1.1   FINE MORPHOLOGY OF THE JUNCTIONAL AREAS

Wherever the membranes of the sarcoplasmic reticulum and T-tubular systems lie side by side, so that the two sets of membranes course alongside each other, specialized 'junctional areas' occur. These junctional areas (*Figure 6.6*) involve (a) the facing membranes of the SR and T-tubules;

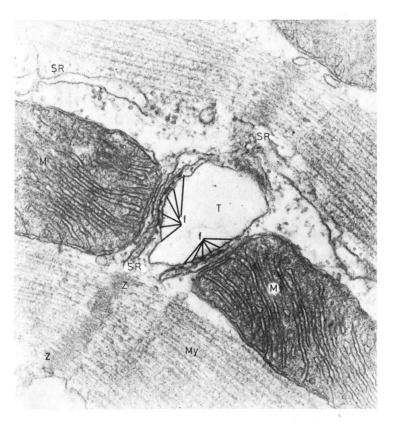

Figure 6.6   Section across a junctional area in a rat heart muscle cell, showing a T-tubular profile, sarcoplasmic reticulum (SR) and small 'feet-like' extensions (f) extending from the facing membrane of the SR into the junctional gap. M.Z. and My as in Figures 6.3 and 6.4. × 42 000

(b) an intervening 'junctional gap', and (c) electron-dense profiles which project from the facing membranes of the SR and extend into the junctional gap. In those muscle cells which have a T-tubular system, for example, mammalian skeletal cells and most mammalian cardiac muscle cells, these junctional areas are readily located along the T-tubular profile. Since the membranes which line the T-tubules are extensions of the plasma membrane, it is not surprising to find that such specialized junctional couplings also occur whenever the membranes of the sr approach the plasma membrane at the level of the sarcolemma as, for example, at the subsarcolemmal cisternae. It is equally important to realize that these specialized junctional areas can also occur in cells which are devoid of a T-tubular system in regions where the membranes of the sr approach those of the plasma membrane. This occurs in the ventricular muscle cells of the frog (Page and Niedergerke, 1972), chicken (Sommer and Johnson, 1969) and lizard (Forbes and Sperelakis, 1971), and some cat atrial muscle cells.

Irrespective of the particular morphology which these various couplings display, and irrespective of whether they occur at the perimeter of the cell adjacent to the sarcolemma, alongside the main T-tubular lumen, or near the fine terminal branches of the T-tubular system, it is generally agreed that these specialized units probably facilitate the distribution of the depolarizing current from the sarcolemma to the sr.

(a)   *The facing membranes of the sarcoplasmic reticulum and T system at the junctional areas.* At the specialized junctions the facing membrane of the

Table 6.3   APPEARANCE OF ELECTRON-DENSE PROFILES WITHIN THE JUNCTIONAL GAP

Species	Muscle	Description of profile	Reference
Sarcoplasmic reticulum–T junctions			
Bat (*Eptesicus* sp.)	Cricothyroid	Regularly spaced electron-dense profiles	Revel (1962)
Crayfish (*Procambarus*)	Walking-leg	Plates, 3.5 nm thick, 6 nm apart	Brandt *et al.* (1965)
Ferret	Cardiac	Electron-dense line	Simpson and Rayns (1968)
Larval newt	Skeletal	Dimples, 32 nm apart, arranged in a regular tetragonal pattern	Kelly and Cahill (1969)
Ascaris lumbricoides	Body wall	Peg-shaped densities, 20 nm wide, centre-to-centre distance 30 nm	Rosenbluth (1969)
Fish (*Molliensia* sp.)	Skeletal	Elongated particles, 10 nm wide	Bertaud, Rayns and Simpson (1970)
Frog (*Rana pipiens*)	Sartorius (twitch)	'Feet', 15 nm wide, 28–30 nm apart	Franzini-Armstrong (1970)
Frog (*Rana pipiens*)	Extraoccular (fast)	'Feet', 30 nm apart, centre-to-centre distance 40 nm	Franzini-Armstrong (1972)
Frog (*Rana pipiens*)	Toe muscle (lumbricalis) (slow)	'Feet', 32.5 nm apart, centre-to-centre distance 27 nm	Franzini-Armstrong (1973)
Barnacle (*Balanus nubilus*)	Depressor scutorum	Electron-dense line or feet	Hoyle, McNeill and Selverston (1973)
Sarcoplasmic reticulum–plasma membrane			
Frog (*Rana pipiens*)	Ventricular	Electron-dense bars, 20–25 nm apart	Page and Niedergerke (1972)
Lizard (*Anolis carolinensis*)	Ventricular	Bars*, 20 nm apart	Forbes and Sperelakis (1971)

* Calculated from published electron micrographs.

T system (or the plasma membrane itself) usually remains flat (Franzini-Armstrong, 1970, 1972, 1973) and apparently without specialization. The facing membranes of the SR, however, may be either flat or scalloped. Thus in the freeze-etched preparations of skeletal muscle cells of the black Mollie fish (Bertaud, Rayns and Simpson, 1970) the facing membranes of the SR at the junctional areas were flat, while in some invertebrate muscle cells (Hoyle, 1965; Rosenbluth, 1969) the facing membranes of the SR seem to be flat, despite the fact that the junctional gap clearly contains periodically repeating electron-dense deposits. However, in many muscle cells the facing membranes of the SR are scalloped, the period of the scallops being regular but their depth variable (Franzini-Armstrong, 1970, 1972). Often electron-dense 'feet-like' extensions can be seen projecting periodically from the facing membrane of the SR into the junctional gap, as described in *Table 6.3*.

(b)   *The junctional gap*. The width of the junctional gap is fairly uniform and without apparent species variation (*Table 6.4*). It is interesting to note,

Table 6.4 DIMENSIONS OF THE JUNCTIONAL GAP

Species	Muscle	Gap width, nm	Reference
Sarcoplasmic reticulum–T junction			
Toad-fish	Swim bladder (fast muscle)	12–13	Franzini-Armstrong (1972)
Frog (*Rana pipiens*)	Extraoccular (fast)	12	Franzini-Armstrong (1972)
Frog (*Rana pipiens*)	Sartorius (twitch)	12	Franzini-Armstrong (1970)
Frog (*Rana pipiens*)	Lumbricalis (slow)	14	Franzini-Armstrong (1973)
Bat (*Eptesicusfucus* or *Myotis lucifugus*)	Cricothyroid (fast)	12	Revel (1962)
Cat	Extraoccular (fast)	12	Franzini-Armstrong (1972)
Barnacle (*Balanus nubilus*)	Depressor scutorum, depressor tegorum	25	Hoyle, McNeill and Selverston (1973)
Ascaris lumbricoides	Body wall	12	Rosenbluth (1969)
Larval newt	Skeletal (slow)	15–20	Kelly and Cahill (1969)
Crayfish (*Procambarus*)	Walking-leg (slow)	15	Brandt *et al.* (1965)
Ferret	Ventricular	25	Simpson and Rayns (1968)
Rat	Ventricular	20	W. G. Nayler, unpublished work
Sarcoplasmic reticulum–plasma membrane junction			
Lizard (*Anolis carolinensis*)	Ventricular	20*	Forbes and Sperelakis (1971)
Frog (*Rana pipiens*)	Ventricular	10–20	Page and Niedergerke (1972)

however, that the gap is slightly wider in the more slowly contracting muscles than it is in the relatively 'faster' muscles, such as the toad-fish swim bladder, and the cat and frog extraoccular muscles. The width of the gap is essentially the same irrespective of whether the facing membrane of the SR is adjacent to a T-tubular or to a plasma membrane. This is shown by the fact that the width of the gap in those muscles which do not have a T system (frog, lizard and chicken ventricular muscle) is essentially the same as that found, for example, in frog sartorius muscle where the T system is well developed (Peachey, 1965; Franzini-Armstrong, 1970). Naturally, if the facing membrane of the SR is scalloped, then the width of the gap will vary.

(c)   *Electron-dense 'feet-like' extensions.* Many investigators have noted the presence of regularly spaced, electron-dense deposits which project from the facing membranes of the SR into the lumen of the junctional gap. Thus Bertaud, Rayns and Simpson (1970) described the occurrence of elongated 'particles' 10 nm wide located in the gap between the facing membranes of the SR and T systems in fish skeletal muscle. Rosenbluth (1969) similarly described the occurrence of a layer of peg-shaped densities extending from the SR membrane towards the membrane of the T system in the body wall cells of *Ascaris lumbricoides*. Kelly and Cahill (1969) refer to similar densities in the junctional gap of the larval newt, and Page and Niedergerke (1972) noted that the narrow gap between the internal plasma membrane surface and SR membrane in frog ventricular muscle cells likewise is bridged by small bars of dense, evenly spaced profiles which repeat at regular intervals of approximately 20–25 nm. Such regularly spaced electron-dense structures undoubtedly occur in rat heart muscle (*Figure 6.6*); presumably it was these electron-dense feet-like projections which were observed by Walker and Schrodt (1965, 1966). For convenience a list of some of those muscle preparations in which the presence of such electron-dense profiles have been described is given in *Table 6.3*. This list is not exhaustive but serves to illustrate the fact that the occurrence of regularly repeating, evenly spaced electron-dense deposits in the gap between the facing membranes of either the SR and T membranes, or the SR and plasma membranes, is widespread.

The number of rows of feet-like projections present in a particular muscle bears little correlation to the contraction pattern (fast, slow or twitch) exhibited by that muscle (Franzini-Armstrong, 1973). For example, in the fast muscles of the toad-fish swim bladder the junctional gap is always occupied by *two* rows of evenly spaced feet, with a centre-to-centre distance of 35–40 nm (Franzini-Armstrong, 1972) while in the fast extraoccular muscles of the cat the corresponding distance is 40 nm (Franzini-Armstrong, 1972). However, in the relatively slowly contracting tail fin muscle of the larval newt (Kelly and Cahill, 1969) and in the tail myotomes of small fish (C. Franzini-Armstrong, personal observation), *three* rows of feet are found; the arrangement of the feet is always periodic and symmetrical. Even in the slow fibres from the slow muscle of the frog *Rana pipiens*, essentially the same distribution of feet is observed (Franzini-Armstrong, 1973). In this muscle the centre-to-centre distance of the feet is 27 nm and there is some evidence of substructure since they seem to be attached to the SR by a small neck and, at a point equidistant from the membrane of the SR and T systems, they apparently expand into a wider zone.

Whether the feet actually link the membranes of the SR and T systems is uncertain. Franzini-Armstrong (1973) has noted that in the junctional areas of the slow muscle fibres of the frog a small gap apparently separates the 'feet-like' projections from the membranes of the T system. The existence of such an apparent gap is not peculiar to the slow muscle fibres of the frog, because a careful examination of many of the published electron micrographs reveals the widespread occurrence of such a gap. Careful high-resolution microscopy is needed to determine if such a gap, which can be no more than 4–5 nm wide, really exists or whether it is an artefact of preparation or due to the absence of electron-dense deposits within that area.

6.3.2   Isolation and characterization

Microsomal fractions which have been prepared from homogenized muscle cells by differential and zonal centrifugation are currently being widely used to investigate the biochemical and biophysical properties of the SR, including its interaction with Ca^{2+} (Harigaya and Schwartz, 1969; Nayler *et al.*, 1970; Repke and Katz, 1972; Katz and Repke, 1973; Martonosi, 1972). In all probability these microsomal fractions are heavily contaminated with mitochondrial and plasma membrane fragments. Many investigators have attempted to define such microsomal preparations further but, in general, they have not been successful (Katz and Repke, 1967; Katz *et al.*, 1970; Wheeldon and Gan, 1971), partly because of the lack of either a specific enzyme marker or a morphological peculiarity. Although the precise interpretation of the results obtained from studies in which impure and undoubtedly heterogeneous microsomal fractions have been used is difficult, it is virtually impossible to escape the conclusion that the microsomal fractions, as commonly prepared, are largely derived from the SR, and that they accumulate Ca^{2+} via an ATP-dependent pathway.

6.3.2.1   ENZYMATIC ACTIVITIES

(a)   *Ca^{2+}-ATPase.* That isolated SR fragments sequester Ca^{2+} is now well documented (Martonosi, 1972). This process of Ca^{2+} accumulation is accompanied by the hydrolysis of ATP (Hasselbach and Makinose, 1961, 1963; Makinose and Hasselbach, 1965) two Ca^{2+} ions being accumulated for each mole of ATP hydrolysed (Hasselbach and Makinose, 1963). The reaction almost certainly involves the formation of a phosphorylated intermediate and probably can be represented as follows:

$$Enzyme + 2\ Ca^{2+}_{outside} + ATP \rightleftarrows Enzyme - P - Ca^{2+}$$

$$\downarrow \qquad\qquad\qquad\qquad \Updownarrow$$

$$ADP \qquad\qquad Enzyme + P_i + 2\ Ca^{2+}_{inside}$$

where the subscripts 'outside' and 'inside' relate to the location of Ca^{2+} with respect to the microsomal membrane, P_i refers to inorganic phosphate, and the enzyme concerned is Ca^{2+}-activated ATPase.

In addition to ATP, other nucleoside triphosphates (Carsten and Mommaerts, 1964) and other phosphate donors such as acetyl phosphate (Pucell and Martonosi, 1969, 1971) can serve as energy donors for this system. A phosphoprotein intermediate for the ATPase involved in this process has been isolated from both skeletal (Makinose, 1969) and cardiac (Fanburg and Matsushita, 1973) microsomal preparations.

Martonosi (1972) has clearly shown that phospholipids are required for the ATPase and Ca transporting activity of fragmental SR fractions. Thus the exposure of isolated microsomes to phospholipase C inhibits their ability both to hydrolyse ATP and to accumulate Ca^{2+} (Martonosi, Donley and Halpin, 1968). The ATPase and Ca^{2+}-transporting activity of preparations treated with phospholipase C can be readily restored, however, by

Table 6.5 SPECIES VARIATION IN THE PHOSPHOLIPID CONTENT OF RAT AND GUINEA-PIG MICROSOMAL FRACTIONS*

Phospholipid	*Phospholipid content*, mg lipid per gram of microsomal protein		
	Rat	*Guinea-pig*	
Phosphatidylserine	16	36	$p < 0.0025$
Phosphatidylethanolamine	130	131	NS
Phosphatidylcholine	162	234	$p = 0.001$

* Each result represents the mean of six separate experiments. Tests of significance were calculated by student's t-test. NS, not significant; p is the probability that the results do not point to a significant difference. The microsomal preparations were isolated by differential centrifugation.

adding exogenous phospholipids. The whole question of the manner in which the phospholipids participate in the ATPase and Ca^{2+}-transporting activity of the SR fragments certainly warrants further investigation, particularly when it is realized that the phospholipid content of the microsomal fractions varies from species to species (*Table 6.5*).

(b)   *Other enzymes*. Recent evidence indicates that the SR may contain other enzymes such as hexokinase (Karpatkin, 1967), glyceraldehyde-3-phosphate dehydrogenase and phosphofructokinase (Aloisi and Margreth, 1967). The possibility that many of the enzymes of the SR are spilled from the reticulum during its isolation, so that they appear in the particulate-free supernatant, cannot be overlooked and, indeed, may explain why enzymes such as lactic dehydrogenase are recovered from the soluble fractions although chemically they are identified with the SR. Even so, isolated SR fragments contain relatively small amounts of the following enzymes (Martonosi, 1972): lactic dehydrogenase, glucose-6-phosphate dehydrogenase, 6-phosphogluconate dehydrogenase, phosphohexose isomerase, aldolase, phosphorylase and glyceraldehyde-3-phosphate dehydrogenase. How many of these enzymes are really associated with the reticulum and how many are absorbed during isolation is unknown. If these enzymes are a functional part of the reticulum then it is possible that the reticulum, as well as the mitochondrial mass, represents a major source of energy supply.

6.3.3 Ca^{2+}-accumulating activity of the sarcoplasmic reticulum

Microsomal fractions, whether they be prepared from skeletal or cardiac muscle cells, are capable of accumulating Ca^{2+}. Studies with isolated microsomal fractions indicate that two distinct processes are involved. One of these processes, designated 'Ca^{2+} binding', involves the ATP-dependent

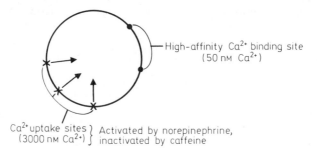

Figure 6.7 Schematic representation of the accumulation of Ca^{2+} by isolated microsomal fractions. Two processes are apparently involved—binding, which probably involves a specific number of binding sites on the SR membrane, and uptake, which involves the precipitation of the Ca^{2+} into the lumen. Uptake is measured in the presence of a precipitating agent, e.g. oxalate or phosphate

association of Ca^{2+} with the microsomal fraction in the absence of precipitating anions while the other process, designated 'Ca^{2+} uptake', relates to the ATP-dependent accumulation of Ca^{2+} in the presence of precipitating anions, for example oxalate or phosphate (Figure 6.7). Under these latter conditions, insoluble products which are presumed to be insoluble calcium salts accumulate within the lumina of the microsomal vesicles. Although significantly more Ca^{2+} is accumulated by the uptake (Figure 6.8) process,

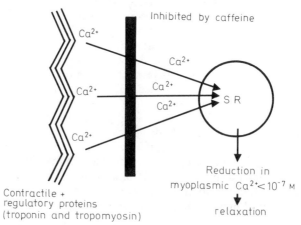

Figure 6.8 Schematic representation of the way in which the SR and Ca^{2+} regulates the transition from systole to diastole. Note that the uptake of Ca^{2+} into the SR (isolated as the microsomal fraction) is inhibited by caffeine

it proceeds so slowly that it may be of little physiological significance in promoting the reduction in the intracellular availability required for relaxation. The rates of Ca^{2+} binding and uptake are known to be influenced by a wide variety of substances and conditions, including pH (Nakamaru and Schwartz, 1972), temperature (Suko, 1973a), thyroid state (Suko, 1973b), *prolonged* ischaemia (Nayler et al., 1971; Lee, Ladinsky and Stuckey, 1967), drugs, including ryanodine (Nayler et al., 1970) and pentobarbital (Nayler and Szeto, 1972), $3',5'$-cAMP (Katz and Repke, 1973; Nayler and Dunnett, 1973a), certain β-adrenoceptor antagonists (Nayler, 1973d) and caffeine. Species variation in the Ca^{2+}-accumulating activity of these microsomal fractions has been described (Nayler and Dunnett, 1973b). Generally microsomal fractions prepared from failing heart muscle accumulate Ca^{2+} more slowly than do similar fractions which have been prepared from non-failing heart muscle (Schwartz, 1971; Gertz et al., 1967; Suko, Vogel and Chidsey, 1970). This list of factors which influence the rate at which isolated microsomal fractions accumulate Ca^{2+} is not meant to be exhaustive. Instead, readers are referred to a review by Martonosi (1972) on this subject. From the data presented in that review it is possible to reach the following conclusions:

1. Cardiac microsomal fractions accumulate Ca^{2+} more slowly than do skeletal microsomal fractions.
2. Microsomal fractions prepared from red skeletal muscle apparently accumulate Ca^{2+} more slowly than do similar fractions prepared from white skeletal muscle (Table 1, Martonosi, 1972).
3. The *yield* of microsomes from red skeletal muscle (0.7–1.4 mg of protein per gram wet weight of tissue) is less than that obtained for white skeletal muscle (2.5–4 mg of protein per gram of tissue). The combination of factors (2) and (3), therefore, may easily explain why it is that relaxation is slower in red than in white muscle (Ranvier, 1874), if it is agreed that the rate of muscle relaxation is determined by the rate of reabsorption of Ca^{2+} into the sarcoplasmic reticulum. Quantitatively this is possible, even if one takes into account only that Ca^{2+} which is accumulated at the specific *binding* sites. Possibly the Ca^{2+}-binding process provides the mechanism whereby Ca^{2+} is withdrawn from the immediate vicinity of the myofibrils, so that relaxation ensues, whereas Ca^{2+} uptake may facilitate the storage of Ca^{2+} at sites within the reticulum and from which it can be displaced during subsequent depolarizations (*Figure 6.1*).
4. The yield of microsomes from cardiac muscle is less than that obtained from skeletal muscle. This, when coupled with observation (1), may mean that the sr in cardiac muscle cells is inefficient with respect to its Ca^{2+}-accumulating activity relative to that found in skeletal muscle cells. Does this mean that more Ca^{2+} is lost to the extracellular space of cardiac muscle cells than is the case for skeletal muscle cells? Could this be why excitation–contraction coupling in cardiac muscle cells depends upon additional Ca^{2+} being made available from the extracellular phase?

6.3.4   Release of Ca^{2+} from microsomal fractions

Isolated fragments of the SR release Ca^{2+} at a rate which approximates to that described for the exchange of Ca^{2+} across resting skeletal (Bianchi, 1961) and cardiac (Winegrad, 1961) cell membranes. This rate is about 10^{-12}–10^{-13} moles Ca^{2+}/cm^2 surface area per second (Martonosi, 1964, 1972). Two distinct processes are involved: (a) a carrier-mediated efflux of Ca^{2+}, and (b) a passive release which probably reflects a diffusion of Ca^{2+} across the vesicular wall. Martonosi (1972) has discussed in detail the significance of the roles played by these two release processes. It is sufficient for our purposes to point out that the carrier-mediated efflux of Ca^{2+} represents a reversal of the Ca^{2+} uptake process, that it is induced by ADP and orthophosphate, and that it is Mg^{2+}-dependent (Bargolie, Hasselbach and Makinose, 1971). This type of Ca^{2+} release is accompanied by the synthesis of one mole of ATP for each two Ca^{2+} ions which are released (Makinose and Hasselbach, 1971). The physiological significance of the carrier-mediated Ca^{2+} efflux is unknown but it has been suggested that its principal role is to regulate the sarcoplasmic concentration of Ca^{2+} in resting muscle cells. Probably the carrier-mediated efflux process is not important with respect to the excitation-induced displacement of Ca^{2+} from the SR into the immediate vicinity of the myofibrils, because the process is inhibited by Ca^{2+} (Martonosi, 1972). The carrier-mediated release of Ca^{2+} from the SR probably does not account for the occurrence of an excitation-induced increase in the rate at which Ca^{2+} is released from intracellular binding sites (*Figure 6.1*). However, it seems probable that membrane depolarization can induce an increase in the passive release process, perhaps by causing a change in the conformational state of the electron-dense feet-like projections present in the junctional gap separating the SR from the plasma membrane and T-tubular membranes, thereby facilitating distribution of the depolarizing current to the SR.

6.3.5   The sarcoplasmic reticulum *in situ*

Although most of the data which are currently available relating to the biochemical and biophysical properties of the SR have originated from studies in which isolated microsomal fractions have been used, additional and valuable information can be obtained by using either skinned muscle fibres (Natori, 1954; Hellam and Podolsky, 1969; Ford and Podolsky, 1972; Fabiato and Fabiato, 1972) or fibres which have been treated with ethylenediaminetetraacetic acid (Winegrad, 1971; Nayler, 1973b) to render them freely permeable to Ca^{2+}. In these preparations the limiting effect of the plasma membrane has been removed, so that the functional behaviour of the SR with respect to the myofibrils and Ca^{2+} can be studied *in situ*. It is, perhaps, interesting to note that the positive inotropic effect of the cardiac glycosides is abolished under these conditions (Nayler, 1973b), an observation which substantiates the conclusion that the positive inotropic response which is normally provoked by this drug cannot be explained simply in terms of an interaction with the SR (Nayler, 1973c) but instead results from an interaction with one of the constituents of the sarcolemma. By contrast, the positive inotropic effect of norepinephrine probably can be accounted for in terms of its interaction with the SR in heart muscle cells (Katz and Repke, 1973).

6.4   EXCITATION–CONTRACTION COUPLING

Although the mechanism of contraction in cardiac and skeletal muscle cells has many points in common there are some important differences.

In mammalian skeletal muscle the action potential at $37\,^{\circ}C$ is terminated before the mechanical response begins. Currents generated at the surface of the cell are probably distributed via the transverse tubular system to the SR, possibly via the feet-like projections described above. How the intra-cellular distribution of the surface action potential initiates a mechanical response is not known for certain, but there is evidence (Ashley and Ridgway, 1970) that the activation of contraction is associated with a rise in the intra-cellular concentration of free Ca^{2+}. In skeletal muscle cells this rise in the intracellular concentration of free Ca^{2+} presumably involves the release of Ca^{2+} from storage sites within the reticulum. It must be this quantum of Ca^{2+} which determines the magnitude of the contractile response, for in skeletal muscle the external concentration of Ca^{2+} has comparatively little effect on the magnitude of the response. By contrast, in cardiac muscle cells the magnitude of the mechanical response is determined largely by the extracellular Ca^{2+} (see Nayler, 1967). It must be recalled that in cardiac muscle cells the membrane remains depolarized during most of the contrac-tion, and that the current which flows across the cell membrane during the plateau phase is, in contrast to the early Na^+ current, carried predominantly by Ca^{2+} (Reuter, 1967, 1973; Beeler and Reuter, 1970; Bassingthwaighte and Reuter, 1972). The data listed in *Table 6.6* show that it is possible to

Table 6.6   CATION- AND DRUG-INDUCED CHANGES IN THE Na^+ AND Ca^{2+} CURRENTS OF MAMMALIAN HEART MUSCLE
(From data of Kohlhardt *et al.*, 1972, 1973)

Active substance	Ca^{2+} *current, % change caused by active substance*	Na^+ *current, % change caused by active substance*
Verapamil (2 mg l^{-1})	−67	0
Lignocaine (50 mg l^{-1})	0	−25
Procaine (1 g l^{-1})	0	−50
Ni^{2+} (2 mM)	−90	−10
Co^{2+} (2 mM)	−58	−10
Mn^{2+} (2 mM)	−60	−10
Ca^{2+} (8.8 mM)	+270	0

differentiate between the transmembrane Na^+ and Ca^{2+} currents in mam-malian muscle cells, by using specific inhibitors—for example, lignocaine to block the Na^+ current, and verapamil to block the Ca^{2+} current (Kohl-hardt *et al.*, 1972, 1973). It is interesting to note that cations such as Ni^{2+}, Co^{2+} and Mn^{2+} (*Table 6.6*) and La^{3+} (Langer and Frank, 1972) which block the inward Ca^{2+} current also uncouple excitation–contraction coupling in cardiac muscle cells (Nayler, 1965). In the case of La^{3+} this must represent a surface effect of the drug, for it is well known that this compound does not penetrate the cell membrane. Although La^{3+} uncouples excitation–contraction coupling in cardiac muscle cells it has little, if any, effect on skeletal muscle (Sanborn and Langer, 1970). When La^{3+} interacts with cardiac cell membranes it displaces Ca^{2+} into the extracellular phase, and the amount of Ca^{2+} which is displaced varies according to the contractile state (Nayler and Szeto, 1972; Nayler, 1973c). It is possible, therefore, that in cardiac, but not in skeletal, muscle cells, the amount of Ca^{2+} which is available at the level of the cell membrane is the prime determinant of the amount of Ca^{2+} which finally is made available for interaction with the myofilaments. Possibly this Ca^{2+} is associated with specific binding sites, either within the plasma membrane or at the level of the basement coat, and is released in response to membrane depolarization. Presumably this Ca^{2+} is displaced into the myoplasm where it may either interact with the myofilaments directly, or alternatively may accumulate as 'activator Ca^{2+}' to be released in response to a subsequent depolarization. This release of 'activator Ca^{2+}' may be a Ca^{2+}-dependent phenomenon, and would resemble other Ca^{2+}-dependent excitation phenomena (Baker, 1972). Isotope studies have already shown that, in cardiac muscle cells, the 'activator Ca^{2+}' appears to originate from two pools (Bailey and Dresel, 1968); perhaps one of these pools is associated with the SR and the other with the sarcolemmal complex.

In skeletal muscle it seems certain that the bulk of the 'activator Ca^{2+}' originates from storage sites within the SR (Inesi, 1972). It seems equally certain that in both skeletal and cardiac muscle cells relaxation results from a reduction in the myoplasmic concentration of Ca^{2+} which is achieved via the accumulation of these ions within the SR.

We are left, then, with the as yet unanswered problem as to why heart muscle has developed in such a way as to require a supply of Ca^{2+} which is external to the cell if excitation–contraction coupling is to take place (Nayler, 1973a). Inevitably, one is left with the conclusion that more data relating to the differences between the biochemical and biophysical properties exhibited by skeletal and cardiac cell membranes are urgently needed if we are to understand the detailed events involved in excitation–contraction coupling.

Acknowledgement

This work was carried out during the tenure of grants in aid from the Medical Research Council of Great Britain and the British Heart Foundation.

REFERENCES

ALLEN, R. D. and ECKERT, R. (1969). *J. Cell Biol.*, **43**:4a.

ALOISI, M. and MARGRETH, A. (1967). *Exploratory Concepts in Muscular Dystrophy and Related Disorders*, pp. 305–317. Ed. A. MILHORAT. International Congress Series No. 147. Amsterdam; Excerpta Medica.

ANDERSON-CEDERGREN, E. (1959). *J. Ultrastruct. Res.*, **1**, Suppl.:1.

ASHLEY, C. C. and RIDGWAY, E. B. (1970). *J. Physiol., Lond.*, **209**:105.

BAILEY, L. E. and DRESEL, P. E. (1968). *J. gen. Physiol.*, **52**:969.

BAKER, P. F. (1972). *Prog. Biophys. molec. Biol.*, **24**:177.

BARGOLIE, B., HASSELBACH, W. and MAKINOSE, M. (1971). *FEBS Lett.*, **12**:267.

BARR, L., DEWEY, M. M. and BERGER, W. (1965). *J. gen. Physiol.*, **48**:797.

BASSINGTHWAIGHTE, J. B. and REUTER, H. (1972). *Electrical Phenomena in the Heart*, pp. 353–395. Ed. W. C. DE MELLO. New York; Academic Press.

BEELER, G. W., JR. and REUTER, H. (1970). *J. Physiol., Lond.*, **207**:191.

BERTAUD, W. S., RAYNS, D. G. and SIMPSON, F. O. (1970). *J. Cell Sci.*, **6**:537.

BIANCHI, C. P. (1961). *Circulation*, **24**:518.

BIRKS, R. I. (1965). *Muscle*, pp. 199–216. Ed. W. M. PAUL, E. E. DANIEL, C. M. KAY and G. MONCKTON. Oxford; Pergamon Press.

BRANDT, P. W., REUBEN, J. P., GIRARDIER, L. and GRUNDFEST, H. (1965). *J. Cell Biol.*, **25**:233.

CARSTEN, M. E. and MOMMAERTS, W. F. H. M. (1964). *J. gen. Physiol.*, **48**:183.

DIETZE, G. and HEPP, K. D. (1971). *Biochem. biophys. Res. Commun.*, **44**:1041.

EBASHI, S. and ENDO, M. (1968). *Prog. Biophys. molec. Biol.*, **18**:123.

FABIATO, A. and FABIATO, F. (1972). *Circulation Res.*, **31**:293.

FANBURG, B. L. and MATSUSHITA, S. (1973). *J. molec. cell. Cardiol.*, **5**:111.

FAWCETT, D. W. and MCNUTT, N. S. (1969). *J. Cell Biol.*, **42**:1.

FAWCETT, D. W. and SELBY, C. C. (1958). *J. biophys. biochem. Cytol.*, **4**:63.

FORBES, M. S. and SPERELAKIS, N. (1971). *J. Ultrastruct. Res.*, **34**:439.

FORD, L. E. and PODOLSKY, R. J. (1972). *J. Physiol. Lond.*, **223**:1.

FORSSMANN, W. G. and GIRARDIER, L. (1970). *J. Cell Biol.*, **44**:1.

FRANZINI-ARMSTRONG, C. (1970). *J. Cell Biol.*, **47**:488.

FRANZINI-ARMSTRONG, C. (1972). *Tissue Cell*, **4**:469.

FRANZINI-ARMSTRONG, C. (1973). *J. Cell Biol.*, **56**:120.

FRANZINI-ARMSTRONG, C. and PORTER, K. R. (1964). *Z. Zellforsch. mikrosk. Anat.*, **61**:661.

GERTZ, E. W., HESS, M. L., CAIN, R. F. and BRIGGS, F. N. (1967). *Circulation Res.*, **20**:477.

HARIGAYA, S. and SCHWARTZ, A. (1969). *Circulation Res.*, **25**:781.

HASSELBACH, W. and MAKINOSE, M. (1961). *Biochem. Z.*, **333**:519.

HASSELBACH, W. and MAKINOSE, M. (1963). *Biochem. Z.*, **339**:94.

HELLAM, D. C. and PODOLSKY, R. J. (1969). *J. Physiol., Lond.*, **200**:807.

HEMMINKI, K. (1973). *Biochim. biophys. Acta*, **298**:810.

HOYLE, G. (1965). *Science, N.Y.*, **149**:70.

HOYLE, G., MCNEILL, P. A. and SELVERSTON, A. I. (1973). *J. Cell Biol.*, **56**:74.

HUXLEY, H. E. (1964). *Nature, Lond.*, **202**:1067.

HUXLEY, H. E. (1965). *Muscle*, pp. 3–28. Ed. W. P. PAUL, E. E. DANIEL, C. M. KAY and G. MONCKTON. Oxford; Pergamon Press.

INESI, G. (1972). *A. Rev. Biophys. Bioengng*, **1**:191.

KARPATKIN, S. (1967). *J. biol. Chem.*, **242**:3525.

KATZ, A. M. (1970). *Physiol. Rev.* **50**:63.

KATZ, A. M. and REPKE, D. I. (1967). *Circulation Res.*, **21**:153.

KATZ, A. M. and REPKE, D. I. (1973). *Am. J. Cardiol.*, **31**:193.

KATZ, A. M., REPKE, D. I., UPSHAW, J. E. and POLASCIK, M. A. (1970). *Biochim. biophys. Acta*, **205**:473.

KELLY, D. E. and CAHILL, M. A. (1969). *J. Cell Biol.*, **43**:66a.

KIDWAI, A. M., RADCLIFFE, M. A., LEE, E. Y. and DANIEL, E. E. (1973). *Biochim. biophys. Acta*, **298**:593.

KOHLHARDT, M., BAUER, B., KRAUSE, H. and FLECKENSTEIN, A. (1972). *Pflügers Arch. ges. Physiol.*, **335**:309.

KOHLHARDT, M., BAUER, B., KRAUSE, H. and FLECKENSTEIN, A. (1973). *Pflügers Arch. ges. Physiol.*, **338**:115.

LANGER, G. A. and FRANK, J. S. (1972). *J. Cell Biol.*, **54**:441.

LEAK, L. V. (1967). *Am. J. Anat.*, **120**:553.

LEE, K. S., LADINSKY, H. and STUCKEY, J. H. (1967). *Circulation Res.*, **21**:439.

MAKINOSE, M. (1969). *Eur. J. Biochem.*, **10**:74.

MAKINOSE, M. and HASSELBACH, W. (1965). *Biochem. Z.*, **343**:360.

MAKINOSE, M. and HASSELBACH, W. (1971). *FEBS Lett.*, **12**:271.

MARTONOSI, A. (1964). *Fedn Proc. Fedn Am. Socs exp. Biol.*, **23**:Suppl. 913.

MARTONOSI, A. (1972). *Current Topics in Membranes and Transport*, pp. 83–197. Ed. F. BRONNER and A. KLEINZELLER. New York; Academic Press.

MARTONOSI, A., DONLEY, J. R. and HALPIN, R. A. (1968). *J. biol. Chem.*, **243**:61.

NAKAMARU, Y. and SCHWARTZ, A. (1972). *J. gen. Physiol.*, **59**:22.

NATORI, R. (1954). *Jikeikai med. J.*, **1**:119.

NAYLER, W. G. (1965). *Muscle*, pp. 167–184. Ed. W. M. PAUL, E. E. DANIEL, C. M. KAY and G. MONCKTON. Oxford; Pergamon Press.
NAYLER, W. G. (1967). *Am. Heart J.*, **73**:379.
NAYLER, W. G. (1973a). *J. molec. cell. Cardiol.*, **5**:214.
NAYLER, W. G. (1973b). *Am. J. Physiol.*, **225**:918.
NAYLER, W. G. (1973c). *J. molec. cell. Cardiol.*, **5**:101.
NAYLER, W. G. (1973d). *New Perspectives in Beta-Blockade*, pp. 41–62. Ed. M. MEIER, S. TAYLOR and R. RONDELL. CIBA Publ., England.
NAYLER, W. G. and DUNNETT, J. (1973a). *Advances in Cardiology*, Vol. 12, pp. 45–58. Ed. R. READER. Basle; S. G. Karger.
NAYLER, W. G. and DUNNETT, J. (1975). *J. molec. cell. Cardiol.*, **7**:663.
NAYLER, W. G. and MERRILLEES, N. C. R. (1964). *J. Cell Biol.*, **22**:533.
NAYLER, W. G. and MERRILLEES, N. C. R. (1971). *Calcium and the Heart*, pp. 24–56. Ed. P. HARRIS and L. H. OPIE. New York; Academic Press.
NAYLER, W. G. and SZETO, J. (1972). *Am. J. Physiol.*, **222**:339.
NAYLER, W. G., DAILE, P., CHIPPERFIELD, D. and GAN, K. (1970). *Am. J. Physiol.*, **219**:1620.
NAYLER, W. G., STONE, J., CARSON, V. and CHIPPERFIELD, D. (1971). *J. molec. cell. Cardiol.*, **2**:125.
NELSON, D. A. and BENSON, E. S. (1963). *J. Cell Biol.*, **16**:297.
PAGE, S. G. (1965). *J. Cell Biol.*, **26**:477.
PAGE, S. G. and NIEDERGERKE, R. (1972). *J. Cell Sci.*, **11**:179.
PEACHEY, L. D. (1965). *J. Cell Biol.*, **25**:209.
PORTER, K. R. and PALADE, G. E. (1957). *J. biophys. biochem. Cytol.*, **3**:269.
PORTIUS, H. J. and REPKE, K. R. H. (1967). *Acta biol. med. germ.*, **19**:876.
PUCELL, A. G. and MARTONOSI, A. (1969). *Abstracts 3rd International Congress Biophysics*, p. 241.
PUCELL, A. G. and MARTONOSI, A. (1971). *J. biol. Chem.*, **246**:3389.
RANVIER, L. A. (1874). *Arch. Physiol.*, **6**:1.
RAYNS, D. G., SIMPSON, F. O. and BERTAUD, W. S. (1968). *J. Cell Sci.*, **3**:467.
REGER, J. F. and COOPER, D. P. (1967). *J. Cell Biol.*, **33**:531.
REPKE, D. I. and KATZ, A. M. (1972). *J. molec. cell. Cardiol.*, **14**:401.
REUTER, H. (1967). *J. Physiol., Lond.*, **192**:479.
REUTER, H. (1973). *Prog. Biophys. molec. Biol.*, **26**:1.
REVEL, J. P. (1962). *J. Cell Biol.*, **12**:571.
REVEL, J. P. and KARNOVSKY, M. J. (1967). *J. Cell Biol.*, **33**:C.7.
ROSENBLUTH, J. (1969). *J. Cell Biol.*, **42**:817.
SANBORN, W. G. and LANGER, G. A. (1970). *J. gen. Physiol.*, **56**:191.
SCHWARTZ, A. (1971). *Cardiac Hypertrophy*, pp. 549–565. Ed. N. R. ALPERT. New York; Academic Press.
SEIMIYA, T. and OHKI, S. (1973). *Biochim. biophys. Acta*, **298**:546.
SELVERSTON, A. (1967). *Am. Zool.*, **7**:515.
SEVERSON, D. L., DRUMMOND, G. I. and SULAKHE, P. V. (1972). *J. biol. Chem.*, **247**:2949.
SIMPSON, F. O. (1965). *Am. J. Anat.*, **117**:1.
SIMPSON, F. O. and OERTELIS, S. J. (1962). *J. Cell Biol.*, **12**:91.
SIMPSON, F. O. and RAYNS, D. G. (1968). *Am. J. Anat.*, **122**:193.
SOMMER, J. R. and JOHNSON, E. A. (1969). *Z. Zellforsch. mikrosk. Anat.*, **98**:437.
SOMMER, J. R., STEERS, R. L., JOHNSON, E. A. and JEWETT, P. H. (1972). *Hibernation and Hypothermia: Perspectives and Challenges*, pp. 291–316. Ed. F. E. SOUTH, J. P. HANNON, J. R. WILLIS, E. T. PENGELLEY and N. R. ALPERT. Amsterdam, Elsevier.
STALEY, N. A. and BENSON, E. S. (1966). *J. Cell Biol.*, **38**:99.
STAM, A. C., WEGLICKI, W. B., FELDMAN, D., SHELBURNE, J. C. and SONNENBLICK, E. H. (1970). *J. molec. cell. Cardiol.*, **1**:117.
STAM, A. C., WEGLICKI, W. B., GERTZ, E. W. and SONNENBLICK, E. H. (1973). *Biochim. biophys. Acta*, **298**:927.
STOECKENIUS, W. and ENGELMAN, D. M. (1969). *J. Cell Biol.*, **42**:613.
STRETER, F. A. (1969). *Archs Biochem. Biophys.*, **134**:25.
SUKO, J. (1973a). *J. Physiol., Lond.*, **228**:563.
SUKO, J. (1973b). *Experientia*, **29**:396.
SUKO, J., VOGEL, J. H. K. and CHIDSEY, C. A. (1970). *Circulation Res.*, **27**:235.
SULAKHE, P. V., DRUMMOND, G. I. and NG, D. C. (1973a). *J. biol. Chem.*, **248**:4150.
SULAKHE, P. V., DRUMMOND, G. I. and NG, D. C. (1973b). *J. biol. Chem.*, **248**:4158.
SUTHERLAND, E. W., ROBINSON, G. A. and BUTCHER, R. W. (1968). *Circulation*, **37**:279.

TADA, M., FINNEY, J. O., SWARTZ, M. H. and KATZ, A. M. (1972). *J. molec. cell. Cardiol.*, **4**:417.
WALKER, S. M. and SCHRODT, G. R. (1965). *J. Cell Biol.*, **27**:671.
WALKER, S. M. and SCHRODT, G. R. (1966). *Anat. Rec.*, **155**:1.
WEBER, A., HERZ, R. and REISS, I. (1967). *Biochim. biophys. Acta*, **131**:188.
WHEELDON, L. W. and GAN, C. (1971). *Biochim. biophys. Acta*, **233**:37.
WINEGRAD, S. (1961). *Circulation*, **24**(2):2523.
WINEGRAD, S. (1971). *J. gen. Physiol.*, **58**:71.

7

The composition and structure of excitable nerve membrane

G. R. Strichartz*

Department of Physiology and Biophysics, State University of New York at Stony Brook

7.1   INTRODUCTION

This chapter describes the plasma membrane of nerve cells. It is a comparative essay dealing primarily with the properties which make nerve membranes excitable. The chapter contains information about the lipid and protein composition of nerve membranes from mammalian nervous systems, those of other vertebrates and some invertebrates as well.

I shall consider the major enzymatic activities which are localized in excitable nerve membranes and associated with nervous activity. The aim of the chapter is to present a view of nerve membranes as dynamic structures which are in constant motion and whose chemical constituents are continually reacting and being degraded and replaced. I hope to show that mammalian nerves are similar in function and composition to other vertebrate and invertebrate nerves and that all these nerves rely on similar excitability mechanisms. The membrane of the specialized nerve ending, the *synapse*, will not be discussed in detail here.

Since the presentation is a comparison of nerve properties from different species, it seems appropriate first to describe the electrophysiological properties of these nerves. A detailed consideration of the excitability of mammalian nerves is presented by D. O. Carpenter in Vol. 4, Chapter 8 of this series.

7.1.1   Morphologic diversity and functional similarity of nerves

The nerve cells of mammals are found in the central nervous system (CNS), which comprises the brain and spinal cord, and in the peripheral nervous

* This review was initiated when the author was at the Department of Pharmacology, Yale University School of Medicine, New Haven, Connecticut.

system, which extends throughout the body. Mammalian nerves differ markedly in appearance depending on their function (Bodian, 1962). Motor neurones of the peripheral nervous system, which activate muscles or glands often located at bodily extremities, may be several meters in length. Neurones of the central nervous system are usually quite restricted in length, but can have extensively branching processes, as in the Purkinje cells of the cerebellum (Ramon y Cajal, 1909).

Despite their morphological differences, neurones behave in very similar ways and a generalized neurone can be described (*see* Grundfest, 1957;

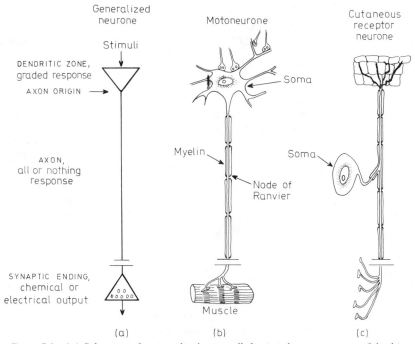

Figure 7.1   (a) Schematic of a generalized nerve cell showing the receptor zone of dendrites and the cell body, the axon origin or initial segment, where action potentials are generated by a summation of signals from the receptor zone, and the axon itself, which conducts action potentials to the synaptic ending. (From Grundfest, 1957). (b), (c) Two mammalian nerve cells, indicating the real structures which correspond to the idealized regions in (a). (After Bodian, 1967)

Bodian, 1962). Such a cell is schematized in *Figure 7.1a*. The nerve is characterized by one long process, the *axon*, many shorter processes, the *dendrites*, and a cell body, or *soma*, which may lie on the path between dendrites and axon (*Figure 7.1b*) or may be displaced to the side (*Figure 7.1c*).

In nerve cells an electrical impulse of short duration, the *action potential*, once initiated at a point on the axon will propagate undiminished to the end of the axon. Action potentials are all-or-nothing responses, thus, if an action potential is generated, it has a constant amplitude, usually changing the potential across the nerve membrane by *ca.* 100 mV so that it becomes positive inside. Since action potentials maintain a constant amplitude with distance traveled, they are called *regenerative* electrical impulses.

The role of the dendrites is to receive electrical signals from other nerves which may also have inputs directly on the soma. These signals are summed, both spatially and temporally, on the nerve membrane. If the sum of these graded electrical inputs is of sufficient size to exceed the threshold, the nerve responds by generating an action potential, which is conducted away from the receiving zone by the axon.

In the vertebrates the large nerve axons are usually enclosed by a multi-lamellar structure called *myelin*. Myelin is formed by the membrane of an oligodendroglial or *Schwann* cell which wraps itself around and around the axon (Geren, 1954; *see also* Peters and Vaughn, 1970) producing an insulator of very high resistance. At regular intervals along the axon, usually about every 1–2 mm, there are narrow ($<$ 1 μm wide) spacings where no myelin exists and the axon membrane is exposed to the external medium. Ionic currents which flow during a nerve impulse enter the axon membrane at these points, which are called *nodes of Ranvier* (Huxley and Stämpfli, 1949).

Membranes displaying regenerative action potentials are referred to as 'excitable' membranes. Axon membranes are always excitable, but cell bodies are frequently inexcitable (Frank, Fuortes and Nelson, 1959; Araki and Terzuolo, 1962) and dendritic processes of mammalian nerves are usually inexcitable or have very high thresholds (Eyzaguirre and Kuffler, 1955; *see also* Purpura, 1967; Shepherd, 1974). There is some evidence of nonpropagated impulses in 'inexcitable' regions of nerve cells which are nonetheless sensitive to specific inhibitors of axonal action potentials (Grundfest, 1965, 1967; Blankenship, 1968). Cell bodies of invertebrates (Geduldig and Gruener, 1970; Wald, 1972) and vertebrates (Koketsu and Nishi, 1969), including some mammalian cells, simultaneously display brief and long regenerative signals (Hirst and Spence, 1973). The shorter of these signals is like an action potential in the axon and requires Na^+ while the slower signal usually depends on Ca^{2+}. Calcium ion influx occurs in the soma of peripheral (Hirst and Spence, 1973) and central (Krnjevic and Lisiewicz, 1972) mammalian neurones and may be a means of controlling excitability during repetitive stimulation (Krnjevic and Lisiewicz, 1972; Jansen and Nicholls, 1973). Calcium-dependent spikes will not be further discussed here since the phenomenon has been thoroughly reviewed by Reuter (1973) and Hagiwara (1973). The variations in impulse behavior in different regions of a single nerve cell demonstrate that the division between excitable and inexcitable membranes is not always well defined electrophysiologically. Nor is the distinction always obvious by the pharmacological criteria tested so far.

7.1.2   **Intrinsic nature of nerve membrane excitability**

The properties which result in excitability are intrinsic to the axon membrane. When the axoplasm of the squid giant axon is extruded and replaced by an artificial medium containing only small ions the action potential remains almost identical to the one observed in intact axons (Baker, Hodgkin and Shaw, 1962a, b). The squid giant axon was the subject of the original, complete description of the electrical changes of a nerve membrane accompanying the action potential (Hodgkin and Huxley, 1952). Subsequent studies have shown that the mechanisms of axonal excitability are very similar among invertebrates and vertebrates, including mammals. In fact, knowledge of excitability in mammalian nerves is not extensive; most of what is known agrees well with, and is best understood in light of, experiments with other vertebrate nerves and with invertebrates.

In resting nerves the electrical potential inside the cell is negative with respect to the outside. This potential difference arises indirectly from the activity of the sodium plus potassium ion pump which transports Na^+ out of the cell and K^+ into the cell and requires energy from ATP to move these ions against their concentration gradients; the intracellular Na^+ concentration is much less than the extracellular concentration and the intracellular K^+ concentration much greater than the extracellular. At rest, the cell membrane is much more permeable to K^+ than to Na^+, hence K^+ ions tend to leak out through the membrane leaving the interior of the cell negatively charged compared with the exterior. Thus, the relatively high permeability to K^+ ions of the resting cell membrane results in a negative resting potential. During an action potential the relative permeabilities of the membrane reverse themselves. The nerve becomes much more permeable to Na^+ ions (Hodgkin and Katz, 1949) which then enter the cell, constituting an inward current, and render the cell interior more positive, or depolarize the cell. Near the peak of the action potential the cell is so permeable to Na^+ that the size of the action potential is very near to that of the Nernst potential calculated from the Na^+ concentrations inside and outside the cell (Hodgkin and Katz, 1949).

The unique properties of axon membranes which account for the appearance of regenerative action potentials are the voltage- and time-dependent ion permeabilities. Both Na^+ and K^+ permeabilities increase when the nerve membrane is depolarized. The increase in Na^+ permeability develops relatively rapidly and permits the *entry* of Na^+, which corresponds to the rising phase of the action potential. However, this permeability is transient and spontaneously decreases, even when the membrane is artificially depolarized for long times. This spontaneous reduction of Na^+ permeability, called *inactivation*, decreases Na^+ current and tends to return the membrane potential to the resting level. A slower-appearing increase in K^+ permeability allows K^+ ions to diffuse *out* of the cell and this also tends to return the membrane potential to its resting value.

Although the total membrane ionic permeability increases several hundred-fold during the action potential, the actual net ion flux is relatively small. The ion gradients produced by the sodium–potassium pump are dissipated very slowly by continuous nerve activity, and are rapidly restored when stimulation ceases (Connelly, 1959; Hodgkin and Keynes, 1955a). In a squid axon of diameter 500 μm the result of one action potential is to increase the internal Na^+ concentration by about 10^{-5} of its concentration at rest (Keynes, 1951), while for a very small nonmyelinated axon (diam. 0.5 μm) with the same ion flux density the corresponding increment of intracellular Na^+ concentration would be about 10^{-2} of the normal concentration.

The ion pump contributes to the maintenance of excitability by providing a concentration difference of ions across the nerve membrane. The pump appears to transport more Na^+ ions out of the cell than K^+ ions into it

(Thomas, 1969), and so produces a net current across the membrane (Rang and Ritchie, 1968). This outward current tends to make the membrane potential more negative, or hyperpolarize the cell. In small fibers and in some nerve cell bodies this hyperpolarization is significantly large and may have profound effects on the excitability of the nerve (Ritchie, 1971; Jansen and Nicholls, 1973).

A knowledge of the voltage dependence and the kinetics of Na^+ and K^+ permeability has been sufficient to explain all the action potentials observed in axons. The original Hodgkin–Huxley formalism describing the regenerative electrical behavior of squid axons has been modified and extended to the properties of other invertebrate axons such as lobster (Julian, Moore and Goldman, 1962) and *Myxicola* (Goldman and Schauf, 1973) giant axons; to myelinated nerves of frog (Dodge and Frankenhaeuser, 1958) and toad (Frankenhaeuser, 1965), as well as to the myelinated axons of mammalian peripheral nerve (Hořackova, Nonner and Stämpfli, 1968; Nonner, 1969) and their cell bodies (Barrett and Crill, 1972). The responses of these nerves, as well as a variety of other excitable membranes (Noble, 1966), are basically similar to those of the squid axon. In all cases there is a Na^+-dependent action potential arising from a membrane permeability change.

The mammalian nervous system includes many bundles of small (diam. < 1 μm), nonmyelinated nerves, the *C-fibers* (*see* Douglas and Ritchie, 1962). Because of their small size and because of the associated Schwann cells, which enclose several fibers intimately and make isolation of single fibers impossible, these nerves cannot be studied by the same techniques as are applicable to larger nerves. However, the action potentials of C-fibers are Na$^+$-dependent, have similar velocities in nerves from fish (Easton, 1965), amphibians (Erlanger and Gasser, 1937; Gasser, Richards and Grundfest, 1938) and mammals (Howarth, Keynes and Ritchie, 1968), although the last-mentioned have a greater temperature dependence than the others. C-fibers also show pharmacological responses which are very similar to those of larger axons. In particular, action potentials in C-fibers are blocked by local anesthetics (Ritchie and Greengard, 1961) and by the poison tetrodotoxin (Colquhoun and Ritchie, 1972), both drugs being selective inhibitors of Na^+ permeability in squid axons (Taylor, 1959; Moore and Narahashi, 1967) and in frog myelinated nerves (Hille, 1966).

In summary, axons in different nerve fibers from a wide variety of animals have Na^+-dependent, regenerative action potentials. The ionic currents which flow during the action potentials must cross the membrane with the aid of some pore or carrier mechanism, because the hydrocarbon region of a lipid bilayer presents a thermodynamically unfavorable milieu even for a hydrated ion, and ion-pair combinations provide insufficient energy to account for observed ion fluxes (Parseghian, 1969). The voltage-dependent changes of Na^+ and K^+ permeability could arise from a macroscopic 'phase transition' of the entire membrane (Delbrück, 1970) or from changes of discrete microscopic sites in the membrane. The evidence presented in this chapter strongly favors the second alternative.

Most investigators believe that there are separate mechanisms for the passage of Na^+ and K^+ ions across the membrane. While these mechanisms are less than perfectly selective, they do discriminate at least tenfold between Na^+ and K^+ ions, respond differently to various voltage patterns imposed across the membrane, and they can be selectively inhibited by different drugs. In addition to discriminating among ions, Na^+ and K^+ permeabilities must respond rapidly to changes in the electric field within the nerve membrane. This response depends on the charged or dipolar properties of the permeability site, which are called 'gating' properties. The selectivity of the Na^+ permeability is independent of time, voltage and pharmacological agents which modify the gating properties (Hille, 1968a, 1971, 1972; Ulbricht, 1969) and may depend on a structure which is different from that of the 'gates'. A review by Hille (1970) concerning these 'ionic channels' in nerves presents the evidence for their existence, separability, membrane surface density and the calculated ion flux through individual channels.

The membrane surface density of channels determined from both theoretical calculations and studies of toxin binding (see Section 7.3) is quite low, being of the order of $10-100$ per μm^2 of axon membrane. For channels embedded in a lipid bilayer this density is equivalent to having one sodium channel for every 10^4-10^5 lipid molecules. Any attempt to measure physical or chemical changes in nerve that are specifically associated with sodium channels must be capable of discriminating between very small signals imposed on the rather overwhelming background presented by the inactive regions of the membrane.

7.2 COMPOSITION OF NERVE MEMBRANES

7.2.1 Lipid bilayer character

The plasma membrane of stained, sectioned nerve cells viewed by the electron microscope appears to be a typical unit membrane with densely staining outer surfaces spaced 7.5–10 nm apart, and a relatively transparent medium between. Nerves from molluscs (Villegas and Villegas, 1968), crustaceans (Simpkins, Panko and Tay, 1971), amphibians (Lentz, 1967), fish (Easton, 1971), and mammalian myelinated (Peters and Vaughn, 1970) and unmyelinated fibers (Keynes and Ritchie, 1965) have the same general morphology although the membrane has characteristic features in specialized regions, such as sensory nerve endings and, notably, at the synapse.

Like other animal membranes, those of nerve are composed of lipids, proteins and some carbohydrates (see Table 7.1, below). The passive electrical properties of nerves give some clue to their organization. First, all cell membranes are capacitors, and have the capability of storing electrical charge, at least transiently. Capacitance is a function of both the width of the membrane and the dielectric constant of the material in the membrane interior. The capacity in terms of membrane area for almost all cells is about $1 \mu F cm^{-2}$ (Cole, 1968), the same value as the capacitance of a phospholipid bilayer formed in the absence of hydrocarbon solvents (Montal and Mueller, 1972). X-ray analyses of model phospholipid bilayers indicate a hydrocarbon interior having a thickness of 3–4 nm (Levine and Wilkins, 1971); similar studies of intact garfish olfactory nerve show a hydrocarbon thickness of 3.4–3.6 nm (J. K. Blasie, personal communication; Blasie et al., 1972). During a nerve impulse the capacitance varies by only 2 percent (Cole and Curtis, 1939). This is evidence that the nerve membrane is largely a lipid bilayer in character and that it undergoes no *major* structural changes during nervous activity.

A second passive electrical property of nerves is the resting membrane ionic permeability. Resting ionic permeability is expressed in the unit of electrical conductance, namely the mho ($= \text{ohm}^{-1}$), and varies from nerve to nerve and also between different regions of the same nerve. The squid giant axon has a resting conductance of *ca.* 10^{-3} mho cm^{-2} (calculated from data of Hodgkin, Huxley and Katz, 1952) while the much smaller toad myelinated axon has a higher nodal resting conductance, 1–3×10^{-2} mho cm^{-2} (Dodge and Frankenhaeuser, 1959). Mammalian motor-neurone dendrites have specific membrane conductances estimated at $< 5 \times 10^{-4}$ mho cm^{-2} (J. N. Barrett, personal communication); an average value for the motor-neurone soma may be 2.5×10^{-3} mho cm^{-2} (Eccles, 1968)*. In contrast, model bilayers formed of phospholipids have much lower resting conductances, 10^{-8} and 10^{-7} mho cm^{-2} in membranes from phosphatidylethanolamine and phosphatidylserine, respectively (McLaughlin, Szabo and Eisenman, 1971), and 10^{-8} mho cm^{-2} in membranes from sphingomyelin (Kauffman and Mead, 1970). Lipids extracted from squid retinal axolemma can form bilayers of 10^{-7} mho cm^{-2} conductance, which is unchanged when the membrane potential varies from $+120$ to -120 mV (Wolff *et al.*, 1971).

The model bilayer values imply that neither resting nor active membrane conductance in nerves can be ascribed to pure phospholipid regions. When proteins are included in model phospholipid bilayers the resting conductances increase from 10^{-8} mho cm^{-2} to 10^{-3}–10^{-5} mho cm^{-2} (Mueller and Rudin, 1963, 1968) and, under the appropriate conditions, regenerative potential changes similar to the nerve action potential are observed (Mueller and Rudin, 1968; *see* Ehrenstein and Lecar, 1972). There is accumulating evidence that the voltage-dependent Na$^+$ permeability in nerve membranes is at least partly mediated by protein (*see below*). Thus, morphological and electrical studies together suggest that nerve membranes are mainly lipid bilayers with some proteins which permit ions to pass through a hydrocarbon interior of otherwise high resistance.

7.2.2   Membrane preparation

In general, plasma membranes are isolated from homogenates of whole cells and are rarely, if ever, free of contamination from organelle membranes. When several cell types exist intimately in the same tissue, as neurones and glial cells often do, they are difficult to separate. Identification of the various fractions is usually made by assaying some enzymatic marker intrinsic to a particular membrane, such as succinic dehydrogenase for mitochondrial inner membranes (DePierre and Karnovsky, 1973). In whole-cell isolations, neurones can be differentiated from glia by recognized differences in morphology or by the presence of enzymes of a biosynthetic

---

* Somatic specific membrane conductances are very difficult to determine, since (a) the soma membrane surface area is only one-quarter of the dendritic membrane surface area, and (b) the state of activity of synapses on the cell body can affect measured membrane conductances quite dramatically. The number presented is, at best, an order of magnitude estimation.

pathway whose product is found uniquely in that cell type (Raine, Poduslo and Norton, 1971; Norton and Poduslo, 1971). In nerve homogenates the criteria for identifying plasma membranes are usually the presence of a ouabain-sensitive Na^+,K^+-ATPase, the ability to bind cholinomimetic compounds or antagonists of acetylcholine binding, or the presence of acetylcholinesterase, although the last-mentioned marker may be rather weakly bound to membranes and is partially solubilized by incubation in a medium of low ionic strength (Hollunger and Niklasson, 1973; Henn, Hansson and Hamberger, 1972). A much better marker for excitable membranes with Na^+-dependent action potentials appears to be the specific binding of neurotoxins. This property of nerve membranes is discussed extensively in Section 7.3.

The axonal preparations included in *Table 7.1* are representative of membranes which support Na^+-dependent action potentials. However, both unmyelinated and myelinated axonal membranes are surrounded by glial cells and the purity of such preparations (70–80 percent) is usually less than that of isolated cell bodies. One exception is the garfish olfactory nerve, a tissue of known excitability and purity whose composition has been extensively analyzed. Garfish olfactory nerve is composed almost exclusively of small (diam. 0.24 µm), nonmyelinated nerve axons; the ratio of axon membrane area to Schwann cell area is about 30:1. The action potentials in these nerves are Na^+-dependent, tetrodotoxin-sensitive (author's unpublished observation), and have a group velocity (0.1–0.2 m s^{-1}) comparable to those of nonmyelinated amphibian and mammalian C-fibers, which are 0.7 and 0.4 m s^{-1}, respectively. The velocity differences are consistent with the differences in average axon diameters in the nerve bundles (Rushton, 1951).

The garfish olfactory nerve appears to be representative of vertebrate nonmyelinated nerves both in electrophysiological properties and lipid composition (*Table 7.1*). The garfish membranes characterized in *Table 7.1* showed 5 percent (lipid phosphorus) contamination with mitochondrial membranes, indicated by the presence of cardiolipin, which is found exclusively in mitochondrial inner membranes (Rouser *et al.*, 1968), and with Schwann cells. A purified axolemma preparation from garfish has been characterized roughly (Chacko *et al.*, 1973) and is not very different from the intact garfish nerve.

Table 7.1 presents the lipid composition of membranes from mammalian nerve and glial cell bodies and peripheral nerve axons. It also includes, for comparison, the lipid analyses of axonal membranes from fish, crustacean and molluscan nerves as well as myelinated mammalian nerves. Among the nonmyelinated nerves the lipid accounts for between 24 and 70 percent of the dry weight. This variation probably reflects differences in analyses and preparative techniques more than species differences. For example, it is always difficult to determine the amount of protein which is solubilized during tissue preparation. Purified garfish olfactory axolemma preparations have *ca.* 65 percent lipid by weight, but mammalian nerves contain considerably less, 35–40 percent (*Table 7.1*). Myelin and preparations of myelinated nerve have a lipid content of about 70 percent dry weight.

The similarity in lipid composition of nonmyelinated nerves is remarkable. The major phospholipid is usually phosphatidylcholine (PC), accounting

Table 7.1a LIPID COMPOSITION

Lipid composition,

Type of lipid	Rat brain neurones (1)*†	Rabbit brain neurones (2)‡	Pig brain neurones (3)†	Rat brain astrocytes (1)†	Ox splenic nerve (4)‡	Human brain gray matter (5)†	Garfish olfactory nerve (6)†	Garfish olfactory nerve (7)‡
Overall lipid content	24	38		39		38		60‖
Polar lipid	84	61		73	62	68	70	73
Cholesterol	10.8	35.5	9	14	38	32	(14)	27
Polar lipids								
PC	48	44	55	50	41	29	45	42
PE	21	31	33	28	21	32	28	32
PS	4.4	7	10	7	14	8.5	16	11
PI	7.1	5.3		5	4.5		1	4.5
SphM	3.6	7.4		5	18	9.4	11	8.3
Cer						2.2	nd	
Cereb	3.0	{4.3	nd}	2.4	1.6	{6.1	nd	nd}
Sulf		{0.9	nd}			{1.2	nd	nd}
Uncharacterized				13		11	10	

Footnotes and abbreviations referring to this table are to be found below *Table 7.1b*.

Table 7.1b LIPID COMPOSITION

Lipid composition,

Type of lipid	Human brain white matter (5)*†	Human brain myelin (5)†	Bovine brain myelin (11)†	Rabbit sciatic nerve (4)‡	Sheep sciatic nerve (4)‡	Monkey sciatic nerve (4)‡
Overall lipid content	66	78	75			
Polar lipid	66	61	72	56	51	40
Cholesterol	34	39	28	44	49	60
Polar lipids						
PC	17	20.9	15	16	12	14
PE	24	25.0	24	38	29	32
PS	9.3	8.9	9	15	20	33
PI			1	5	6	3
SphM	9.5	7.8	10	25	29	32
Cer	1.2	2.9				
Cereb	19.2	22.5	33}	20	25	27
Sulf	6.5	7.5	5}			
Uncharacterized	13.5	4.4				

Abbreviations—PC: phosphatidylcholine; PE: phosphatidylethanolamine; PS: phosphatidylserine; PI: phosphatidylinositol; SphM: sphingomyelin; Cer: ceramide; Cereb: cerebroside; Sulf: sulfatide; nd: none detected.

References—(1) Norton and Poduslo (1971); (2) Hamberger and Svennerholm (1971); (3) Tamai, Matsukawa and Satake (1971a); (4) Sheltawy and Dawson (1966); (5) O'Brien and Sampson (1965a); (6) Holton and Easton (1971); (7) Chacko, Goldman and Pennock (1972); (8) Zambrano, Cellino and Canessa-Fischer (1971); (9) Camejo *et al.* (1969); (10) Keesey, Salle and Adams (1972) (whole nerve); (11) Norton and Autilio (1966); (12) Poduslo and Norton (1972); (13) Rouser *et al.* (1968).

%, in:

Squid retinal axon (8)‡	Squid stellar nerve axolemma (9)‡	Lobster walking-leg nerve (10)§	Lobster leg nerve (4)‡	Lobster circum-esophageal nerve trunk (10)§	Crab claw nerve (4)‡	Crab leg nerve (4)‡	Lobster claw nerve (4)‡	Average values	
								Verte-brate	Inverte-brate
45.7	70.5								
67	58		70		63	62	67	71	65
22	28		30		37	38	33	29	35
40	46	37	37	43	37	40	39	46	40
37	34	38	28	30	28	28	18	28	30
10	10	nd	8	2.0	10	8	11	9.3	7.4
		5	2	4	2	4	3	4.3	2.5
4	10	15.5	18	14	16	14	23	7.5	14.3
								2.0	nd

OF MYELINATED NERVE

%, in:

Domestic fowl sciatic nerve (4)‡	Garfish trigeminal nerve (6)§	Garfish trigeminal nerve (7)‡	Lobster circum-esophageal nerve sheath (10)§	Bovine oligodend-roglia (12)†	Human skeletal muscle (13)‡	Rat liver (13)‡
				30		
50		64		80		
50	(22)	27		14		
17	28	32	43	41	53.9	52.2
40	28	41	31	19	26.6	25.3
13	15	12	3	6	3.1	3.7
2	0.3	1.2	5	6	8.4	9.0
25	9.6	12.6	13	10	4.0	4.5
				nd		
25	⎰ 1.3	2		10		
	⎱ 6.4	6		2		
	6.4	nd		11		

* The numerals in parentheses refer to the references as set out opposite.
† Percentage of lipid weight.
‡ Percentage of lipid phosphorus.
§ Densitometric reading of thin-layer chromatography spot.
‖ This value taken from Chacko et al. (1973).

for *ca.* 42 percent of the polar lipids, while phosphatidylethanolamine (PE) accounts for *ca.* 30 percent. The negatively charged phospholipids, namely phosphatidylserine (PS) and phosphatidylinositol (PI), together constitute *ca.* 15 percent of the polar lipids in vertebrate nonmyelinated nerve, and *ca.* 10 percent in invertebrate nerves. The turnover of polyphosphoinositides in nerve tissue is very rapid, making the amounts of the di- and triphosphorylated compounds *in vivo* difficult to determine following tissue removal and homogenization. Most of the phosphoinositide analyzed is, therefore, PI itself (Birnberger *et al.*, 1971; Sheltawy and Dawson, 1966). Only a small amount (*ca.* 2 percent) of the total phospholipid of nonmyelinated vertebrate nerves is cerebroside or sulfatide. These lipids are characteristic of myelin membrane and their appearance probably indicates contamination by myelin membranes.

In myelinated nerves the phospholipid profile is quite different. Mammalian myelinated nerves have much lower PC and higher sphingomyelin and particularly cerebroside and sulfatide contents than their nonmyelinated counterparts. The phospholipid pattern of myelin is similar to that of the soma membrane of oligodendroglial cells, the cells which wrap themselves around CNS nerve axons many times to form the myelin sheath (Geren, 1954).

Metabolism of lipids in nerve and myelin also differs. In 2-week-old rats, an age of rapid brain myelination (Agrawal *et al.*, 1970), the rate of incorporation of radiophosphorus into cerebral sphingomyelin is about four times as high as it is in adult rats, while there is almost no change in the rate of incorporation of radiophosphorus into phosphatidylcholine with age (Ansell and Spanner, 1961). The systematic changes with age in lipid composition of whole mammalian brain, which have been elaborately analyzed by Rouser and his colleagues, may be ascribed completely to increases in brain myelination (Rouser *et al.*, 1972).

Organelle membranes isolated from whole brain resemble the average plasma membrane in phospholipid composition, but not in cholesterol content. Both the phospholipid types and their fatty acid moieties occur to the same extent in the brain mitochondrial and microsomal fractions as they do in whole-brain homogenates (Biran and Bartley, 1961; Hamberger and Svennerholm, 1971). The plasma membrane, mitochondrial outer membrane and nuclear membrane of one cell may have a gross phospholipid composition which is determined by common factors, for example synthesizing systems of the endoplasmic reticulum, whereas the enzyme composition varies greatly between plasma and organelle membranes and may be determined by the membrane environment, cholesterol content and by the distribution and control of the genome in the nucleus and the organelles of the cell.

7.2.3   Fatty acid composition

The fatty acid composition of the lipids of nerve tissues is represented graphically in *Figure 7.2*. As in most animal cell membranes, there is an abundance of palmitic (C16:0) and stearic (C18:0) acids, and smaller contributions from the even-numbered, longer-chain fatty acids. Compared with the gray matter of human brain, axons from garfish olfactory nerve and squid retinal and stellar nerves have a relatively high content of long-chain, polyunsaturated fatty acids (C20:4–C22:6). Although long-chain, polyunsaturated fatty acids are more prominent components of tissues from

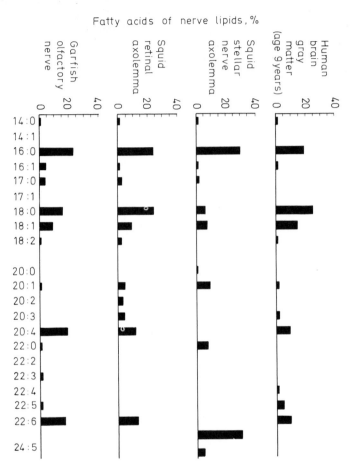

Figure 7.2   The profiles of total fatty acids from the lipids of membranes of nerve tissue. An analysis of the fatty acid composition of each lipid class is given in Tables 7.2–7.4

marine organisms in general (Gray and MacFarlane, 1961; Brockerhoff, Ackman and Hoyle, 1963), they also occur in relatively high proportion in the neurone fraction of mammalian brain (*Tables 7.2–7.5*). It is not clear to what extent these components are specific to nerve tissue or result from a species-specific difference (*see* Chacko *et al.*, 1972).

The fatty acid distribution in other mammalian organs differs from that of brain. For example, in rat liver the C18 fatty acids are mostly unsaturated in the phosphatide fraction and partly saturated in the neutral fraction (Veerkamp, Mulder and van Deenen, 1962). The fatty acid composition of brain actually tells much more about myelin than about nerve membrane.

Analyses of whole-brain tissue are, therefore, of limited value for an understanding of neural excitability.

In general, the fatty acid patterns of phosphatides from the same organs of different animals are similar, while fatty acid analyses of neutral lipids are similar if the lipids come from different organs of the same animal and are probably determined by diet and metabolism (Veerkamp, Mulder and van Deenen, 1962; van Deenen, 1965). Like the distributions of lipid types themselves, the fatty acids of organelle membranes are very similar to those of the plasma membrane of the parent cell (Veerkamp, Mulder and van Deenen, 1962; Hamberger and Svennerholm, 1971).

Table 7.2 FATTY ACID DISTRIBUTION IN NERVE PHOSPHATIDYLCHOLINE*

Fatty acid	Rabbit brain neurones†	Garfish olfactory nerve‡	Human brain gray matter§	Rabbit brain glia†	Rabbit brain myelin†	Human brain myelin§
	(44)‖	(42)	(29)	(39)	(21)	(15–21)
14:0	nd	1.1	0.5	nd	nd	0.1
16:0	38.1	40.3	42.5	45.8	32.9	32.0
16:1	1.9	7.6	3.5	1.4	2.3	0.7
17:0	nd	4.2	nd	nd	nd	nd
18:0	10.7	9.4	11.7	10.7	18.6	17.3
18:1	30.0	12.0	30.6	27.8	34.7	49.8
18:2	2.9	0.3	tr	2.2	3.6	tr
20:0	nd	nd	nd	nd	nd	nd
20:1	1.1	0.8	1.2	1.2	1.7	tr
20:2	nd		tr	nd	nd	nd
20:3	0.7		3.2	0.3	0.6	nd
20:4	8.9	11.6	5.3	6.1	2.9	tr
22:0	nd	0.2		nd	nd	
22:3	nd	2.0		nd	nd	
22:4	1.2	nd	nd	1.1	0.6	nd
22:5	1.7	0.7	nd	1.5	0.8	nd
22:6	2.5	8.9	1.0	1.9	0.8	tr
Total polyunsaturated	17.9	23.5	9.5	13.1	9.3	tr
Percentage sum in C14–18	83.6	74.9	88.8	87.9	92.1	99.9
Percentage sum in C19–22	16.1	24.2	10.7	12.1	7.4	tr

Abbreviations—nd: none detected; tr: trace.
* Data listed are the wt % of total fatty acid in phosphatidylcholine.
† From Hamberger and Svennerholm (1971).
‡ From Chacko, Goldman and Pennock (1972).
§ 6-year-old male; from O'Brien and Sampson (1965b).
‖ Numbers in parentheses are the percentage of total polar lipid in each tissue which is phosphatidylcholine.

How are the fatty acids distributed among the lipid classes present in nerve membranes? *Tables 7.2–7.5* list the distribution of fatty acid moieties in the major phosphoglycerides of mammalian neurones, garfish olfactory nerve, human brain (6-year-old) gray matter and mammalian glial cells and myelin. In axonal membranes, PC (*Table 7.2*) has a fatty acid content primarily composed of palmitate (C16:0), which accounts for 40 percent of the fatty acid, and oleate (C18:1), which accounts for about 25 percent of the fatty acid, with about 10 percent coming from stearate (C18:0) and a trace from linoleic acid (C18:2). Arachidonic acid (C20:4) and C22:6 are the prominent long-chain, polyunsaturated fatty acids in PC of nerve membranes (*ca.* 12 percent) but occur at much lower levels (*ca.* 4 percent) in the PC fraction of glia and myelin. At least 75 percent of the fatty acids in PC of nerve are 18 carbon atoms or less in length and only about 20 percent are polyunsaturated. This trend is even more extreme in PC from myelin; there, more than 90 percent of the fatty acids are C18 or shorter and only 5 percent, on average, are polyunsaturated.

Table 7.3   FATTY ACID DISTRIBUTION IN NERVE PHOSPHATIDYLETHANOLAMINE*

Fatty acid	Rabbit brain neurones	Garfish olfactory nerve	Human brain gray matter	Rabbit brain glia	Rabbit brain myelin	Human brain myelin
	(31)†	(32)	(32)	(35)	(29)	(24–25)
14:0	nd	0.4	0.2	nd	nd	0.8
16:0	6.1	4.6	10.2	6.8	6.2	13.8
16:1	0.7	0.6	0.9	0.8	0.7	1.5
17:0	nd	3.8	nd	nd	nd	nd
18:0	23.3	13.6	26.7	24.9	16.3	11.9
18:1	13.3	3.8	10.9	11.7	27.9	57.1
18:2	3.2	0.3	0.2	2.8	1.8	tr
20:0	nd	0.1	nd	nd	nd	nd
20:1	0.6	1.0	0.8	1.1	4.8	1.7
20:2	nd	nd	tr	nd	nd	tr
20:3	1.2	nd	1.6	1.3	3.3	tr
20:4	20.9	38.2	18.2	18.6	15.2	2.8
22:0	nd	0.2	nd	nd	nd	nd
22:3	nd	3.0	nd	nd	nd	nd
22:4	0.4	0.8	0.6	6.8	8.9	tr
22:5	9.7	3.1	11.4	11.0	7.2	tr
22:6	14.3	26.4	16.7	14.3	6.2	1.3
Total polyunsaturated	55.7	71.8	48.7	54.8	42.6	13.3
Percentage sum in C14–18	46.6	27.1	49.1	47.0	52.9	85.1
Percentage sum in C19–22	53.1	72.8	49.3	53.1	45.6	5.8

Abbreviations and sources of data are as in *Table 7.2.*
* Data listed are the wt % of total fatty acid in phosphatidylethanolamine.
† Numbers in parentheses are the percentage of total polar lipid in each tissue which is phosphatidylethanolamine.

In the axonal PE fraction of nerve a much larger proportion of the fatty acids is C20 and C22 polyunsaturated acids (50–70 percent). Most of the C18 hydrocarbons are saturated, as compared with the myelin lipids, where most have one double bond. The PE of garfish olfactory nerve is particularly high in plasmalogen (58 percent of the hydrocarbon), and C18:0 plus C18:1 (65.5 percent) account for more than 80 percent of the plasmalogen fatty aldehyde moiety. Plasmalogen is usually associated with the PE or PC fraction of membrane lipids (van Deenen, 1965).

Of the negatively charged phospholipids, PS has half of its fatty acid as C18, mostly saturated C18:0, in both 'nerve' membrane and myelin, and in

Table 7.4 FATTY ACID DISTRIBUTION IN NERVE PHOSPHATIDYLSERINE AND PHOSPHATIDYLINOSITOL

Fatty acid	Phosphatidylserine						Phosphatidylinositol			
	Rabbit brain neurones	Garfish olfactory nerve	Human brain gray matter	Rabbit brain glia	Rabbit brain myelin	Human brain myelin	Rabbit brain neurones	Garfish olfactory nerve	Rabbit brain glia	Rabbit brain myelin
	(7.0)†	(11)	(8.5)	(7.2)	(8.1)	(9)	(5.2)	(4.5)	(2.9)	(1.0)
14:0	nd	nd	0.3	nd	nd	0.3	nd	0.3	nd	nd
16:0	7.9	1.7	2.8	5.8	5.3	2.7	6.0	17.6	9.3	13.9
16:1	1.5	nd	0.6	2.4	0.6	0.4	1.1	1.3	0.7	1.0
17:0	nd	2.7	nd	nd	nd	nd	nd	1.9	nd	nd
18:0	37.6	41.7	46.0	38.6	38.6	44.4	34.0	45.5	34.6	36.7
18:1	12.8	4.8	7.7	12.5	30.1	36.9	10.3	5.3	9.4	12.2
18:2	1.5	0.6	tr	1.6	0.9	tr	0.5	nd	0.6	1.3
20:0	nd	0.4	nd	nd	nd	nd	nd	nd	nd	nd
20:1	1.0	0.3	0.8	0.8	3.1	3.7	0.5	nd	0.5	1.3
20:2	nd	nd	1.4	nd	nd	1.7	nd		nd	nd
20:3	1.1	nd	1.5	0.4	2.5	0.8	0.9		1.1	1.2
20:4	6.2	9.2	5.4	4.2	5.8	1.4	43.4	10.2	40.1	29.2
22:0	nd	2.6	nd	nd	nd	nd	nd	nd	nd	nd
22:3	nd	nd	nd	nd	nd	nd	nd	nd	nd	nd
22:4	4.8	1.5	tr	4.8	3.9	tr	0.8	nd	0.9	1.0
22:5	10.7	4.3	10.0	13.7	4.5	2.6	0.7	nd	0.8	0.6
22:6	14.3	28.3	23.5	14.9	3.5	1.1	0.8	17.0	1.4	1.0
Total polyunsaturated	31.3	43.9	41.8	39.6	21.1	7.6	47.1	27.2	44.9	34.3
Percentage sum in C14–18	61.9	51.5	57.4	60.9	75.5	84.7	51.9	71.9	54.6	65.1
Percentage sum in C19–22	38.1	46.6	42.6	23.3	23.3	11.3	47.1	28.1	44.8	34.3

Abbreviations and sources of data are as in *Table 7.2*.

* Data listed are the wt % of total fatty acid in phosphatidylserine and phosphatidylinositol respectively.

† Numbers in parentheses are the percentage of total polar lipid in each tissue which is phosphatidylserine or phosphatidylinositol respectively.

nerve about one-third of the fatty acid of PS is polyunsaturated C22:5 plus C22:6, but less than 10 percent is C20:4. In myelin, which contains the same percentage of PS as nerve membrane, the contribution from poly-unsaturated fatty acids is minute; almost all the fatty acid of PS exists as C18:0 and C18:1.

PI is a minor constituent of mammalian nerve membranes (5 percent of the polar lipid), but its polyphosphate derivatives exchange rapidly and the degree of PI phosphorylation is affected by nervous activity. The PI of garfish olfactory nerve differs markedly in fatty acid content from mammalian PI.

Table 7.5   FATTY ACID DISTRIBUTION IN SPHINGOLIPIDS OF HUMAN BRAIN*

Fatty acid	Sphingomyelin		Cerebroside		Sulfatide	
	Gray matter	Myelin	Gray matter	Myelin	Gray matter	Myelin
	(9.4)†	(7.8–10)	(6.1)	(22–33)	(1.2)	(5–7.5)
14:0	0.7	0.3	3.0	0.9	2.5	0.5
16:0	12.5	3.7	33	8.2	40	4.4
16:1	nd	nd	nd	nd	nd	nd
17:0	nd	nd	nd	nd	nd	nd
18:0	77.0	41.1	35.6	10.1	11.6	2.9
18:1	2.8	0.4	20	2.3	15.5	0.8
18:2	nd	nd	nd	nd	nd	nd
20:0	0.7	0.9	0.1	0.4	0.6	0.2
20:1	nd	nd	nd	nd	nd	nd
22:0	tr	0.6	0.5	1.7	1.6	1.0
22:1	0.5	1.6	tr	tr	tr	0.2
23:0	0.7	1.6	0.2	2.0	1.2	2.1
23:1	0.3	0.9	tr	tr	1.2	0.3
24:0	1.1	9.9	2.0	13.8	5.4	20.9
24:1	2.1	32.1	4.2	38.3	10.2	43.7
25:0	tr	1.8	tr	7.1	2.2	4.8
25:1	tr	0.4	tr	3.6	3.0	7.1
26:0	tr	tr	tr	0.9	tr	1.3
26:1	tr	3.1	tr	7.2	2.1	9.5
Total saturated	92.7	59.9	74.4	45.1	65.1	38.1
Percentage sum in C14–18	94.2	45.5	93.0	24.0	72.5	8.9
Percentage sum in C19–26	5.8	54.5	7.0	76.0	27.5	91.1

Abbreviations and sources of data are as in *Table 7.2.*
* All data are from O'Brien and Sampson (1965b).
† Number in parentheses are the percentage of total polar lipid in each tissue which is the particular sphingolipid concerned.

Unlike the other phosphoglycerides, PI tends to be shorter in length and lower in number of double bonds in the garfish nerve than in mammalian membranes. About one-quarter of the fatty acids of PI in garfish nerve and nearly one-half in mammalian nerve are polyunsaturated hydrocarbons, C20:4 and C22:6.

Several generalizations may be stated about the fatty acid pattern in these different tissues and lipids. The phosphoglycerides of nerve membranes are significantly more polyunsaturated and the proportion of long-chain fatty acids above C18 is usually greater than in myelin membranes. In phosphatidylethanolamine of nerve, the fatty acids are highly polyunsaturated (*ca.* 50 percent), while in the negatively charged lipids, PS and PI, from 30–40 percent of fatty acids are polyunsaturated, and in phosphatidylcholine less than 20 percent are polyunsaturated. The average chain lengths for each phosphoglyceride class are in the order PE > PI > PS > PC.

An analysis of the fatty acids of neural sphingolipids is not available. At best there are the data on human brain gray matter given in *Table 7.5*. The overall lipid composition of gray matter compared with that of vertebrate nerves and myelin suggests that between one-quarter and one-third of the tissue membrane may be myelin. The fatty acid profile in the sphingomyelin from gray matter is like the profile of fatty acids of PC from myelin membranes; almost 80 percent are C18:0. The sum of the two choline-containing phospholipids is about the same in both myelinated and nonmyelinated nerve. It may be that these phospholipids are interchangeable in membranes in general but this is not inevitable (Dervichian, 1964). In gray matter, less than 10 percent of the sphingomyelin fatty acids are unsaturated and only about 5 percent exceed 18 carbons in length (*Table 7.5*). All the sphingosine-derived lipids (sphingomyelin, cerebrosides, sulfatides) of gray matter contain mostly shorter-chain (C14–C18) unsaturated fatty acids, a pattern quite distinct from that of the fatty acid distribution in phosphoglycerides of gray matter (*Tables 7.2–7.5*). The cerebroside and sulfatide of gray matter are exclusively associated with oligodendrocytes and myelin, plus an uncharacterized variety of nerve endings. They are not characteristic of nerve soma or axon membrane (*Table 7.1*), but in gray matter their fatty acid distribution also differs from that of pure myelin. This apparent discrepancy probably arises from the difference between somatic membranes of oligodendrocytes found in gray matter and membranes of mature myelin (Norton, 1971).

The gangliosides are glycosphingolipids containing one or more sialic acid residues. They comprise less than 1 percent of the total lipid of nerve membranes from brain tissue of rabbit (Hamberger and Svennerholm, 1971), rat (Poduslo and Norton, 1972) and cat (Dekirmenjian *et al.*, 1969), and are found in both neuronal and glial fractions. Gangliosides appear to be concentrated at the synaptic endings from nerves (Hamberger and Svennerholm, 1971; Bosmann, 1972), particularly in the outer membranes of synaptosomes (Avrova, Chenykaeva and Obukhova, 1973). Interestingly, the enzymes which metabolize gangliosides are found in the synaptic fraction (Dicesare and Rapport, 1973) and other brain fractions, but the ganglioside-synthesizing enzymes are concentrated in the microsomal fraction which originates primarily from endoplasmic reticulum (Neskovic, Sarlieve and Mandel, 1973). The fatty acid content of synaptosomal gangliosides is, like that of neuronal glycosphingolipids (*Table 7.5*), largely composed of stearic acid (C18:0 > 80%) with minute fractions of polyunsaturated hydrocarbons; this pattern is identical in brain homogenate, microsomal, and synaptosomal fractions (Avrova, Chenykaeva and Obukhova, 1973).

7.2.4   Lipid composition and membrane dynamics

The lipid composition of the nerve membrane influences its structure and dynamic characteristics, and modulates the activities of some of the intrinsically bound enzymes. Earlier we have seen that the morphological and electrical properties of nerve membranes are quite similar to those of lipid bilayers which contain some protein or other ion-transporting agent. Direct structural determinations of nerve membranes, by X-ray diffraction, for example, further support the bilayer model. Studies of bilayers of lipid extracted from nerve are comparable with those of pure lipid bilayers and, when combined with spin-label studies of nerve, provide us with an average picture of structure and motion in nerve membranes.

The low-angle X-ray diffraction pattern of garfish olfactory nerve has been analyzed by Blasie et al. (1972). The electron density profile of the axolemma (after calculated corrections) is very similar to that of a stack of bilayers made from the extracted lipid of the garfish nerve and has the features of bilayers from single phospholipids (Levine and Wilkins, 1971). The difference may be accounted for by the presence of a thin layer of protein localized at the phospholipid headgroups in whole nerve (Blasie et al., 1972). Protein accounts for more than 30 percent of garfish olfactory nerve dry weight, yet most of it is localized at the membrane surface and not within the hydrocarbon region. When X-ray patterns of bilayers made only from phospholipids are compared with those made from total lipid extract, it appears that the inclusion of cholesterol increases the separation of polar lipid headgroups to that seen in the axolemma and causes an increase in average degree of orientation of fatty acid chains. This finding is consistent with those of Wolff et al. (1971), who noted that monolayers of phospholipids from squid retinal axons are condensed by the inclusion of cholesterol; that is, cholesterol increases the total membrane area but decreases the area per molecule at the surface, as it does in monolayers formed from unsaturated lipids (van Deenen et al., 1962; Ladbrooke, Williams and Chapman, 1968). Inclusion of cholesterol in bilayers containing cis-unsaturated lipids decreases the fluidity of the hydrocarbon interior and increases the degree of orientation (Long et al., 1970).

Melting studies and monolayer packing both support the notion that lipids with longer or more saturated hydrocarbon chains pack more closely and interact more strongly in membranes than do lipids with short-chain or polyunsaturated hydrocarbons, particularly of cis configuration (Williams and Chapman, 1970; Overath and Traüble, 1973). Wolff et al. found that the phospholipids of squid nerves were separated by about 0.9 nm in monolayers at a pressure of 10 dyn cm^{-2}. This is the intermolecular distance in model monolayers formed at the same pressure from phosphatidylcholine molecules with two saturated fatty acids, each only 10 carbon atoms long (Phillips and Chapman, 1968). Since most of the fatty acids in squid retinal axons are longer than C16 (Zambrano, Cellino and Canessa-Fischer, 1971), their weak interaction is an indication of the high degree of polyunsaturation of these lipids.

The motion in nerves can be detected by electron spin resonance 'spin labels' which are dissolved in the hydrocarbon phase or covalently bound to molecules which become oriented in the membrane. Spin-label studies show that the hydrocarbon region of nerves is very fluid. Lipid fatty acids are in a semi-'liquid' state and are rotating and bending at frequencies up to 10^9 s^{-1} in membranes of lobster walking-leg and rabbit vagus nerves (Hubbell and McConnell, 1968; McConnell, 1970). Signals from stearic acid derivatives (C18:0) spin-labeled at different positions along the hydrocarbon chain indicate that the nerve membranes in which they are inserted are most fluid and have the least degree of orientation near the center of the hydrocarbon region. The fluidity decreases while the degree of orientation increases as the label is moved closer to the polar headgroup region of the lipids (Hubbell and McConnell, 1969; Simpkins, Tay and Panko, 1971). Removal of 70–80 percent of the zwitterionic lipid headgroups by treatment with phospholipase C makes the environment of the stearic-acid-linked spin label in lobster nerve even more fluid (Simpkins, Panko and Tay, 1971). In contrast, cleaving the fatty acids from the 2-position of phosphoglycerides was found to decrease spin-label fluidity in lobster nerve (Simpkins, Tay and Panko, 1971), possibly by permitting tighter packing of the hydrocarbons from the released fatty acids.

Lipids diffuse laterally in the plane of the membrane, and transversely from one face of the membrane to another. Lateral diffusion of lipids has been detected by electron and nuclear spin resonance studies of model membranes (Devaux and McConnell, 1972; Traüble and Sackmann, 1972; Lee, Birdsall and Metcalfe, 1973) and biological membranes from muscle (Davis and Inesi, 1971; Robinson et al., 1972), nerve (Lee, Birdsall and Metcalfe, 1973) and kidney medulla (Barnett and Grisham, 1972). Proteins on muscle membranes also diffuse laterally, with a calculated diffusion constant of 1–3×10^{-9} cm^2 s^{-1}, about one order of magnitude smaller than that for lipids (Edidin and Fambrough, 1973).

Phospholipids with spin labels attached to the headgroups are able to switch from one side of a vesicular bilayer to the other, a motion referred to as 'flip-flop' (Kornberg and McConnell, 1971). In vesicles of pure phosphatidylcholine an asymmetric distribution of spin-labeled phospholipid is reduced by half in 6.5 h at 30 °C (Kornberg and McConnell, 1971), while the same spin-label in biological membranes has a flip-flop half-time of about 5 min at 15 °C (McNamee and McConnell, 1973).

The orientation of the polar headgroups of lipids at the membrane surface is not known. Theoretical arguments and studies of model phospholipid–water systems (Hanai, Haydon and Taylor, 1965) indicate that the direction of the phosphate–nitrogen dipole in PC and PE is parallel to the membrane surface, but phospholipid headgroup orientation in membranes containing both dipolar and charged phospholipids interacting with proteins is an unresolved problem.

Many studies have shown that fixed charges near 'sodium channels' in nerve can influence excitability. Such fixed charges create local surface potentials which contribute to the electric field in the membrane, and thus influence the voltage sensors which control Na$^+$ permeability (McLaughlin, Szabo and Eisenman, 1971). A membrane with fixed negative charges tends to accumulate cations and repel anions in the adjacent solution known as the 'double layer'. In turn, cations, particularly multivalent cations, decrease the local surface potential as they accumulate in the double layer (McLaughlin, Szabo and Eisenman, 1971), and in this way ions can influence excitability. For example, increasing external Ca^{2+} raises the firing threshold of nerves (Brink, 1954), and this is explained by a change in the electric field within the membrane which is detected by voltage sensors as a hyperpolarization of the nerve (*Figure 7.3*). In squid giant axons (Frankenhaeuser

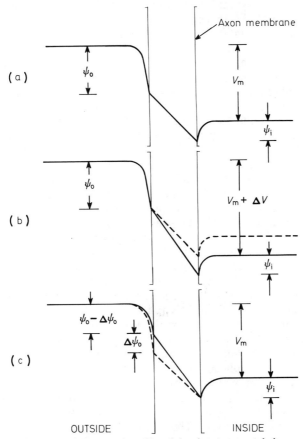

Figure 7.3   Schematized profiles of the electric potential changes across the nerve membrane. (a) The total membrane potential, V_m, is the sum of the potential change within the membrane plus the surface potentials at the outer and inner surfaces, ψ_o and ψ_i, respectively. Surface potentials arise from fixed negative charges on the nerve membrane. (After McLaughlin and Harary, 1974.) (b) When the membrane potential becomes more negative inside $(+\Delta V_m)$, the slope of the electric potential within the membrane becomes steeper, i.e. the electric field in the membrane increases. (c) If the outer surface potential $\Delta\psi_o$ becomes more positive, the electric field is changed to the same extent as in (b) but without changing the total membrane potential, V_m. This can occur when sufficient divalent cation is present outside the nerve

and Hodgkin, 1957) and in frog myelinated nerve (Hille, 1968b) a tenfold increase in external Ca^{2+} is equivalent to a 20 mV hyperpolarization detected by the Na^+ permeability voltage sensor. Changes of ionic strength (Chandler, Hodgkin and Meves, 1965) and divalent cation concentration (T. Begenisich, personal communication) inside squid giant axons produce much smaller changes in Na^+ permeability, indicating a lower density of fixed negative charges on the inside than the outside surface of the nerve membrane.

The difference between internal and external charge density could arise from charges specifically bound to 'ionic channel' protein at each membrane surface or from an asymmetric distribution of charge across the membrane as a whole. It is probable that cationic toxins which block Na^+ permeability, such as tetrodotoxin and saxitoxin, are influenced by the same fixed charges at their binding site. In whole nerves, divalent cations do affect toxin binding in a manner consistent with some screening of fixed charges (see Section 7.3); however, nerve membranes which have been solubilized by nonionic detergents bind toxins with no influence from fixed charges. This is the only evidence we have to suggest that fixed charges near ionic channels may reside on loosely bound (i.e. detergent-soluble) phospholipids.

If the fixed charges are lipid-associated and able to move laterally in the membrane, from where does the apparent asymmetry of charge density at each end of the sodium channel voltage sensor arise? As seen in *Table 7.1*, only about 15 percent of the total lipids are negatively charged, yet near the outside of the sodium channel it appears that there is a negative charge every $0.75-0.8$ nm^2, which requires that approximately 50 percent of the phospholipids have a net negative charge [using the monolayer density calculated by Wolff *et al.* (1971)]. Either phospholipids are not uniformly distributed on each membrane surface or they are distributed unequally between the inner and outer membrane surfaces. If phospholipid molecules can flip-flop across the membrane (see above), then a resting potential which is negative inside could provide the driving force to produce an asymmetric charge distribution across the membrane. McLaughlin and Harary (1974) have calculated this asymmetry using the equations for the potential in the double layer, the Boltzmann distribution for charged particles at an electric potential, a resting potential of -70 mV, and an average density of negative charges equal to 15 percent of the total phospholipid. The values which they derived for the surface potential from theoretical calculations, -50 mV at the outer surface and -15 mV at the inner, are reasonably close to those for squid axons, -62 mV (Gilbert and Ehrenstein, 1969) and -17 mV (Chandler, Hodgkin and Meves, 1965), respectively. This means that the actual potential change across the membrane (35 mV) is only one-half of the potential difference between internal and external media (*Figure 7.3*).

As noted above, the negative phospholipids PS and PI have hydrocarbon chains of intermediate length with multiple double bonds, and probably interact weakly with neighboring phospholipid molecules. Their motion is, therefore, less restricted and they may be able to flip-flop rapidly. So, the difference between inner and outer surface charge density can be accounted for by the flip-flopping of lipid molecules into an equilibrium distribution dependent on the membrane potential. But the charge asymmetry can also arise from specific interactions of charged lipids and sodium channel proteins which are themselves asymmetric across the nerve membrane.

7.2.5   Lipid–protein interactions in membranes

Certain enzyme activities in membranes are intimately dependent on the nature of surrounding lipid. For example, the sodium–potassium pump in the plasma membrane requires phospholipid for activity and the rate of coupled ATP hydrolysis by the pump depends on the fluid character of the surrounding lipids (Grisham and Barnett, 1973). When membrane lipids are in the crystalline state the enzyme rate is only one-third of that when they are in the liquid-crystalline state with fluid hydrocarbon motion. The change in rates occurs abruptly, at a temperature within the range that brings about the crystalline–liquid crystal phase transition of the lipids, and is accompanied by a change of about 15 kcal mol^{-1} in the activation energy of the enzyme (Grisham and Barnett, 1973).

The rate constants for Na$^+$ permeability changes in squid nerve also show a large temperature dependence, increasing about threefold for each 10 °C increase in temperature (Hodgkin and Huxley, 1952). This may be interpreted as an activation energy of about 20 kcal mol^{-1} according to the transition state theory. There is a uniform change of rate with temperature for Na$^+$ permeability in the squid (Hodgkin, Huxley and Katz, 1952), and of conduction velocity with temperature in the rabbit vagus (Howarth, Keynes and Ritchie, 1968), making it unlikely that lipid phase transitions occur or significantly affect Na$^+$ permeability changes in nerve.

The Na$^+$,K$^+$-ATPase appears to function as a 'carrier' of ions across the cell membrane (Sen and Post, 1964). This enzyme hydrolyzes no more than 600 molecules of ATP per second (Baker and Willis, 1969), coupled to the transport of 1.8×10^3 Na$^+$ ions per second and 1.2×10^3 K$^+$ ions per second (Sen and Post, 1964). In contrast, the voltage-dependent 'sodium channel' of nerve has a maximum Na$^+$ flux estimated at 10^8 Na$^+$ ions per second (Hille, 1970). This rate is considerably faster than the rate of carrier-mediated transport (Eigen and Winkler, 1970; Pressman, 1971), and supports the notion that Na$^+$ conductance is mediated by a pore, or channel.

7.2.6   Phospholipids and excitation: phospholipid metabolism

Experimental evidence reveals no direct role for phospholipids in nerve excitation. Extensive removal of fatty acids from the lipid fraction using phospholipase A produces lysophosphatidyl compounds whose detergent properties can solubilize nerve membranes (Condrea and Rosenberg, 1968; Rosenberg and Condrea, 1968). This effect accounts for the conduction block following treatment of crudely dissected squid axons with snake venom or purified phospholipase A. But if squid axons are carefully dissected, very little soluble lysophosphatide is produced by venom or phospholipase A treatment and conduction is not blocked, even though axon phospholipids are hydrolyzed to the same extent (Rosenberg, 1973). Similarly, removal of 85–100 percent of the phosphorylated bases by phospholipase C does not change resting permeability or conduction in the squid giant axon (Rosenberg and Condrea, 1968; Rosenberg, 1970). These results do not preclude the possibility that phospholipids specifically bound to proteins are important for excitation. Scant evidence from toxin binding studies suggests that fixed negative charges near ionic channels may be from lipid headgroups (Section 7.3).

There is no evidence that gangliosides are critical for excitability. The distribution of ganglioside types in neurones is very similar to the distribution in glia, although glia usually contain twice the weight percentage of gangliosides (Poduslo and Norton, 1972; Hamberger and Svennerholm, 1971). The ganglioside and the neutral glycosphingolipid compositions of neuroblastoma are not appreciably changed when the cells differentiate from an inexcitable to an electrically excitable form (Yogeeswaran et al., 1973). Treatment with neuraminidase, which cleaves the sialic acid group from gangliosides, has no effect on the binding to nerve membranes of tetrodotoxin (Benzer and Raftery, 1972), a molecule which blocks Na^+ permeability (see Section 7.1.2), nor on the action potential in squid giant axons (Rosenberg, 1965). In contrast, the potential importance of gangliosides as drug receptors at synapses is suggested by experimental inactivation of botulinus toxin by exogenous gangliosides present in the solution (Simpson and Rapport, 1971a, b). Protection against botulinus toxin by gangliosides may be a nonspecific effect of these lipids and surely is not strong evidence that they are the physiological receptors for toxin.

Labeling of phospholipids in nerve is affected by excitation. Exposure to [^{32}P]orthophosphate of resting axons from squid (Larabee and Brinley, 1968), lobster (Birnberger et al., 1971), rat and rabbit vagus (White and Larabee, 1973; Salway and Hughes, 1972) and hen sciatic nerves (Dawson, 1969) results in incorporation of the label into phospholipid. Most of the radioactivity is located in the phosphatidylinositols, some in phosphatidylethanolamine and phosphatidic acid, and a smaller portion in phosphatidylcholine, sphingomyelin and phosphatidylserine. The phosphatidylinositols contained ^{32}P in the order tri- > di- > mono-. Stimulation of the nerves while they are being exposed to ^{32}P$_i$ increases labeling of the phosphoinositides dramatically (Larabee and Brinley, 1968; Salway and Hughes, 1972). If nerves are preincubated in ^{32}P$_i$ and then stimulated in nonradioactive solution, the labeling of the tri- and diphosphoinositides decreases (Birnberger et al., 1971; Hawthorne, 1972). Changes in phosphoinositide labeling from stimulation of rat vagus and phrenic nerves and rabbit vagus nerve are inhibited by tetrodotoxin, which blocks Na^+ permeability (Salway and Hughes, 1972), and by ouabain, which inhibits the sodium–potassium pump (Hawthorne, 1972). Since the increase in internal Na^+ concentration which results from nerve impulses activates the sodium–potassium pump in rabbit vagus nerves (Rang and Ritchie, 1968), it follows that both tetrodotoxin and ouabain would inhibit stimulation-induced pump activity. Tetrodotoxin blocks Na^+ permeability but has no effect on the small capacitative currents associated with movement of the Na^+ permeability 'gating molecules' (Armstrong and Bezanilla, 1973a); hence, with tetrodotoxin present at least the molecular conformational changes which produce charge movement can occur, although other molecular rearrangements or associations may be inhibited. Treatment with cinchocaine (a local anesthetic) of nerves exposed to ^{32}P$_i$ increases the labeling of the phosphoinositides greatly, especially monophosphoinositide, regardless of whether the nerves are stimulated or not (Salway and Hughes, 1972). Clearly, an understanding of the phosphoinositide effect in axonal conduction requires more careful study. Labeling of polyphosphoinositides is very active in synaptic preparations and is modified by a variety of drugs. The reader is referred to Burt and Larabee (1973) for a discussion of these effects.

7.3   PROTEINS AND NERVE CONDUCTION

From the previous sections it should be clear that the lipids of nerve cannot account for excitability exclusively. Neurone membrane lipids are identical to those of inexcitable astroglia (Kuffler, Nicholls and Orkand, 1966; Kuffler, 1967) (*Table 7.1*), bilayers made from nerve lipid alone are inexcitable, and model bilayers of lipids have resistances much higher than nerve membranes unless protein is present. The analysis of neuronal membrane proteins is just beginning. Protein accounts for 30–60 percent of the dry weight of nonmyelinated nerve (*Table 7.1*). Gel electrophoresis of detergent-treated nerve membranes shows 10–15 detectable bands of protein, with five major bands from proteins of garfish olfactory membranes (Chacko *et al.*, 1973) and four from proteins of rabbit brain neurones (Karlsson *et al.*, 1973). None of the bands has been functionally identified.

There are several enzymes of nerve membrane whose function is related to excitability. The sodium–potassium pump has already been discussed with regard to its electrogenic effect and the modulation of excitability (*see* Section 7.1), and the pump turnover number and dependence on lipid fatty acid fluidity. In C-fibers of the rabbit vagus nerves the pump is present at a density of 750 per μm^2 (Landowne and Ritchie, 1970), and at 10^2–10^3 per μm^2 in snail giant neurone soma (calculated from Thomas, 1969).

What evidence is there for a direct role of proteins in nerve excitation? Nonmyelinated nerves contain receptors for acetylcholine, the application of which causes depolarization of C-fibers in rabbit vagus nerves (Armett and Ritchie, 1961), and an acetylcholine-binding protein has been isolated from lobster nerve which also binds a variety of drugs that act at post-synaptic membranes (Denburg, Eldefrawi and O'Brien, 1972). However, reagents which irreversibly block enzymes of the acetylcholine system in nerve by almost 100 percent do not affect axonal conduction irreversibly (Kremzner and Rosenberg, 1971). Many other observations (Ritchie and Armett, 1963; Rosenberg, 1973) are inconsistent with a functional role for acetylcholine in nerve conduction.

Proteolytic enzymes do alter excitability, but the effects are usually rather nonspecific (Tasaki and Takenaka, 1964; Narahashi and Tobias, 1964), and many of the excitability parameters change simultaneously. An interesting exception to this occurs when the interior of the squid giant axon is treated with the mixture of enzymes called 'pronase'. Pronase changes the transient nature of the voltage-dependent Na^+ permeability; when the axon membrane is depolarized after pronase treatment the Na^+ permeability increases to a plateau value but does not spontaneously decrease (inactivate), as it does in untreated axons (Rojas and Armstrong, 1971; Armstrong and Bezanilla, 1973b). Since the voltage-dependent K^+ permeability is not affected by pronase, these experiments show that part of the Na^+ permeability mechanism involves protein, accessible from the inner axon membrane surface, and is separate from the K^+ permeability mechanism. The inactivation of Na^+

permeability is also sensitive to proteins from scorpion venom applied to the outer surface of axons (Cahalan, 1973), so the structure controlling Na^+ permeability 'gating' may span the membrane.

7.3.1 Pharmacological probes of excitability

Nerve activity can be modified by many drugs. Local anesthetics are drugs which selectively block Na^+ permeability. The potency of a local anesthetic in blocking a de-sheathed nerve depends on the lipid solubility of the drug (Skou, 1954a, b). Local anesthetics are usually tertiary amines and such molecules appear to act primarily from inside the nerve when in the cationic form (Greengard and Ritchie, 1966; Narahashi, Frazier and Yamada, 1970). Quaternary forms of local anesthetics, which are permanently cationic, have a much greater blocking effect from inside the nerve than from outside (Frazier, Narahashi and Yamada, 1970; Strichartz, 1973). Different stereo-isomers of certain local anesthetics show differential nerve potency, although their uptakes are identical (Akerman, Camougis and Sandberg, 1969; Akerman, 1973), suggesting that local anesthetic molecules form specific complexes rather than being merely dissolved in the membrane. Quaternary local anesthetics can block Na^+ permeability from inside the nerve to an extent controlled by the membrane potential; the development of such a block depends on the opening of sodium channel 'gates' (Strichartz, 1973). Local anesthetics may act like their quaternary derivatives, plugging up sodium channels from the axoplasmic opening when the 'gates' are open. Other compounds such as veratridine (Ulbricht, 1969) and batrachotoxin (Albuquerque, Daly and Witkop, 1971) are lipid-soluble compounds which change the 'gating' characteristics of Na^+ permeability. Apparently the sodium channel 'gating' molecules are in contact with a lipid matrix, probably in the nerve membrane interior.

In nerve membranes the slowly developing K^+ currents can be separated from the ionic currents passing through sodium channels, indicating that different structures mediate these two ion permeabilities. Potassium ion permeability increases with membrane depolarization but does not spontaneously diminish during the short periods over which Na^+ permeability does (Hodgkin and Huxley, 1952). Potassium ion currents are completely insensitive to tetrodotoxin and are relatively resistant to local anesthetics in frog myelinated nerve (Hille, 1966, 1968a) and squid giant axons (Narahashi, Anderson and Moore, 1967; Taylor, 1959). But K^+ currents are blocked by the cation tetraethylammonium (TEA) when it is present inside squid axons and myelinated nerves (Armstrong and Binstock, 1965; Armstrong and Hille, 1972). TEA and its derivatives appear to be swept into potassium 'channels' when they are opened by depolarization, and the inhibition from these drugs may be reversed by membrane hyperpolarization and the subsequent influx of K^+ ions (Armstrong, 1969, 1971). Derivatives of TEA where one ethyl group has been replaced by a very hydrophobic group are much more effective inhibitors of K^+ currents, suggesting that a region near the TEA receptor is lipophilic (Armstrong, 1971; Armstrong and Hille, 1972).

Armstrong (1971) has emphasized the ability of hyperpolarization and elevated concentrations of external K^+ to aid in reversing TEA-induced inhibition in stressing that K^+ permeability is probably mediated by a pore or channel mechanism rather than a carrier. One-way K^+ fluxes through these channels are quite responsive to K^+ concentrations on both sides of the axon membrane in squid, exhibiting 'long pore' effects where K^+ ions occupy plural sites in a channel which is too narrow to allow individual ions to pass each other (Hodgkin and Keynes, 1955b). A few ions other than K^+ (Rb^+, NH_4^+) can permeate these channels, but alkali metal cations such as Na^+ and Cs^+ block potassium channels from the axoplasmic side. They appear capable of entering the channel but become 'stuck' within it (Bezanilla and Armstrong, 1972; Hille, 1973). In addition, K^+ currents in myelinated nerve are inhibited by external TEA, independently of membrane potential or the size or direction of K^+ currents (Hille, 1967).

In summary, the voltage-dependent K^+ permeability in nerves is mediated by very selective channels which several ions can occupy simultaneously. The channels have a wider opening at the axoplasmic end, allowing some cations to enter and partially traverse them, and near this opening there is a lipophilic binding site. Myelinated nerves uniquely possess a binding site for TEA at the outer opening of their potassium channels. One estimate for the density of potassium channels in squid axons is about 70 per μm^2 (Armstrong, 1966) but this calculation assumes a particular model for the TEA–channel interaction. Unfortunately, there is no available drug which inhibits K^+ currents selectively enough and at concentrations low enough to permit counting of potassium channels by binding experiments.

Studies of the binding of neurotoxins have provided information about the properties of sodium 'channels' in nerve membranes. The Na^+ permeability of nerve is selectively inhibited by very low concentrations of both tetrodotoxin and saxitoxin. The actions of these toxins are almost identical

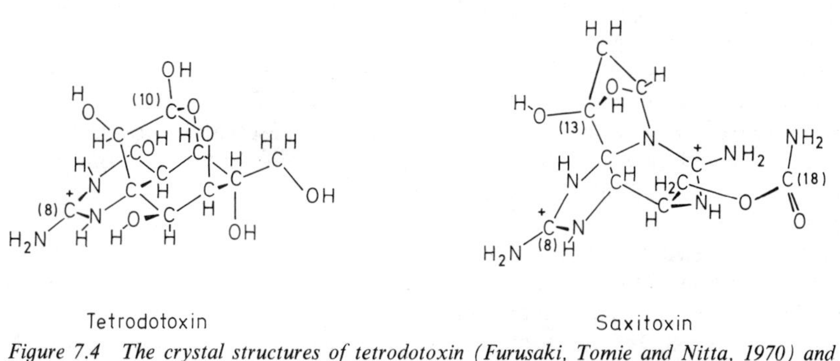

Tetrodotoxin   Saxitoxin

Figure 7.4   The crystal structures of tetrodotoxin (Furusaki, Tomie and Nitta, 1970) and saxitoxin (Schantz et al., 1975), showing their similar dimensions and the presence of cationic guanidinium groups

(Evans, 1972); both have a molecular weight of about 300 and have positively charged guanidinium groups (Figure 7.4). Sodium ion permeability is halved by toxin concentrations of 2–10 nM applied to squid axons (Cuervo and Adelman, 1970), frog and toad nodes of Ranvier (Hille, 1968a; Scwarz, Ulbricht and Wagner, 1973) and C-fibres of the rabbit vagus nerve (Colquhoun and Ritchie, 1972). Tetrodotoxin also inhibits the early inward current in cat motor-neurone cell bodies (J. Barrett, personal communication), the chemically induced activity in cat central nervous cells (Zieglgansbërger and Puil, 1972), and the early Na^+ current in *Aplysia* giant neurones (Geduldig and Gruener, 1970). The inhibition is reversible, and appears only when tetrodotoxin is applied outside the nerve, not when it is present internally (Narahashi, Anderson and Moore, 1966). Application of toxin to unstimulated nerve causes a small hyperpolarization of the membrane potential consistent with the blockage of resting Na^+ permeability (Narahashi, 1964; Freeman, 1969). Furthermore, the rate of appearance of toxin inhibition is not influenced by stimulating the nerve during toxin application (Schwarz, Ulbricht and Wagner, 1973). Therefore, the effects of tetrodotoxin are not modulated by electrical activity in the nerve.

The binding of radioactively labeled tetrodotoxin and saxitoxin to various nerves has been investigated. Both toxins show Langmuir-type binding to rabbit vagus, lobster walking-leg and garfish olfactory nerves (Colquhoun, Henderson and Ritchie, 1972; Benzer and Raftery, 1972)

$$\text{Toxin (T)} + \text{Receptor (R)} \rightleftharpoons \text{Toxin . Receptor (TR)}$$

with equilibrium dissociation constants of 3–10 nM. The density of saturable binding sites is relatively low, of the order of 20–30 sites per μm^2 in rabbit and lobster nerve, and only 3 sites per μm^2 in the small-diameter garfish olfactory nerve (Colquhoun, Henderson and Ritchie, 1972; Moore, Narahashi and Shaw, 1967).

Tetrodotoxin-binding activity can be solubilized by treating nerve membranes with nonionic detergents (1–2% Triton X-100) for several hours at $0\,^\circ C$ (Henderson and Wang, 1972); the binding is measured by equilibrating radioactive toxin with the solubilized binding material as it passes through a gel filtration column. About 80 percent of the original binding capacity per milligram of protein is solubilized by this procedure, while the apparent affinity for the toxin is slightly increased by solubilization (Henderson and Wang, 1972; Benzer and Raftery, 1973). The dissociation rates of the solubilized receptor–tetrodotoxin complex and receptor–saxitoxin complex (*Table 7.6*) are identical to the rate of reversal of the inhibition of Na^+ currents by toxins in intact myelinated nerves: these values are $0.95\ min^{-1}$ and $3.2\ min^{-1}$ respectively for the dissociation of the two complexes and $0.82\ min^{-1}$ and $1.5\ min^{-1}$ for the respective reversals of inhibition by these toxins. The constants noted were all measured at $20\,^\circ C$, except for the last which was measured at $12\,^\circ C$, and this accounts for the discrepancy between binding and physiological rate values (Schwarz, Ulbricht and Wagner, 1973). The first-order *binding* rate constant, calculated as the equilibrium binding constant divided by the dissociation rate constant, is also identical to the physiologically determined rate constant for the appearance of tetrodotoxin inhibition of Na^+ currents (Henderson and Wang, 1972; Henderson, Ritchie and Strichartz, 1973; B. Hille, personal communication; Schwarz, Ulbricht and Wagner, 1973).

Toxin binding to intact nerves is unaffected by local anesthetics or by agents such as veratrine alkaloids or batrachotoxin which alter the kinetics or membrane potential dependence of Na^+ permeability (Colquhoun, Henderson and Ritchie, 1972; Benzer and Raftery, 1972; Henderson, Ritchie and Strichartz, 1973). Metal cations inhibit the binding of toxins to intact and solubilized nerves. Unpublished studies from the laboratory of J. M. Ritchie have shown that the binding of trivalent, divalent and monovalent cations competes reversibly with toxin binding. On the assumption that cation binding and toxin binding are mutually exclusive, dissociation constants for cations can be calculated as inhibitory constants of toxin binding

Table 7.6 BINDING PARAMETERS FOR TETRODOTOXIN AND SAXITOXIN

	Binding to intact nerve	Binding to solubilized receptor from garfish nerve	Physiological effects
Tetrodotoxin			
K_{diss} (nM)	3.0 ± 0.4 (1)*†	6 ± 1.5 (3)	3–5 (6)†
	10.1 ± 1.3 (1)‡	2.5 (4)	3.60 ± 17 (7)§
	8.3 (2)		12 ± 6 (8)†
K_{off}		0.95 min^{-1} (3)	0.35 min^{-1} (12 °C) (9)‖
			0.82 min^{-1} (7)§
K_{on}		1.6×10^8 M^{-1} min^{-1} (5)	1.5×10^8 M^{-1} min^{-1} (12 °C) (8)‖
			1.7×10^8 M^{-1} min^{-1} (7)§
Saxitoxin			
K_{diss}	6.7 ± 3.0*	6.3 (5)	6 ± 3 (8)‡
K_{off}		3.2 min^{-1} (5)	1.5 min^{-1} (12 °C) (9)‖
K_{on}		5×10^8 M^{-1} min^{-1} (5)	5.7×10^8 M^{-1} min^{-1} (12 °C) (9)‖

All measurements were performed at 20 °C unless otherwise noted.
References—(1) Colquhoun, Henderson and Ritchie (1972); (2) Benzer and Raftery (1972); (3) Henderson and Wang (1972); (4) Benzer and Raftery (1973); (5) Henderson, Ritchie and Strichartz (1973); (6) Colquhoun and Ritchie (1972); (7) Schwarz, Ulbricht and Wagner (1973); (8) R. Henderson and G. Strichartz, unpublished observation on the inhibition of ^{22}Na efflux from garfish nerves; (9) B. Hille, personal communication.

* The numerals in parentheses refer to the references as set out above.
† Rabbit vagus nerve.
‡ Garfish olfactory nerve.
§ *Xenopus laevis*, node of Ranvier; inhibition of I_{Na} during voltage clamp.
‖ *Rana pipiens*, node of Ranvier; inhibition of I_{Na} during voltage clamp.

(Henderson, Ritchie and Strichartz, 1973). Trivalent cations have K_{diss} of about 10^{-3} M, for divalent the value is $(1–3) \times 10^{-2}$ M, and monovalent ions usually have $K_{diss} \geqslant 0.5$ M. Exceptions to the usual behavior of monovalent cations are Tl$^+$ and Li$^+$, which have apparent dissociation constants of 2×10^{-2} M and $(2–3) \times 10^{-1}$ M, respectively. Protons also reversibly inhibit the binding of tetrodotoxin and saxitoxin to intact nonmyelinated nerves; inhibition is consistent with a toxin binding site having a single acidic group with $pK_a = 5.9$ (Henderson, Ritchie and Strichartz, 1973, and unpublished observations).

These cation-binding properties of the saturable binding site for tetrodotoxin and saxitoxin are very similar to the properties of the sodium channel in nerve. Sodium ion permeability is inhibited by Tl$^+$ with K_{diss} of about 2×10^{-2} M [inferred from measurements of Na$^+$ currents in nerves bathed by a thallous-containing medium (B. Hille, personal communication)]. Both H$^+$ and Ca^{2+} ions block axonal Na$^+$ permeability, and to a degree which depends on the membrane potential (Woodhull, 1973). In resting C-fibers, which have a membrane potential of -40 mV (Keynes and Ritchie, 1965), the calculated dissociation constants for Ca^{2+} ions and protons would be

2.3×10^{-2} M and 1.2×10^{-6} M, corresponding to $pK_a = 5.9$ (Henderson, Ritchie and Strichartz, 1973). Woodhull (1973) found the voltage dependence of the inhibition by Ca^{2+} to be twice that of the inhibition by H^+; she concluded that both ions were bound to the same site, within the sodium channel, in the pathway of Na^+ ions as they crossed the axon membrane during an action potential. From the similarities between the kinetics of toxin blockage of Na^+ currents and the kinetics of toxin binding, and between the ionic (H^+, Ca^{2+}, Tl^+) blockage of Na^+ currents and of toxin binding at equilibrium, Henderson, Ritchie and Strichartz (1973) concluded that tetrodotoxin and saxitoxin were bound to the external opening of the sodium channels and blocked Na^+ currents by 'plugging' the channels, a hypothesis previously suggested by Kao and Nishiyama (1965) and by Hille (1971).

Hille (1968b) had shown earlier that Na^+ currents were inhibited at low pH, and later (Hille, 1971, 1972) he described a small-molecule filter for the sodium channel, lined with hydrogen bond acceptors and including an acidic group, to account for the selectivity of the channel towards organic and metal cations. If this selective structure is the site of the voltage-dependent acidic group described by Woodhull (1973), then its voltage dependence locates it towards the external opening of the sodium channel. The good fit between molecular models to toxins and this ion filter indicates that it could serve as the receptor for tetrodotoxin and saxitoxin molecules.

The biochemical properties of the tetrodotoxin receptor have been studied in nerve homogenates and in the solubilized membrane preparation. Homogenates of garfish olfactory nerve bind toxins with affinities like that of intact nerve, and the binding is relatively insensitive to treatment with various nucleases, proteolytic enzymes, and phospholipases C and D (Benzer and Raftery, 1973). Phospholipases C and D cleave the headgroup of phosphoglycerides, leaving or removing the phosphate moiety respectively. Neuraminidase treatment, which cleaves sialic acid residues from gangliosides, actually enhances the amount of toxin bound, although this may be a nonspecific effect unrelated to sodium channel sites, and treatment with phospholipase A removes about 30 percent of the binding activity. If the homogenate preparation is pretreated with phospholipase A, the actions of proteolytic enzymes are greatly enhanced and up to 90 per cent of the binding is abolished (Benzer and Raftery, 1973). Phospholipase A hydrolyzes fatty acids from phosphoglycerides at the 2-position, and such free fatty acids, which are often polyunsaturated, may act as ionic detergents to disrupt the bilayer character of the membrane. Ionic detergent treatment of intact nerve is known to destroy toxin binding (Henderson and Wang, 1972).

When the membrane binding activity has been solubilized with nonionic detergents, it becomes immediately vulnerable to the action of proteolytic enzymes (Benzer and Raftery, 1973). Neuraminidase has no effect on the binding of tetrodotoxin to solubilized receptors; however, binding is reduced drastically by a reagent which breaks disulfide bonds (dithiothreitol) but not by one which reacts with free sulfhydryl groups (iodoacetamide) (Benzer and Raftery, 1973).

Taken together, these studies indicate that tetrodotoxin and saxitoxin bind to receptors in the nerve membrane which are probably protein embedded in a lipid matrix, which have an acidic site that must be negatively charged to bind toxin molecules and which can bind metal cations. Neither the polar headgroups nor the phosphoryl moiety of phosphoglycerides seem necessary for toxin binding, but disulfide bonds appear to be critically involved, perhaps by maintaining the conformation of the binding protein. There is accumulating evidence that tetrodotoxin binds to the sodium channel at its external opening. If this is so, then the properties of the toxin receptor must also be properties of the sodium channels in nerve.

Studies of toxin binding also indicate that fixed charges on the membrane have little effect on toxin binding. In experiments in J. M. Ritchie's laboratory, we have investigated the ability of divalent cations to displace saxitoxin relative to tetrodotoxin. Divalent cations are known to decrease the electric potential arising from fixed charges at the surface of negatively charged membranes (McLaughlin, Szabo and Eisenman, 1971). This effect has been manifested electrophysiologically as a shift of the Na^+ permeability of nerve along the voltage axis, probably due to the screening of fixed charges by divalent cations (*see Figure 7.4*). The Na^+ permeability function of frog myelinated fibers and squid giant axons is shifted by 20 mV for each tenfold increase in the external Ca^{2+} concentration (Hille, 1968b; Frankenhaeuser and Hodgkin, 1957). The surface charges near the toxin binding site will affect the relative concentrations of monovalent tetrodotoxin and divalent saxitoxin, and so changes in surface charge will alter their apparent affinities. Thus, the ratio of the apparent dissociation constant of saxitoxin–receptor to that of tetrodotoxin–receptor can be used as a measure of the surface charge at the binding site. Our experiments show that binding of toxin to intact garfish olfactory nerve detects a surface potential change of only 6 mV per tenfold increase of external Ca^{2+}, much lower than that which affects Na^+ permeability. In solubilized nerve membranes the binding of tetrodotoxin and saxitoxin is affected equally by changes in external Ca^{2+}. The results are consistent with the idea that fixed charges at the outer surface of the sodium channel are on negatively charged lipids that are removed by detergent treatment and that they are closer to the molecules which control the voltage sensitivity of Na^+ permeability than they are to the mouth of the channel. However, these regions may be separated by as little as 0.8 nm (if we assume that the potential from fixed surface charges decreases with distance like that from a point charge in 0.1 M ionic solution (Robinson and Stokes, 1959).

Radioactively labeled tetrodotoxin promises to be valuable for the study of the development of excitability in nerves. For example, increasing amounts of toxin are bound by homogenates of chick and mouse brain as neural development proceeds (Hafemann and Unsworth, 1973), tetrodotoxin sensitivity of skeletal muscle changes following denervation (Redfern and Thesleff, 1971), and studies in cultured cardiac muscle cells indicate that tetrodotoxin desensitization may be accompanied by the synthesis *de novo* of a different kind of ionic channel (McDonald, Sachs and DeHaan, 1973). By studying differentiating neuroblastoma cells, muscle cells and denervated muscle membrane with radioactive toxins we may begin to determine the factors which control the synthesis and insertion of the structures necessary for excitability.

This review was completed in October 1974. The author acknowledges the kind support of Prof. J. M. Ritchie. Supported by USPHS Grants NS-08304 and 5 S07RR0573604.

REFERENCES

AGRAWAL, H. C., BANIK, N. L., BONE, A. H., DAVISON, A. N., MITCHELL, R. F. and SPOHN, M. (1970). *Biochem. J.*, **120**:635.

AKERMAN, B. (1973). *Acta pharmac. tox.*, **32**:225.

AKERMAN, B., CAMOUGIS, G. and SANDBERG, R. V. (1969). *Eur. J. Pharmac.*, **8**:337.

ALBUQUERQUE, E. X., DALY, J. W. and WITKOP, B. (1971). *Science, N.Y.*, **172**:995.

ANSELL, G. B. and SPANNER, S. (1961). *Biochem. J.*, **79**:176.

ARAKI, T. and TERZUOLO, C. A. (1962). *J. Neurophysiol.*, **25**:772.

ARMETT, C. J. and RITCHIE, J. M. (1961). *J. Physiol., Lond.*, **155**:372.

ARMSTRONG, C. M. (1966). *J. gen. Physiol.*, **50**:491.

ARMSTRONG, C. M. (1969). *J. gen. Physiol.*, **54**:553.

ARMSTRONG, C. M. (1971). *J. gen. Physiol.*, **58**:413.

ARMSTRONG, C. M. and BEZANILLA, F. (1973a). *Nature, Lond.*, **242**:459.

ARMSTRONG, C. M. and BEZANILLA, F. (1973b). *Abstracts 17th Meeting Biophys. Soc.*, No. 242a.

ARMSTRONG, C. M. and BINSTOCK, L. (1965). *J. gen Physiol.*, **48**:859.

ARMSTRONG, C. M. and HILLE, B. (1972). *J. gen. Physiol.*, **59**:388.

AVROVA, N. F., CHENYKAEVA, E. Y. and OBUKHOVA, E. L. (1973). *J. Neurochem.*, **20**:997.

BAKER, P. F., HODGKIN, A. L. and SHAW, T. I. (1962a). *J. Physiol., Lond.*, **164**:330.

BAKER, P. F., HODGKIN, A. L. and SHAW, T. I. (1962b). *J. Physiol., Lond.*, **164**:355.

BAKER, P. F. and WILLIS, J. S. (1969). *Biochim. biophys. Acta*, **183**:646.

BARNETT, R. E. and GRISHAM, C. M. (1972). *Biochem. biophys. Res. Commun.*, **48**:1362.

BARRETT, J. N. and CRILL, W. E. (1972). *Fedn Proc. Fedn Am. Socs exp. Biol.*, **31**:305.

BENZER, T. I. and RAFTERY, M. A. (1972). *Proc. natn Acad. Sci. U.S.A.*, **69**:3634.

BENZER, T. I. and RAFTERY, M. A. (1973). *Biochem. biophys. Res. Commun.*, **51**:939.

BEZANILLA, F. and ARMSTRONG, C. M. (1972). *J. gen. Physiol.*, **60**:588.

BIRAN, L. A. and BARTLEY, W. (1961). *Biochem. J.*, **79**:159.

BIRNBERGER, A., BIRNBERGER, K., ELIASSON, S. and SIMPSON, P. (1971). *J. Neurochem.*, **18**:1291.

BLANKENSHIP, J. E. (1968). *J. Neurophysiol.*, **31**:186.

BLASIE, J. K., GOLDMAN, D. E., CHACKO, G. and DEWEY, M. (1972). *Abstracts 16th Meeting Biophys. Soc.*, No. 253a.

BODIAN, D. (1962). *Science, N.Y.*, **137**:323.

BODIAN, D. (1967). *The Neurosciences*, pp. 6–24. Ed. G. C. QUARTON, T. MELNECHUK and F. O. SCHMITT. New York; Rockefeller University Press.

BOSMANN, H. B. (1972). *FEBS Lett.*, **22**:97.

BRINK, F. (1954). *Pharmac. Rev.*, **6**:243.

BROCKERHOFF, H., ACKMAN, R. G. and HOYLE, R. J. (1963). *Archs Biochem. Biophys.*, **100**:9.

BURT, D. R. and LARABEE, M. G. (1973). *J. Neurochem.*, **21**:255.

CAHALAN, M. (1973). *Abstracts 17th Meeting Biophys. Soc.*, No. 242a.

CAMEJO, B., VILLEGAS, G. M., BARNOLA, F. V. and VILLEGAS, R. (1969). *Biochim. biophys. Acta*, **193**:247.

CHACKO, G. K., GOLDMAN, D. E. and PENNOCK, B. E. (1972). *Biochim. biophys. Acta*, **280**:1.

CHACKO, G., GOLDMAN, D. E., MALHOTRA, H. C. and DEWEY, M. M. (1973). *Abstracts 15th Meeting Biophys. Soc.*, No. 240a.

CHANDLER, W. K., HODGKIN, A. L. and MEVES, H. (1965). *J. Physiol., Lond.*, **180**:821.

COLE, K. S. (1968). *Membranes, Ions and Impulses*. Berkeley, California; University of California Press.

COLE, K. S. and CURTIS, H. J. (1939). *J. gen. Physiol.*, **22**:649.

COLQUHOUN, D., HENDERSON, R. and RITCHIE, J. M. (1972). *J. Physiol., Lond.*, **227**:95.

COLQUHOUN, D. and RITCHIE, J. M. (1972). *J. Physiol., Lond.*, **221**:533.

CONDREA, E. and ROSENBERG, P. (1968). *Biochim. biophys. Acta*, **150**:271.

CONNELLY, C. M. (1959). *Rev. mod. Phys.*, **31**:475.

CUERVO, L. A. and ADELMAN, W. J. (1970). *J. gen. Physiol.*, **55**:309.

DAVIS, D. G. and INESI, G. (1971). *Biochim. biophys. Acta*, **241**:1.

DAWSON, R. M. C. (1969). *Ann. N.Y. Acad. Sci.*, **165**:774.

DEKIRMENJIAN, H., BRUNNGRABER, E. G., LEMKEY JOHNSTON, N. and LARRAMENDI, L. M. H. (1969). *Expl Brain Res.*, **8**:97.

DELBRÜCK, M. (1970). *The Neurosciences Second Study Program*, pp. 677–684. Ed. F. O. SCHMITT. New York; Rockefeller University Press.

DENBURG, J. L., ELDEFRAWI, M. E. and O'BRIEN, R. D. (1972). *Proc. natn. Acad. Sci. U.S.A.*, **69**:177.

DEPIERRE, J. W. and KARNOVSKY, M. L. (1973). *J. Cell Biol.*, **56**:275.

DERVICHIAN, D. G. (1964). *Prog. Biophys.*, **14**:263.
DEVAUX, P. and MCCONNELL, H. M. (1972). *J. Am. chem. Soc.*, **94**:4475.
DICESARE, J. L. and RAPPORT, M. M. (1973). *J. Neurochem.*, **20**:1781.
DODGE, F. A. and FRANKENHAEUSER, B. (1958). *J. Physiol., Lond.*, **143**:76.
DODGE, F. A. and FRANKENHAEUSER, B. (1959). *J. Physiol., Lond.*, **148**:188.
DOUGLAS, W. W. and RITCHIE, J. M. (1962). *Physiol. Rev.*, **42**:297.
EASTON, D. M. (1965). *Cold Spring Harb. Symp. quant. Biol.*, **30**:15.
EASTON, D. M. (1971). *Science, N.Y.*, **172**:952.
ECCLES, J. C. (1968). *The Physiology of Nerve Cells.* Baltimore, Maryland; Johns Hopkins Press.
EDIDIN, M. and FAMBROUGH, D. (1973). *J. Cell Biol.*, **57**:27.
EHRENSTEIN, G. and LECAR, H. (1972). *A. Rev. Biophys. Bioengng*, **1**:347.
EIGEN, M. and WINKLER, R. (1970). *The Neurosciences Second Study Program*, pp. 685–696. Ed. F. O. SCHMITT. New York; Rockefeller University Press.
ERLANGER, J. and GASSER, H. S. (1937). *Electrical Signs of Nervous Activity.* Philadelphia; University of Pennsylvania Press.
EVANS, M. H. (1972). *Int. Rev. Neurobiol.*, **15**:83.
EYZAGUIRRE, C. and KUFFLER, S. W. (1955). *J. gen. Physiol.*, **39**:87.
FRANK, K., FUORTES, M. G. F. and NELSON, P. G. (1959). *Science, N.Y.*, **130**:38.
FRANKENHAEUSER, B. (1965). *J. Physiol., Lond.*, **180**:780.
FRANKENHAEUSER, B. and HODGKIN, A. L. (1957). *J. Physiol., Lond.*, **137**:218.
FRAZIER, D., NARAHASHI, T. and YAMADA, M. (1970). *J. Pharmac. exp. Ther.*, **171**:45.
FREEMAN, A. R. (1969). *Fedn Proc. Fedn Am. Socs exp. Biol.*, **28**:333.
FURUSAKI, A., TOMIE, Y. and NITTA, I. (1970). *Bull. chem. Soc. Japan*, **43**:3332.
GASSER, H. S., RICHARDS, C. H. and GRUNDFEST, H. (1938). *Am. J. Physiol.*, **123**:299.
GEDULDIG, D. and GRUENER, R. (1970). *J. Physiol., Lond.*, **211**:217.
GEREN, B. B. (1954). *Expl Cell Res.*, **7**:558.
GILBERT, D. L. and EHRENSTEIN, G. (1969). *Biophys. J.*, **9**:447.
GOLDMAN, L. and SCHAUF, C. L. (1973). *J. gen. Physiol.*, **61**:361.
GRAY, G. M. and MACFARLANE, M. G. (1961). *Biochem. J.*, **81**:480.
GREENGARD, P. and RITCHIE, J. M. (1966). *A. Rev. Pharmacol.*, **6**:405.
GRISHAM, C. M. and BARNETT, R. E. (1973). *Biochemistry*, **12**:2635.
GRUNDFEST, H. (1957). *Physiol. Rev.*, **37**:337.
GRUNDFEST, H. (1965). *Cold Spring Harb. Symp. quant. Biol.*, **30**:1.
GRUNDFEST, H. (1967). *Science, N.Y.*, **156**:1771.
HAFEMANN, D. R. and UNSWORTH, B. R. (1973). *J. Neurochem.*, **20**:613.
HAGIWARA, S. (1973). *Adv. Biophys.*, **4**:71.
HAMBERGER, A. and SVENNERHOLM, L. (1971). *J. Neurochem.*, **18**:1821.
HANAI, T., HAYDON, D. A. and TAYLOR, J. (1965). *J. theor. Biol.*, **9**:278.
HAWTHORNE, J. N. (1972). *Biochem. Soc. Symp.*, **35**:383.
HENDERSON, R., RITCHIE, J. M. and STRICHARTZ, G. R. (1973). *J. Physiol., Lond.*, **235**:783.
HENDERSON, R. and WANG, J. H. (1972). *Biochemistry*, **11**:4565.
HENN, F. A., HANSSON, H. and HAMBERGER, A. (1972). *J. Cell Biol.*, **53**:654.
HILLE, B. (1966). *Nature, Lond.*, **210**:1220.
HILLE, B. (1967). *J. gen. Physiol.*, **50**:1287.
HILLE, B. (1968a). *J. gen. Physiol.*, **51**:199.
HILLE, B. (1968b). *J. gen. Physiol.*, **51**:221.
HILLE, B. (1970). *Prog. Biophys. molec. Biol.*, **21**:1.
HILLE, B. (1971). *J. gen. Physiol.*, **58**:599.
HILLE, B. (1972). *J. gen. Physiol.*, **59**:637.
HILLE, B. (1973). *J. gen. Physiol.*, **61**:669.
HIRST, G. D. S. and SPENCE, I. (1973). *Nature, New Biol.*, **243**:54.
HODGKIN, A. L. and HUXLEY, A. F. (1952). *J. Physiol., Lond.*, **117**:500.
HODGKIN, A. L., HUXLEY, A. F. and KATZ, B. (1952). *J. Physiol., Lond.*, **116**:424.
HODGKIN, A. L. and KATZ, B. (1949). *J. Physiol., Lond.*, **108**:37.
HODGKIN, A. L. and KEYNES, R. D. (1955a). *J. Physiol., Lond.*, **128**:28.
HODGKIN, A. L. and KEYNES, R. D. (1955b). *J. Physiol., Lond.*, **128**:61.
HOLLUNGER, E. G. and NIKLASSON, B. H. (1973). *J. Neurochem.*, **20**:821.
HOLTON, J. B. and EASTON, D. M. (1971). *Biochim. biophys. Acta*, **239**:61.
HOŘACKOVA, M., NONNER, W. and STÄMPFLI, R. (1968). *Proc. int. Union physiol. Sci.*, **7**:198.
HOWARTH, J. V., KEYNES, R. D. and RITCHIE, J. M. (1968). *J. Physiol., Lond.*, **194**:745.
HUBBELL, W. L. and MCCONNELL, H. M. (1968). *Proc. natn. Acad. Sci. U.S.A.*, **61**:12.

HUBBELL, W. L. and MCCONNELL, H. M. (1969). *Proc. natn. Acad. Sci. U.S.A.*, **64**:20.

HUXLEY, A. F. and STÄMPFLI, R. (1949). *J. Physiol., Lond.*, **108**:315.

JANSEN, J. K. S. and NICHOLLS, J. G. (1973). *J. Physiol., Lond.*, **229**:635.

JULIAN, F. J., MOORE, J. W. and GOLDMAN, D. E. (1962). *J. gen. Physiol.*, **45**:1217.

KAO, C. Y. and NISHIYAMA, A. (1965). *J. Physiol., Lond.*, **180**:50.

KARLSSON, J.-O., HAMBERGER, A. and HENN, F. A. (1973). *Biochim. biophys. Acta*, **298**:219.

KAUFFMAN, J. W. and MEAD, C. A. (1970). *Biophys. J.*, **10**:1084.

KEESEY, J. C., SALLEE, T. L. and ADAMS, G. M. (1972). *J. Neurochem.*, **19**:2225.

KEYNES, R. D. (1951). *J. Physiol., Lond.*, **114**:119.

KEYNES, R. D. and RITCHIE, J. M. (1965). *J. Physiol., Lond.*, **179**:333.

KOKETSU, K. and NISHI, S. (1969). *J. gen. Physiol.*, **53**:608.

KORNBERG, R. D. and MCCONNELL, H. M. (1971). *Biochemistry*, **10**:1111.

KREMZNER, L. T. and ROSENBERG, P. (1971). *Biochem. Pharmac.*, **20**:2953.

KRNJEVIC, K. and LISIEWICZ, A. (1972). *J. Physiol., Lond.*, **225**:363.

KUFFLER, S. W. (1967). *Proc. R. Soc., Ser. B*, **168**:1.

KUFFLER, S. W., NICHOLLS, J. G. and ORKAND, R. K. (1966). *J. Neurophysiol.*, **29**:768.

LADBROOKE, B. D., WILLIAMS, R. M. and CHAPMAN, D. (1968). *Biochim. biophys. Acta*, **150**:333.

LANDOWNE, D. and RITCHIE, J. M. (1970). *J. Physiol., Lond.*, **207**:529.

LARABEE, M. G. and BRINLEY, F. J., JR. (1968). *J. Neurochem.*, **15**:533.

LEE, A. G., BIRDSALL, N. J. M. and METCALFE, J. C. (1973). *Biochemistry*, **12**:1650.

LENTZ, T. L. (1967). *Am. J. Anat.*, **121**:647.

LEVINE, Y. K. and WILKINS, M. H. F. (1971). *Nature, New Biol.*, **230**:69.

LONG, R. A., HRUSKA, F., GESSER, H. D., HSIA, J. C. and WILLIAMS, R. (1970). *Biochem. biophys. Res. Commun.*, **41**:321.

MCCONNELL, H. M. (1970). *The Neurosciences Second Study Program*, pp. 697–706. Ed. F. O. SCHMITT. New York; Rockefeller University Press.

MCDONALD, T. F., SACHS, H. G. and DEHAAN, R. L. (1973). *J. gen. Physiol.*, **62**:286.

MCLAUGHLIN, S. G. A. and HARARY, H. (1974). *Biophys. J.*, **14**:200.

MCLAUGHLIN, S. G. A., SZABO, G. and EISENMAN, G. (1971). *J. gen. Physiol.*, **58**:667.

MCNAMEE, M. G. and MCCONNELL, H. M. (1973). *Biochemistry*, **12**:2951.

MONTAL, M. and MUELLER, P. (1972). *Proc. natn. Acad. Sci. U.S.A.*, **69**:3561.

MOORE, J. W. and NARAHASHI, T. (1967). *Fedn Proc. Fedn Am. Socs exp. Biol.*, **26**:1655.

MOORE, J. W., NARAHASHI, T. and SHAW, T. I. (1967). *J. Physiol., Lond.*, **188**:99.

MUELLER, P. and RUDIN, D. O. (1963). *J. theor. Biol.*, **4**:268.

MUELLER, P. and RUDIN, D. O. (1968). *J. theor. Biol.*, **18**:222.

NARAHASHI, T. (1964). *J. cell. comp. Physiol.*, **64**:73.

NARAHASHI, T., ANDERSON, N. C. and MOORE, J. W. (1966). *Science, N.Y.*, **153**:765.

NARAHASHI, T., ANDERSON, N. C. and MOORE, J. W. (1967). *J. gen. Physiol.*, **50**:1413.

NARAHASHI, T., FRAZIER, D. and YAMADA, M. (1970). *J. Pharmac. exp. Ther.*, **171**:32.

NARAHASHI, T. and TOBIAS, J. M. (1964). *Am. J. Physiol.*, **207**:1441.

NESKOVIC, N. M., SARLIEVE, L. L. and MANDEL, P. (1973). *J. Neurochem.*, **20**:1419.

NOBLE, D. (1966). *Physiol. Rev.*, **46**:1.

NONNER, W. (1969). *Pflügers Arch. ges. Physiol.*, **309**:176.

NORTON, W. T. (1971). *Chemistry and Brain Development*, pp. 327–337. Ed. R. PAOLETTI and A. N. DAVISON. New York; Plenum Press.

NORTON, W. T. and AUTILIO, L. A. (1966). *J. Neurochem.*, **13**:213.

NORTON, W. T. and PODUSLO, S. E. (1971). *J. Lipid Res.*, **12**:84.

O'BRIEN, J. S. and SAMPSON, E. L. (1965a). *J. Lipid Res.*, **6**:537.

O'BRIEN, J. S. and SAMPSON, E. L. (1965b). *J. Lipid Res.*, **6**:545.

OVERATH, P. and TRAÜBLE, H. (1973). *Biochemistry*, **12**:2625.

PARSEGHIAN, A. (1969). *Nature, Lond.*, **221**:844.

PETERS, A. and VAUGHN, J. E. (1970). *Myelination*, pp. 3–79. Ed. A. N. DAVISON and A. PETERS. Springfield, Illinois; Charles C. Thomas.

PHILLIPS, M. C. and CHAPMAN, D. (1968). *Biochim. biophys. Acta*, **163**:301.

PODUSLO, S. E. and NORTON, W. T. (1972). *J. Neurochem.*, **19**:727.

PRESSMAN, B. C. (1971). *Neurosci. Res. Prog. Bull.*, **9**:320.

PURPURA, D. P. (1967). *The Neurosciences: A Study Program*, pp. 372–393. Ed. G. C. QUARTON, T. MELNECHUK and F. O. SCHMITT. New York; Rockefeller University Press.

RAINE, C. S., PODUSLO, S. E. and NORTON, W. T. (1971). *Brain Res.*, **27**:11.

RAMON Y CAJAL, S. (1909). *Histologie du système nerveux de l'homme et de vertébrés*. Paris; Maloine.

RANG, H. P. and RITCHIE, J. M. (1968). *J. Physiol., Lond.*, **196**:183.

REDFERN, P. and THESLEFF, S. (1971). *Acta physiol. scand.*, **82**:70.

REUTER, H. (1973). *Prog. Biophys. molec. Biol.*, **26**:1.

RITCHIE, J. M. (1971). *Curr. Topics Bioenergetics*, **4**:327.

RITCHIE, J. M. and ARMETT, C. J. (1963). *J. Pharmac. exp. Ther.*, **139**:201.

RITCHIE, J. M. and GREENGARD, P. (1961). *J. Pharmac. exp. Ther.*, **133**:241.

ROBINSON, J. D., BIRDSALL, N. J. M., LEE, A. G. and METCALFE, J. C. (1972). *Biochemistry*, **11**:2903.

ROBINSON, R. A. and STOKES, R. H. (1959). *Electrolyte Solutions*, 2nd edn. London; Butterworths.

ROJAS, E. and ARMSTRONG, C. M. (1971). *Nature, New Biol.*, **229**:177.

ROSENBERG, P. (1965). *Toxicon*, **3**:125.

ROSENBERG, P. (1970). *Toxicon*, **8**:235.

ROSENBERG, P. (1973). *Meth. Neurochem.*, **4**:97.

ROSENBERG, P. and CONDREA, E. (1968). *Biochem. Pharmac.*, **17**:2033.

ROUSER, G., NELSON, G. J., FLEISCHER, S. and SIMON, G. (1968). *Biological Membranes*, pp. 5–69. Ed. D. CHAPMAN. New York; Academic Press.

ROUSER, G., KRITCHEVSKY, G., YAMAMOTO, A. and BAXTER, C. F. (1972). *Adv. Lipid Res.*, **10**:261.

RUSHTON, W. A. H. (1951). *J. Physiol., Lond.*, **115**:101.

SALWAY, J. G. and HUGHES, I. E. (1972). *J. Neurochem.*, **19**:1233.

SCHANTZ, E. J., GHAZAROSSIAN, V. E., SCHNOES, H. K., STRONG, F. M., SPRINGER, J. P., PEZZANITE, J. O and CLARDY, J. (1975). *J. Am. chem. Soc.*, **97**:1238.

SCHWARZ, J. R., ULBRICHT, W. and WAGNER, H. H. (1973). *J. Physiol., Lond.*, **233**:167.

SEN, A. K. and POST, R. L. (1964). *J. biol. Chem.*, **239**:345.

SHELTAWY, A. and DAWSON, R. M. C. (1966). *Biochem. J.*, **100**:12.

SHEPHERD, G. M. (1974). *The Synaptic Organization of the Brain*. New York; Oxford University Press.

SIMPKINS, H., PANKO, E. and TAY, S. (1971). *Biochemistry*, **10**:3851.

SIMPKINS, H., TAY, S. and PANKO, E. (1971). *Biochemistry*, **10**:3579.

SIMPSON, L. L. and RAPPORT, M. M. (1971a). *J. Neurochem.*, **18**:1751.

SIMPSON, L. L. and RAPPORT, M. M. (1971b). *J. Neurochem.*, **18**:1761.

SKOU, J. C. (1954a). *Acta pharmac. tox.*, **10**:281.

SKOU, J. C. (1954b). *Acta pharmac. tox.*, **10**:305.

STRICHARTZ, G. R. (1973). *J. gen. Physiol.*, **62**:37.

TAMAI, Y., MATSUKAWA, S. and SATAKE, M. (1971). *J. Biochem., Tokyo*, **69**:235.

TASAKI, I. and TAKENAKA, T. (1964). *Proc. natn. Acad. Sci. U.S.A.*, **52**:804.

TAYLOR, R. E. (1959). *Am. J. Physiol.*, **196**:1071.

THOMAS, R. C. (1969). *J. Physiol., Lond.*, **201**:495.

TRAÜBLE, H. and SACKMANN, E. (1972). *J. Am. chem. Soc.*, **94**:4499.

ULBRICHT, W. (1969). *Ergebn. Physiol.*, **61**:18.

VAN DEENEN, L. L. M. (1965). *Prog. Chem. Fats*, **8**:1.

VAN DEENEN, L. L. M., HOUTSMULLER, U. M. T., DE HAAS, G. H. and MULDER, E. (1962). *J. Pharm. Pharmac.*, **14**:429.

VEERKAMP, J. H., MULDER, I. and VAN DEENEN, L. L. M. (1962). *Biochim. biophys. Acta*, **57**:299.

VILLEGAS, G. M. and VILLEGAS, R. (1968). *J. gen. Physiol.*, **51**:44s.

WALD, F. (1972). *J. Physiol., Lond.*, **22**:267.

WHITE, G. L. and LARRABEE, M. G. (1973). *J. Neurochem.*, **20**:783.

WILLIAMS, R. M. and CHAPMAN, D. (1970). *Prog. Chem. Fats*, **11**:3.

WOLFF, D., CANESSA-FISCHER, M., VARGAS, F. and DIAZ, G. (1971). *J. Membrane Biol.*, **6**:304.

WOODHULL, A. M. (1973). *J. gen. Physiol.*, **61**:687.

YOGEESWARAN, G., MURRAY, R. K., PEARSON, M. L., SANWAL, B. D., MCMORRIS, F. A. and RUDDLE, F. H. (1973). *J. biol. Chem.*, **248**:1231.

ZAMBRANO, F., CELLINO, M. and CANESSA-FISCHER, M. (1971). *J. Membrane Biol.*, **6**:289.

ZIEGLGÄNSBERGER, W. and PUIL, E. A. (1972). *Nature, New Biol.*, **239**:204.

8

Spermatozoa and ova: the role of membranes in the fertilization process

C. R. Austin
Physiological Laboratory, University of Cambridge

8.1 MEMBRANOUS COMPONENTS OF MAMMALIAN GAMETES

8.1.1 The spermatozoon

In all mammals so far studied, and a large proportion of non-mammals also, the spermatozoon is a motile cell consisting of a head and a flagellum or tail, and in overall size and proportionality it varies greatly. The general structure of animal spermatozoa has been reviewed by Bishop and Austin (1957) and details of ultrastructure in several different species were ably set out by Fawcett (1970). The sperm head takes many different shapes and, indeed, its precise form appears to be peculiar to each individual species (a point brought out clearly by Friend, 1936) and may even be used to identify strains of animals within a species (Braden, 1958). Among the spermatozoa of mammals the head most often roughly resembles a circular or oval disc, being somewhat more flattened in one transverse direction than the other. Sperm heads in many rodents are characteristically hook-shaped or long and pointed, and among the marsupials bizarre pivoted forms are seen. The sperm tail may be short and thin as in the human spermatozoon, in which the length of the tail is less than the diameter of the egg, or else thick and very long as in the Chinese hamster spermatozoon where the tail is about three times the diameter of the egg.

The major parts of the sperm head are the nucleus and acrosome, and these are enclosed within the expanse of plasma membrane (*Figure 8.1*).

The *acrosome* is a vesicular or sac-like structure moulded over the front half of the nucleus, commonly protruding some distance anteriorly, often in asymmetric shapes. The wall of the acrosome is a typical trilaminar cell membrane and it encloses a mass of dense granular material which is more or less homogeneous in consistency. The acrosomal contents have been found to exhibit reactivity towards periodic acid–Schiff reagent (PAS) (Onuma and Nishikawa, 1963), and consist largely of a lipoglycoprotein complex (Srivastava, Adams and Hartree, 1965) made up of saccharides, including galactose, mannose, fucose, glucosamine, galactosamine and sialic acid, in addition to a variety of hydrolytic enzymes. The most abundant,

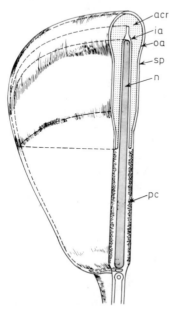

Figure 8.1 Diagram of a mammalian sperm head cut in saggital section. acr, acrosome; ia and oa, inner and outer acrosomal membranes; n, nucleus; pc, postnuclear 'cap'; sp, sperm plasma membrane. (From Austin, 1968, courtesy of Holt, Rinehart & Winston)

and seemingly most significant, of the enzymes are hyaluronidase and a trypsin-like agent, but others have also been identified, namely catalase, carbonic anhydrase, lactic dehydrogenase, acid phosphatase, aryl sulphatase, β-N-acetylglucosaminidase and phospholipase A (Stambaugh and Buckley, 1970; Allison and Hartree, 1970). The acrosome has been identified as a modified lysosome because of its origin in association with the Golgi apparatus in the spermatid, and its bright red fluorescence when treated with acridine orange and irradiated in the near-ultraviolet.

The space between the nucleus and the acrosome (the 'subacrosomal space' of Fawcett, 1965) is occupied by readily demonstrable cytoplasmic material, particularly anteriorly where it fills an invagination in the acrosome referred to as the marginal ridge. In many preparations the cytoplasmic material in the marginal ridge, as seen in sagittal section, takes the form of a relatively dense structure and has been called the 'apical body', though its existence as a real entity is not universally accepted. The apical body seems to undergo a distinct evolution during the course of spermiogenesis (Franklin and Fussell, 1972), when it appears early as a membranous whorl but later exhibits a more homogeneous granular texture. Franklin and Fussell believed that the apical body should not be dismissed as merely entrapped cytoplasmic material but could well have a particular property and function.

By contrast, between the acrosome and the plasma membrane of the sperm head there is a space of the order of 10 nm in which virtually no structural material can be demonstrated; this feature may have special significance in relation to the acrosome reaction, which will be discussed below. Despite the apparent emptiness of the space, R. Yanagimachi (personal communication, 1974) has reported positive staining for ATPase on the opposed surfaces of the plasma and outer acrosome membranes.

In the posterior half of the sperm head, a structure known as the *post-nuclear* or *postacrosomal cap* was formerly accepted as being a distinct entity (Hancock and Trevan, 1957); the idea was based chiefly on appearances presented in histological preparations, especially when silver compounds were used for staining. With more refined techniques, it has become evident that the term 'cap' is inappropriate since continuity is lacking over the posterior pole of the sperm head, but there is no doubt that a relatively large accumulation of cytoplasmic material exists in this region and is closely applied to the plasma membrane. Most electron micrographs show this as an amorphous layer, but sometimes a distinct pattern can be discerned (Fawcett and Ito, 1965); freeze-etching techniques have indeed revealed regular arrays of parallel ridges and grooves, and the general impression is of some structural complexity (Koehler, 1969).

The *sperm nucleus* is bounded by a typical nuclear envelope consisting of two layers of cell membrane. These layers are clearly evident in the developing spermatid and the envelope is also seen at these stages to carry numerous nuclear pores, except in the region immediately beneath the developing acrosome. In the mature spermatozoon, a separate nuclear envelope can still be distinguished from the dense mass of chromatin, though closely applied over most of the nuclear surface. Posteriorly, however, 'surplus' membrane projects from the nucleus down over the butt of the sperm tail and forms scrolls, and nuclear pores can be seen in these regions. The extent of the scrolls varies greatly with species, being well shown in the bat (Fawcett and Ito, 1965), dormouse and Russian hamster (Fawcett, 1970), and, most impressively, in the bushbaby (*Galago senegalensis*), where the membranous mass forms a complete collar about the anterior part of the tail mid-piece (Bedford, 1967a). The development of these curious scrolls of nuclear envelope was followed in the monkey spermatid by Franklin (1968), who maintained that the explanation of their existence as deriving simply from an excess of nuclear envelope following condensation of the nuclear chromatin lacked conviction. Their functional significance, however, remains quite unknown.

The sperm *plasma membrane* has not yet been demonstrated to possess a surface layer or extraneous coat, at least not one visible by electron microscopy. On the other hand, many observations have been made on the surface antigens of spermatozoa, some of which would appear to be attached in the manner of an extraneous coat.

Surface antigens of spermatozoa belong to two types—structural entities that are integral with the plasma membrane, and adsorbed antigens deriving from various parts of the male genital tract. Structural antigens can be further divided into three groups—the species-specific antigens, the alloantigens and the autoantigens. The adsorbed antigens on human spermatozoa include the ABO blood-group substances as well as the HLA antigens. Mouse spermatozoa have been shown to carry transplantation (H-2) and other antigens, including the Y antigen and a determinant specified at the T locus.

Studies on the antigenic characteristics of spermatozoa have been made not only with the classical techniques of immunology but also by the use of plant lectins, or phytoagglutinins, which have the advantage of identifying specific carbohydrate groupings on the antigen molecules. The terminal saccharides on the cell surface in rabbit and hamster spermatozoa have been quantitatively determined by treatment with ^{125}I-labelled plant agglutinins, namely concanavalin A and the agglutinins from *Ricinus communis* and wheat-germ (Edelman and Millette, 1971; Nicolson, Lacorbiere and Yanagimachi, 1972). Rabbit and hamster spermatozoa showed species differences in their reactions to some plant agglutinins but were similar in their reactivity to influenza virus (Nicolson and Yanagimachi, 1972). Treatment with concanavalin A and horseradish peroxidase has been used to demonstrate the distribution of terminal oligosaccharide residues over the sperm surface by electron microscopy (Gordon, Dandekar and Bartoszewicz, 1974). Many experiments have demonstrated distinctive species differences in the distribution and number of surface antigens, and the work has been reviewed by Johnson and Howe (1974).

Recent studies by Fawcett (1975) employing the freeze-etching technique revealed an interesting difference in the structure of the plasma membrane overlying the acrosomal region of the sperm head: fractures passing through the thickness of the plasma membrane displayed very numerous large particles identified as protein molecules arranged in a regular reticular formation. These were notably much less common and more random in the plasma membrane overlying the postacrosomal region of the sperm head. The plasma membrane overlying the acrosome could well have the kind of structure described by Vanderkooi (1972) for retinal rod membrane, where the protein molecules exist in paracrystalline array with their non-polar regions embedded in the hydrophobic zone of the lipid bilayer. If the structure of the outer acrosome membrane should prove to be similar, a plausible explanation might be offered for the normally stable spatial relationship that exists between plasma and outer acrosome membranes, in the absence of 'gap substance' or cytoplasmic material between the two. Such an explanation would be based on the short-range interaction between protein molecules with similar charges on their polar surfaces which, with close packing of the proteins, would confer rigidity on each membrane and also maintain the separation between the two membranes. As Green (1972) has pointed out, the structure of the membrane is well suited to the development of the necessary potential differences on either side of the dielectric provided by the hydrophobic interior zone. According to this model, membranes have asymmetry and it is reasonable to suppose that the plasma and outer acrosome membranes present surfaces of similar charge towards each other.

Both the head and the tail of the spermatozoon carry negative surface charges which appear to be more concentrated in the tail region (Nevo,

Michaeli and Schindler, 1961). The charge density has been studied by observing the surface distribution of colloidal iron hydroxide applied by a labelling technique to spermatozoa of the rabbit, mongoose, rat, guinea-pig and hamster (Yanagimachi *et al.*, 1972). Labelling density indicated a higher concentration of negative charges on the tails than on the heads of the spermatozoa, and the acrosomal region of the sperm head was more strongly charged than the postacrosomal.

The preponderance of charge on the sperm tail is consistent with earlier observations that spermatozoa suspended in an appropriate medium and subjected to an electric field become oriented, though still motile, with the tail directed towards the anode; if the strength of the current is sufficient the spermatozoa can be drawn to the anode. Spermatozoa vary somewhat in the charge carried, and according to Koltzoff and Schröder (1933) it is possible to separate spermatozoa bearing X and Y chromosomes by careful selection of current strength. Despite further similar reports from these workers, the claims have not been supported by other investigators; the matter was critically reviewed by Bishop and Walton (1960). A further result of the uneven distribution of surface charge is seen when spermatozoa are suspended in media in which the osmolarity is due chiefly to non-electrolytes; tail-to-tail agglutination then becomes evident.

The *sperm tail* is morphologically divisible into four regions: the neck, the mid-piece, the main piece and the end piece. The *neck* encloses the centrioles, or such structures as derive from them (a point discussed in general terms by Fawcett, 1970, and specifically for the rat spermatozoon by Woolley and Fawcett, 1973), and around this region there project also, at least in some species, the scrolls of 'surplus' nuclear envelope mentioned above. In the neck the flagellar components of the sperm tail originate, namely the nine outer coarse fibres and the axial complex of 9+2 Hicrotubules. The *mid-piece* of the sperm tail is distinguished by the presence of rod-shaped mitochondria arranged end-to-end in a double spiral around the axial complex. In the *main piece*, the axial complex is surrounded instead by a dense fibrous sheath composed of a large number of annular rings. The only structures continuing into the *end piece* of the sperm tail are the axial complex and enclosing plasma membrane. The posterior end of the mid-piece is marked by the presence of the annulus (formerly called the 'ring centriole'), which may have some significance during the process of spermatogenesis (*see* Fawcett, 1970).

8.1.2   The egg or oocyte

Among the contents of the egg, the structures of principal interest in the present context are the nucleus, the mitochondria, the cortical vesicles and the cytomembranes.

In most mammals, sperm entry, which initiates fertilization, occurs after the egg has undergone the first meiotic division, so that the nucleus of an egg that is ready for fertilization consists of a group of condensed chromosomes in the metaphase of the second meiotic division. The spindle now lies just beneath the egg surface and its pole-to-pole axis is generally oriented paratangentially, or nearly so, though in some species a radial orientation seems to be normal at this stage. The chromosomes form a dense group with irregular outline and cannot be resolved individually in whole mounts of living eggs or sections of fixed uncompressed material. With appropriate treatment, however, adequate separation of chromosomes can be achieved and each unit may then be identified by banding and other properties (Kaufman, 1973).

Egg mitochondria are distinctive in presenting a rounded and somewhat inflated appearance; cristae are not numerous and are often seen to be flattened against the wall of the mitochondrion. Mitochondria are randomly distributed through the cytoplasm in the egg awaiting fertilization. The cortical vesicles (originally, and still perhaps more generally, called 'cortical granules') consist of dense PAS-positive material enclosed by a cell membrane. In the mature egg they are disposed just below the plasma membrane, but achieve this location only in the later phases of maturation. The cortical vesicles vary somewhat in size, being 0.1–0.5 μm in diameter, and exist in very large numbers (50–100 per 100 μm^2). They can be seen by phase-contrast microscopy in the living eggs of the golden hamster (Austin, 1956a), and in the eggs of other mammals by electron microscopy (Szollosi, 1962). Their distribution is not especially uniform throughout the egg cortex and they are noticeably lacking in the region above the second meiotic spindle, before emission of the second polar body.

Cytomembranes are hard to discern in the unfertilized egg or during the early stages of fertilization, but later they become more abundant, evidently testifying to increased metabolic activity in the egg. Such membranes as are encountered are mostly of the smooth variety, the majority of ribosomes lying free in the cytoplasm. Golgi complexes are quite abundant, especially in the cortical regions, and they appear actively to be producing cortical vesicles while the egg lies in the oviduct awaiting fertilization (Zamboni, 1970).

All available evidence shows that the *plasma membrane* of the egg is a standard bimolecular (or trilaminar) lipoprotein leaflet. In the ovary the surface of the developing oocyte is thrown up into numerous microvilli projecting into the zona pellucida, which at that stage is closely adherent to the egg surface. At other points, processes from the surrounding follicle cells extend through the zona pellucida and indent the surface of the egg to varying degrees; in the depths of these indentations the plasma membranes of the egg and the follicle-cell process lie in close apposition and desmosome-like bodies are often visible here (Anderson and Beams, 1960). The implication is that the follicle-cell processes fulfil a nutritive function for the growing oocyte. Evidence for this function was obtained some years ago by Wotton and Village (1951) and has more recently been reviewed and discussed by Schjeide *et al.* (1970).

With the emission of the first and second polar bodies, the surface of the egg is progressively withdrawn from the zona pellucida, and the follicle-cell processes are pulled away from the vitellus or stretched to thin threads and broken. When strands are broken the club-shaped termination may remain within the egg as a double-walled inclusion. These inclusions do not persist for long, evidently degenerating during the process of fertilization. During polar body emission the microvilli are also retracted from the zona pellucida and stand free in the newly formed perivitelline space. Microvilli are particularly numerous in the regions adjacent to the polar bodies and their total number varies rather irregularly at different stages of fertilization.

As with the sperm plasma membrane, a surface layer on the vitelline plasma membrane has not yet been satisfactorily demonstrated, investigations being hampered by the fact that the enzymatic digestion necessary for prior removal of the zona pellucida is likely also to remove the extraneous coat. Studies have been made, however, with the zona pellucida itself, a thick (10–20 μm) mucoprotein coat (Braden, 1952) immediately surrounding the egg, of homogeneous texture and high transparency. Most of this work involves induced changes in sperm permeability of the zona pellucida, and will be considered later in that context. More externally, the freshly ovulated mammalian egg is endowed with the cumulus oophorus, a much broader band, consisting of follicle cells embedded in a gelatinous matrix. The matrix of the cumulus is composed of acidic mucopolysaccharide (Braden, 1952), and in this case hyaluronic acid is the carbohydrate moiety.

8.2   SPERM PENETRATION THROUGH EGG INVESTMENTS

8.2.1   Capacitation

It has been recognized for some years now that the mammalian spermatozoon, in all species so far investigated, is not capable of taking part in fertilization immediately upon its release from the male tract but must first undergo some sort of functional change. This change is termed *capacitation* and its occurrence helps to make possible sperm penetration through the cumulus oophorus and zona pellucida. Normally capacitation takes place during the passage of the spermatozoon through the female tract and the times required for capacitation *in vivo* differ according to species (*Table 8.1*). In recent years it has been possible, with the gametes of several species, to obtain conditions permitting the change to occur also *in vitro* (*Table 8.2*).

Table 8.1   TIME REQUIRED FOR CAPACITATION OF SPERMATOZOA
IN VIVO

Species	Time, h	Species	Time, h
Mouse	<1	Pig	3–6
Sheep	$1\frac{1}{2}$	Ferret	$3\frac{1}{2}$–$11\frac{1}{2}$
Rat	2–3	Rabbit	5
Hamster	2–4	Rhesus monkey, man	~5 or 6

Work on capacitation *in vitro* began with the observation by Yanagimachi and Chang (1964) that hamster eggs could be fertilized *in vitro* by incubating them with epididymal spermatozoa. Clearly, residence of the spermatozoa in the female tract was not necessary for capacitation, but since the eggs had been obtained from the oviduct the possibility remained that interaction had occurred between the spermatozoa and components of female tract secretions. The need for tract secretions was eliminated when Barros and Austin (1967) obtained fertilization of follicular eggs matured *in vitro* with spermatozoa recovered from the epididymis. Next Bavister (1971) showed that capacitation of hamster spermatozoa could be brought about by prior incubation in an artificial medium (containing bovine serum albumin) and later (Bavister, 1974) that fertilization would take place in a fully defined medium* (*see also* Austin, Bavister and Edwards, 1973, which describes much of the work leading up to this point). These observations supported the view that capacitation is not a change induced by a specific external agent but rather is an endogenous process in the spermatozoon promoted by maintenance in an appropriate environment (Austin and Bavister, 1975).

Table 8.2 MAMMALIAN EGGS FERTILIZED *IN VITRO*

Animal	Source of spermatozoa	Stage reached with culture
Rabbit	M	Blastocyst
Golden hamster	M	Two-cell
Mouse	M	Blastocyst
Chinese hamster	M	Two-cell
Cat	U	Cleavage
Man	M	Blastocyst
Sheep	U	Cleavage
Pig	U	Cleavage
Guinea-pig	M	Cleavage
Rat	M	Cleavage

Abbreviations—U: uterus; M: male tract.

Human spermatozoa were shown to achieve capacitation *in vitro* when they were incubated in a medium containing diluted follicular fluid, and in the presence of eggs (Bavister, Edwards and Steptoe, 1969); in this environment the process could be inferred to require about 7 hours.

Investigations on the rabbit have shown that spermatozoa recovered after residence in the uterus for a time sufficient to have undergone capacitation can be rendered incapable of immediate fertilization if they are resuspended in seminal plasma (Chang, 1957; Bedford and Chang, 1962). Washing will not restore fertilizing ability; for this to happen the spermatozoa must spend a further period of time in the female tract. Evidently some substance from the seminal plasma becomes adherent to the sperm surface and must be removed if the spermatozoon is to achieve a fertile state. The substance is referred to as the 'decapacitation factor'; in the crude state it behaves as a protein but purification produces a lipid with a molecular weight of 300–500 (Williams, Robertson and Dukelow, 1970). Observations on the decapacitation factor and its subsequent removal have suggested that capacitation itself may consist in the elimination of an inhibitory factor from the sperm surface. The evidence does not favour this as a complete explanation, but the possibility remains that removal of adsorbed materials represents the first stage of capacitation in the rabbit. Such materials are not limited to lipoproteins, but evidently include carbohydrates, to judge from the effects produced by treating spermatozoa with enzymes such as β-amylase (Kirton and Hafs, 1965; Hunter and Nornes, 1969) and β-glucuronidase (Gwatkin

* The medium consists of Tyrode's solution (including glucose), pH 7.6, also containing sodium pyruvate (0.5 mM), dextran (10 mg ml^{-1}) and penicillin (300 mosmol ml^{-1}).

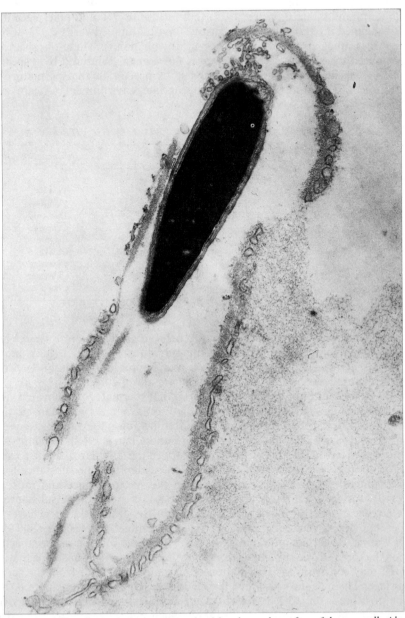

Figure 8.2   Photograph of a hamster sperm head found near the surface of the zona pellucida, and displaying vesiculation of plasma and outer acrosomal membranes. (From Barros et al., 1967, courtesy of The Rockefeller University Press)

and Hutchinson, 1971; Johnson and Hunter, 1972). In species other than the rabbit, such as the mouse, hamster and man, capacitation appears to be more of a one-step phenomenon.

Although conditions required for capacitation *in vitro* are especially exacting and the nature of the change remains a mystery, the one thing that seems certain is that capacitation unblocks the next step in the process of sperm penetration, namely the acrosome reaction.

8.2.2   Acrosome reaction

The often dramatic change in the morphology of the sperm head provoked by close approach to the egg surface was first described in marine invertebrate spermatozoa [for an account of the earlier work *see* Dan (1967); this author was herself a pioneer in this field]. In these animals the most distinctive change is the formation of a filament, often of remarkable length and fineness, which projects from the sperm head through a jelly coat to the egg surface. The mammalian equivalent was recognized a few years later as a change in the shape of the sperm head, which clearly involved loss of at least part of the acrosome (Austin and Bishop, 1958).

Ultrastructural studies showed that no filament is produced in the mammalian acrosome reaction but rather an area of vesiculated membrane becomes detached from the sperm head (*Figure 8.2*) (Barros *et al.*, 1967; Bedford, 1967b). A similar change can take place at almost any time, even in the testis or during passage through the female genital tract, but only as a rare event, and the vast majority of reacted spermatozoa are found in close relation to the surface of the cumulus oophorus or within that investment (these points are further described and discussed by Bedford, 1970). Analysis of the configurations seen has led to the conclusion that the acrosome reaction involves multiple fusion between the outer acrosome membrane and the overlying plasma membrane. Such a process would produce a number of apertures through which the enzymatic contents of the acrosome could escape to the exterior (*Figure 8.3*). The most extensive fusion evidently occurs near the equator of the sperm head, for around this region the membrane complex that is to be lost becomes detached from the sperm head [detailed descriptions of the hamster sperm acrosome reaction are given by Franklin, Barros and Fussell (1970) and Yanagimachi and Noda (1970a)].

The mechanism of the acrosome reaction is such that the continuity of the membrane enclosing the spermatozoon is maintained (an essential requirement for the cell's survival), though the membrane in question is actually a mosaic of plasma and inner acrosomal membranes. The multiplicity of the fusions commonly observed between the two membranes may conceivably be to some degree a fixation artefact, so that under natural conditions the acrosome reaction could perhaps consist merely in a ring of fusion in the equatorial region of the sperm head, resulting in the detachment of an intact cap consisting of acrosomal and plasma membranes. However, in view of the frequency with which vesiculation has been encountered with differing experimental procedures, this seems an improbable explanation. In any event, the main effect of the reaction is to permit the escape of the acrosomal contents, including hydrolytic enzymes whose function would seem clearly to be to enable spermatozoa to make their way through the matrix of the cumulus (Austin, 1961).

Unlike capacitation, the acrosome reaction does appear normally to be induced by an external agent. Initially, induction was thought to be peculiar to the ovarian follicle, and follicular fluid is certainly a highly potent inducer [as first observed by Barros and Austin (1967), and studied in greater detail by Yanagimachi (1969)]. The follicle, however, is not the only source, since oviduct fluid is also effective, even after removal of the ipselateral ovary.

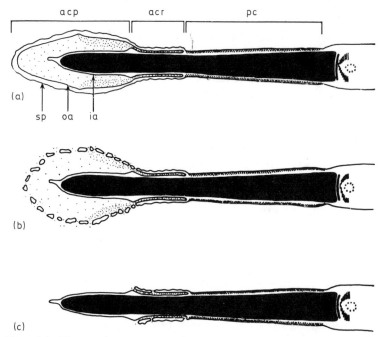

Figure 8.3   Diagram showing stages in the acrosome reaction in the hamster sperma-tozoon. sp, sperm plasma membrane; ia and oa, inner and outer acrosomal membranes; acp, acrosomal cap; acr, acrosome; pc, postnuclear cap. (From Yanagimachi and Noda, 1970a, courtesy of the Wistar Institute of Anatomy and Biology)

Moreover, incubation of hamster spermatozoa in medium containing serum which has been heated to destroy complement will favour the progressive changes that culminate 3–4 hours later in the acrosome reaction (Barros and Garavagno, 1970), and media containing heated serum are, of course, appropriate for preparing spermatozoa for fertilization *in vitro*. By contrast, incubation for a similar period in media in which whole serum is replaced by crystalline bovine serum albumin does not result in spermatozoa showing the acrosome reaction (Bavister, 1971). Treating guinea-pig spermatozoa with unheated serum, which contains complement (M. H. Johnson, personal communication, 1970), or holding them on a glass slide under a coverslip (Barros, Berrios and Herrera, 1973), can produce the acrosome reaction within 10–15 minutes. The former result may well depend on the same mechanism as that responsible for the production of 8–10 nm perforations in erythrocyte membranes by treatment with immune serum and complement (Humphrey and Dourmashkin, 1969). Guinea-pig spermatozoa will also develop an acrosome reaction if incubated for about 10 hours in an artificial medium, but only if Ca^{2+} ions are present (Yanagimachi and Usui, 1974).

Clearly species differences exist in the facility with which changes occur leading up to the acrosome reaction, but the normal participation of a specific agent still seems likely. Some of the evidence favours a steroid, possibly progesterone, for this role (Austin, Bavister and Edwards, 1973), but definitive data are lacking. Some observations probably represent false leads, such as the rapid induction by fresh serum or glass surfaces of the acrosome reaction in guinea-pig spermatozoa, but they may eventually help in the understanding of the process. The involvement of Ca^{2+} is of special interest, since divalent ions were earlier invoked as possibly being instrumental in stabilizing the adhesion between cells when the separation between their surfaces is reduced to less than 1 nm (Pethica, 1961).

8.2.3   Attachment to the zona pellucida

As pointed out some years ago (Austin, 1948), sperm penetration *in vivo* through the zona pellucida normally occurs, at least in some species, before discernible breakdown in the cumulus oophorus. Under these circumstances the dense array of follicle cells remaining about the egg tends to preclude observations on sperm contact prior to penetration of the zona pellucida. Accordingly most studies have been made with eggs that have been denuded of cumulus by enzyme treatment and then fertilized *in vitro*; the possibility of aberrations due to artefact must therefore be kept in mind.

Spermatozoa attach readily to denuded eggs, if these are unfertilized and of the same species, but not if they are fertilized or of different species. However, the position may be more complex than this. Capacitated spermatozoa of the hamster have been reported to show two phases of adhesion to the zona pellucida, attributed to the existence of two binding sites on the spermatozoon, identified as B1 and B2 (Hartmann and Gwatkin, 1971; Hartmann and Hutchison, 1974). Adhesion by site B1 occurs within 2 minutes of the mixing of spermatozoa and eggs and is shortly neutralized by a factor emanating from the vitellus. The spermatozoa then remain loosely attached until about 30 minutes later when a second firm adhesion becomes evident, this time involving the B2 site. Shortly afterwards the spermatozoon penetrates the zona pellucida. The occurrence of such a sequence of events has yet to be confirmed by other investigators, and Barros and Yanagimachi (1972) have in fact reported rather divergent observations; they noted that capacitated spermatozoa became attached to cumulus-free eggs immediately on mixing, and spent at least 5 minutes on the surface, but then started moving through the investment, reaching the perivitelline space about 15 minutes after attachment.

Antibodies raised against zona pellucida have been shown capable of inhibiting fertilization in the hamster. The treatment produced a discernible precipitation layer on the surface of the zona pellucida which was found to be resistant to tryptic digestion (Ownby and Shivers, 1972; Sacco and Shivers, 1973).

Oikawa, Yanagimachi and Nicolson (1973) observed that treating hamster eggs with wheat-germ agglutinin reduced sperm attachment to the zona pellucida and tended to prevent penetration (with lower concentrations of agglutinin sperm attachment became firm but the frequency of penetration was less than normal); in addition such treatment inhibited the dissolution of the zona pellucida by otherwise effective enzymes. As a result of treatment, the outer surface of the zona pellucida showed an increase in its light-scattering power, which could have denoted a change in molecular organization. The effects could be prevented by making available the haptenic determinant of the lectin, N-acetyl-D-glucosamine, and this was interpreted as indicating the existence in the surface of the zona pellucida of oligosaccharides with terminal residues of this kind. Resistance to lytic action could have been due to the lectins forming cross-linkages between oligosaccharide chains of neighbouring glycoprotein molecules; cross-linkage would also tend to make attachment sites for spermatozoa unavailable and oppose the action of the 'zona lysin', thus preventing sperm penetration. A similar blocking action was produced by certain other lectins, including that from *Ricinus communis*, which is specific for D-galactose, that from *Dolichos biflorus*, which is specific for N-acetyl-D-galactosamine, and concanavalin A, which is specific for α-D-mannose and α-D-glucose. It is clear that the nature of a wide variety of specific saccharides in investments and membranes can be studied by these techniques.

8.2.4   Penetration through the zona pellucida

Ultrastructural studies on rabbit and hamster spermatozoa found in the zona pellucida or in the perivitelline space make it abundantly clear that they lack most of the outer acrosome membrane, overlying plasma membrane and acrosomal contents. If sperm passage through the zona pellucida depends on the action of a lytic agent, which is a reasonable assumption, such an agent must be located on the surface of the inner acrosome membrane which now covers the anterior part of the sperm head. Furthermore the agent must be firmly attached to the membrane or even a component of it, since tracks left by spermatozoa that have penetrated the zona pellucida are narrow slits, uniform in width and showing no evidence of any diffusion of lytic activity from the sperm head. The appropriate relationship is known in other cells wherein enzymes exist as structural components of cell membranes (Singer, 1971).

The claim to have extracted the 'zona lysin' from sperm head membranes was made some years ago (Srivastava, Adams and Hartree, 1965), and this might appear inconsistent with the notion of a firmly attached enzyme; to be sure, the investigators used detergents in their procedure but their claim lacks conviction because they failed to exclude from their preparations trypsin-like enzymes known to exist in the acrosomal contents. some of which are capable of exerting a strongly lytic action on the zona pellucida (Zaneveld, Polakowski and Williams, 1972). Recent work by Srivastava (1973) may have overcome this objection, since he used extraction methods that could be shown to remove the inner acrosome membrane from the sperm head, and was able to demonstrate that the 'zona lysin' was present in the fraction containing this element, rather than in earlier fractions deriving from other parts of the acrosome.

8.3   SPERM–EGG UNION

8.3.1   Attachment at the vitelline surface

Soon after passage through the zona pellucida the sperm head comes to lie on the surface of the cytoplasmic body of the cell, which is called the *vitellus*, and attachment is made with the egg surface. With observations on living cells it is difficult to determine how long the spermatozoon remains attached before fusion occurs, but indications are that this can be as short as a few minutes. R. Yanagimachi (personal communication, 1973), watching hamster eggs undergoing fertilization *in vitro*, noticed that spermatozoa quite often became almost immobile virtually as soon as they had come into contact with the surface of the vitellus. In several eggs fixed at this time and examined by electron microscopy, fusion between spermatozoon and vitellus was found to have occurred already. The initial point of attachment appears to be a small area in the equatorial region of the sperm head, since close study of living material shows that residual motility of the sperm tail is capable of turning the head through an angle of about 90° or more. Fine-structural studies also support the idea that fusion occurs first in the postacrosomal region of the sperm plasma membrane. This is really rather surprising. Among marine invertebrates initial fusion between sperm and egg involves the newly exposed inner acrosomal membrane covering the acrosome filament (*see* Dan, 1967), and the implication seemed to be that only the inner acrosome membrane was adapted for fusion. When hamster eggs are denuded of the zona pellucida and placed in sperm suspension, only capacitated (and acrosome-reacted) spermatozoa fuse with the eggs (Yanagimachi and Noda, 1970b). The inference clearly is that some kind of functional change must take place in the plasma membrane covering the postacrosomal region of the sperm head as a result of capacitation, fitting it for fusion, but as yet no concomitant structural change has been detected.

The nature of the pre-fusion attachment is as yet unknown, but there are some observations indicating that the functional state of the vitelline plasma membrane is important. The membrane is certainly refractory to sperm contact after the occurrence of the vitelline block to polyspermy (discussed later), and there may be other circumstances also. Primary oocytes are occasionally ovulated in rats and mice and some of these have been reported to contain spermatozoa only in the perivitelline space (Austin and Braden, 1954a). Secondary oocytes with only perivitelline spermatozoa have been found well after the normal time of fertilization in mice that have received heat treatment (Austin and Braden, 1956) and in mice of certain inbred strains (Krzanowska, 1960).

Earlier reports suggested that the first association between the sperm head and vitellus was at the surface of the microvilli (Piko and Tyler, 1964), and this idea has found support in more recent studies with standard electron

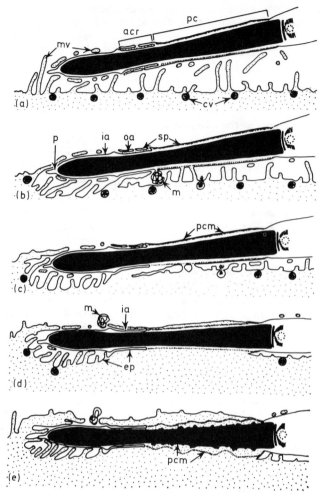

Figure 8.4   Diagram showing stages in the initial contact and fusion of the hamster spermatozoon with the egg surface. acr, acrosome; cv, cortical vesicles; ep, egg plasma membrane; ia and oa, inner and outer acrosomal membranes; m, egg microvilli; mv, vesiculated membrane; p, marginal ridge; pc, postnuclear cap; pcm, postnuclear cap material; sp, sperm plasma membrane. (From Yanagimachi and Noda, 1970a, courtesy of the Wistar Institute of Anatomy and Biology)

microscopy and stereoscan microscopy (*Figure 8.4*) (Yanagimachi and Noda, 1970a, 1972).

8.3.2   Membrane fusion

Steps following initial fusion between the gamete surfaces must take place relatively slowly, as progressive stages have often been described and illustrated. The fused sperm–egg membranes form a distinct continuum and at this stage there is little to suggest any dispersal of membrane by vesiculation in the egg cytoplasm, a process that has been described in certain invertebrates (*Hydroides*—Colwin and Colwin, 1961; *Saccoglossus*—Colwin and Colwin, 1963). The area of fusion spreads posteriorly and soon involves the tail of the spermatozoon, eventually taking in the whole of the structure.

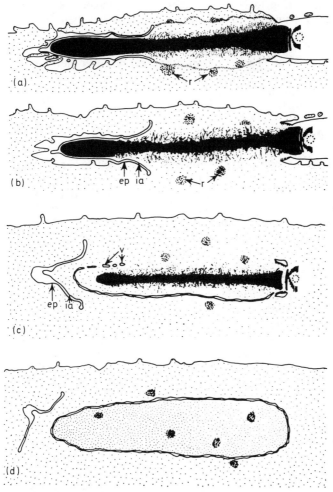

Figure 8.5   Diagram showing stages in the incorporation of the sperm head in the egg and early formation of the male pronucleus. ep, egg plasma membrane; ia, inner acrosomal membrane; r, granular aggregates; v, vesicles. (From Yanagimachi and Noda, 1970a, courtesy of the Wistar Institute of Anatomy and Biology)

This statement is true only for those species in which the tail is normally incorporated into the vitellus; in some animals, such as the field-vole, the sperm tail is commonly relegated to the perivitelline space and in the Chinese hamster it nearly always fails to pass through the zona pellucida, possibly because of its extraordinary girth and length.

Fusion in the anterior direction, on the other hand, halts at about the point of junction between plasma and inner acrosome membranes, so that continuity is not established between the inner acrosome membrane and the main sheet of the egg plasma membrane. The anterior portion of the sperm head lies for a while in a deep pit (*Figures 8.4* and *8.5*). The residual inner acrosomal membrane is then, as it were, lifted away from the sperm nucleus, appearing to become enclosed in a phagocytotic vacuole, and it subsequently disintegrates in the egg cytoplasm. This behaviour of the inner acrosome membrane in the mammalian spermatozoon is quite different from that of the corresponding structure in marine invertebrates; in these creatures it is the inner acrosome membranes that cover the acrosome filament and so fuse with the egg plasma membrane (a point firmly established by Colwin and Colwin, 1967).

8.3.3 Pronuclear development

As the region of sperm–egg fusion increases in extent, egg cytoplasm flows around the sperm nucleus, which then begins to change. The chromatin mass expands, and becomes more diffuse and fibrillar in appearance, the fibrils on its periphery spreading out into the cytoplasm of the egg (*Figure 8.5*) (Yanagimachi and Noda, 1970a). A somewhat similar appearance is presented by the chromatin mass of the rabbit sperm head after treatment *in vitro* with sodium dodecyl sulphate and dithiothreitol (Calvin and Bedford, 1971). These reagents disrupt the disulphide bonds in proteins, and it may be reasonable to infer therefore that similar chemical changes in the sperm head are involved in early pronucleus formation. In the fertilizing spermatozoon, the changes begin in the equatorial region of the nucleus and gradually involve the pole. Eventually the entire sperm nucleus achieves an expanded state as it lies in the cytoplasm of the egg. Conditions permitting this expansion of the sperm nucleus apparently exist in the egg during only a limited period, namely after the breakdown of the germinal vesicle and before the formation of vesicular pronuclei (Yanagimachi and Usui, 1974); spermatozoa may, in fact, enter the egg before or after this phase, even as late as the two-cell state, but nuclear expansion does not then occur.

When sperm penetration occurs at the normal time the sperm nuclear envelope seems to disappear as soon as it comes into contact with egg cytoplasm, the manner of its going being obscure (Yanagimachi and Noda, 1970b). In several marine invertebrates the nuclear envelope breaks up early into vesicles which are dispersed into the egg cytoplasm as the nucleus moves centrally, the vesiculation taking place by multiple fusion in much the same way as occurs in the mammalian acrosome (*see* Longo, 1973). Both in mammals and invertebrates the sperm nuclear envelope persists for a while at the anterior and posterior poles of the sperm head, but eventually disappears from these regions also.

During this time the egg chromosomes also undergo expansion, eventually forming a fibrillar mass not altogether dissimilar in appearance from that produced from the sperm nucleus. More or less synchronously, new nuclear envelopes are now constituted about the developing male and female pronuclei—small cytomembranous vesicles gather about the chromatin masses, become aligned and flattened, and form the characteristic double-membrane structure (*Figures 8.5c* and *8.5d*). Initially the contours are irregular, especially for the early female pronucleus, but as expansion occurs a smooth rounded

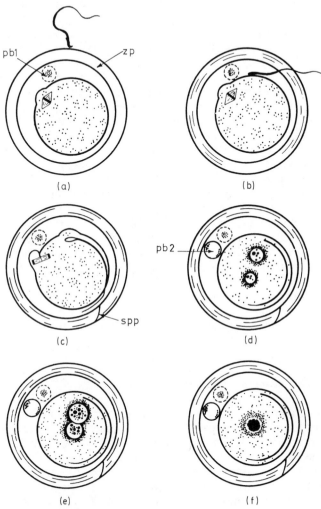

Figure 8.6   Diagram showing stages in the fertilization of the rat egg. (a) Approach and contact of the spermatozoon. (b) Penetration through the zona pellucida; the lines in the zona signify the occurrence of the zona reaction. (c)–(e) Formation and growth of the pronuclei. (f) Prophase of the first cleavage division. pb1, first polar body (which often fragments early; pb2, second polar body; spp, sperm penetration path; zp, zona pellucida. (From Austin and Short, 1973, courtesy of Cambridge University Press)

shape is achieved. Both chromatin masses become highly dispersed, and as this takes place nucleoli form, enlarge and coalesce, so that soon the vesicular nucleus comes to contain from one to several of these distinctive bodies (*Figure 8.6*).

In the rat egg there is a definite pattern in pronuclear development, with two clear-cut generations of nucleoli (Austin, 1952). Those of the first generation arise at different points in the mass of the early pronucleus, grow and coalesce to form usually a single large nucleolus; a second generation of nucleoli then becomes evident in close apposition to the nuclear envelope, while the original single nucleolus enlarges still further. Early in the second phase, appearances by electron microscopy suggest that nucleoli may also be extruded into the cytoplasm of the egg (Szollosi, 1965a). In this process small nucleoli become closely adherent to the inner lamina of the nuclear envelope, which at that point comes to protrude into the interlaminar space. The outer lamina then conforms to the shape of the protrusion and the whole structure assumes a form suggesting that it is being budded off into the cytoplasm. Consistently, bodies can be seen in the neighbouring cytoplasm consisting of small rounded masses of material, resembling that of the nucleoli, bounded by two layers of cell membrane.

The pronuclei grow considerably in size and come into close apposition in the centre of the egg (*Figure 8.6*). Pronuclear fusion does not take place in the mammalian egg as it does in that of animals such as *Arbacia* (*see* Longo, 1973). Instead the contiguous regions of nuclear envelope become highly convoluted and interdigitated. The continuity of the nuclear envelopes is interrupted in a number of places, apparently through the same mechanism of multiple fusion and vesiculation that was seen with the sperm nuclear envelope. Progressively the envelopes of both nuclei disintegrate and the vesicular products become dispersed in the egg cytoplasm. While this is going on chromosomes begin to condense in the regions of the two pronuclei, and the two groups thus formed move together in what represents the prophase of the first cleavage division. The process of fertilization is thus completed.

8.3.4   Fate of sperm components

Observations on fertilization in mammals and other animals tend to support the generalization that the paternal chromosomes represent the sole significant contribution of the spermatozoon to the zygote. The sperm plasma membrane makes a very minor contribution to the egg plasma membrane at sperm entry, and the disposal of the inner acrosome membrane by apparent phagocytosis in the mammalian egg has already been noted. In some invertebrate species a sperm centriole appears to serve an important function in the formation of the first cleavage spindle (Boveri, 1891), but there is no satisfactory evidence that mammalian sperm centrioles have any role to play in development (Austin, 1964), and indeed Woolley and Fawcett (1973) maintain that the mature rat spermatozoon contains only traces left by the degeneration of both proximal and distal centrioles. True centrioles have not as yet been reported in electron-microscopic studies of mammalian eggs undergoing fertilization. In the rat egg, the sperm mitochondria persist virtually unchanged throughout fertilization, but degenerate during early cleavage (Szollosi, 1965b). The fibrillar constituents of the tail disappear during either fertilization or cleavage, though remnants of the coarse fibres often persist to the eight-cell stage in the rat (Austin and Braden, 1953).

The more obvious sperm tail components thus seem to lack further function in the initiation of development, at least as related to their structural features, and this is consistent with the failure of entry of the tail as a normal feature of fertilization in some mammals, notably the Chinese hamster (Austin and Walton, 1960). There remains only the cytoplasmic material of the sperm head, located between the nucleus and inner acrosome membrane, and between the nucleus and the plasma membrane in the postacrosomal region. These could be appropriate regions for plasmagenes or other cytoplasmic determinants capable of influencing development.

8.4   REACTIONS OF THE EGG TO SPERM ENTRY

One of the earliest morphological changes in the egg following upon sperm–egg fusion is the disappearance of the cortical vesicles ('cortical granules'). This involves fusion of the vesicle membrane with the plasma membrane of the egg, and consequent opening of the cavity of the vesicle to the perivitelline space. The contents of the cortical vesicle are then released and soon disperse in the perivitelline fluid. Cortical vesicles also appear to fuse with each other so as to form short tubular structures which communicate with the perivitelline space at one point (Szollosi, 1967). Szollosi used ruthenium red as a dye to make the contents of the cortical vesicles more easily visible. Observations indicated that the cortical vesicle response spreads out from the point of contact of the fused spermatozoon to involve the whole egg cortex (Braden, Austin and David, 1954). Variations in the pattern of response are often seen, even in apparently normal eggs. The reaction does not necessarily involve all cortical vesicles and some can still be found intact in eggs at later stages of fertilization. Some rat eggs have been found (all in one animal) which retained the cortical vesicles even though they were penetrated by spermatozoa (Szollosi, 1967). Activation of hamster eggs (Austin, 1956b) or rabbit eggs (Flechon and Thibault, 1964) by cold shock did not cause a breakdown of the cortical vesicles. On the other hand, stimulation of hamster eggs with an electric square-wave pulse (150 V for 1 ms) was highly effective in inducing breakdown (Gwatkin et al., 1973).

Associated in time with the breakdown of the cortical vesicles are two other reactions on the part of the egg, the relative importance of which differs in different animals. One is the change in the egg plasma membrane whereby it loses its affinity for attachment of further spermatozoa after the entry of the fertilizing spermatozoon; from the nature of its effect this has long been known as the 'block to polyspermy'. Its mechanism of action is obscure and its causal relation with cortical vesicle breakdown is inferential. Barros and Yanagimachi (1972) estimated that the vitelline block to poly-spermy in hamster eggs is established not less than 2 nor more than $3–3\frac{1}{2}$ hours after attachment of the spermatozoon to the vitellus.

The other response, which seems more clearly connected with cortical vesicle breakdown, is the zona reaction whereby the zona pellucida becomes refractory to sperm contact, attachment and penetration. Like the vitelline block to polyspermy, the zona reaction also constitutes a protection against the penetration of supernumerary spermatozoa, and both reactions are of vital importance since polyspermy is almost invariably lethal in mammals (*see* Austin, 1969).

It was long surmised that the zona reaction reflects a change in the physical or chemical properties of the zona pellucida induced by substances released by the breakdown of the cortical vesicles (Braden, Austin and David, 1954). Smithberg (1953) showed that the zona pellucida of the mouse egg was digested by trypsin more readily before, rather than after, sperm penetration, and Chang and Hunt (1956) found evidence for the same change in rabbit and rat (but not hamster) eggs. Gwatkin (1964) detected a difference between penetrated and unpenetrated mouse eggs with crude β-glucuronidase, but not with proteases. A possible mechanism was adduced when Yanagimachi and Chang (1961) observed that the cortical vesicles contained PAS-positive material which was released into the perivitelline space when the vesicles broke down, and later Barros and Yanagimachi (1971) showed that perivitelline fluid from eggs in which the cortical vesicle reaction had occurred was capable of rendering the zona pellucida of other eggs impermeable to spermatozoa. Evidence that the active agent in hamster cortical vesicles was a trypsin-like protease was reported by Gwatkin *et al.* (1973), who showed that the effect of cortical vesicle material on the zona pellucida could be inhibited by inhibitors of trypsin. This observation was in good agreement with the demonstration by Vacquier, Epel and Douglas (1972) that cortical vesicles of the sea-urchin egg contain a trypsin-like protease, but not with the claim by Conrad, Buckley and Stambaugh (1971) that the active agent in rabbit egg cortical vesicles was a protease inhibitor. This may not be a contradictory theory since the rabbit egg has no zona reaction and other masking factors may be released with the protease inhibitor, blocking sites on the sperm head that would normally be involved in initial attachment with the vitelline surface. If this is true, the supposed vitelline block to polyspermy may in reality be due to a change in the sperm surface.

The time required for the hamster cortical vesicle breakdown was estimated to be between $2\frac{1}{2}$ and 5 minutes after attachment of the sperm to the vitellus of the zona-free egg (Barros and Yanagimachi, 1971), and the zona reaction in the intact egg was found to take place in rather less than 15 minutes after penetration of the first spermatozoon (Barros and Yanagimachi, 1972). The latter represents a shorter time than that estimated by Braden, Austin and David (1954) for the zona reaction of the rat egg, namely not less than 10 minutes nor more than about $1\frac{1}{2}$–2 hours. Gwatkin *et al.* (1973) noted that their protease took about 8 minutes to render the zona pellucida impermeable to spermatozoa.

Neither the vitelline block to polyspermy nor the zona reaction is completely efficient and consequently polyspermic fertilization does occur though normally at a low incidence. The incidence is increased under certain circumstances, and these are of special interest in the present context since the mechanism probably involves disturbance in membrane properties and functions (*Table 8.3*). The following factors have been identified: (a) *Ageing.* Delay in the time of coitus or artificial insemination relative to oestrus in rats and pigs results in quite a large increase in the incidence of polyspermy, which is more evident in some colonies of animals than in others. (b) *Genetic.* Some stocks of rats show a higher spontaneous incidence than others, as well as a greater response to ageing. Braden (1958) also observed significantly different frequencies after delayed mating in different rat strains. (c) *Temperature*. Application of heat to the oviduct, or induction of generalized hyperthermia, at the time of fertilization, produces a large increase in the frequency of polyspermy in the rat. (d) *Other environmental conditions*. When fertilization of rat and hamster eggs occurs *in vitro*, the incidence of polyspermy varies greatly with pH and other (unidentified) properties of the medium.

Table 8.3 INCIDENCE OF POLYSPERMY OR POLYGYNY IN THREE MAMMALS UNDER DIFFERENT CONDITIONS

| Animal | Incidence, % | | | Reference |
	Controls	Aged eggs	In vitro	
Rat				
Group 1	1.2 (s)	8.2 (s)		Austin and Braden (1953)
Group 1	1.2 (s)	16.0 (s)*		Austin and Braden (1954b)
Group 2	1.8 (s)	3.3 (s)		Austin (1956c)
		34.0 (s)†		
Group 3	0.3 (s)	3.3 (s)		Odor and Blandau (1956)
Group 4	0	4.5 (s)		Piko (1958)
Group 5	0.9 (s)	7.1 (s)		Piko (1958)
Group 6	3.2 (s)	6.6 (s)		Piko (1958)
Group 7			44 (s)	Toyoda and Chang (1974)
Hamster				
Group 1	3.8 (s)	2.3 (s)		Chang and Fernandez-Cano
		34.0 (g)		(1958)
Group 2			20 (s)	Yanagimachi and Chang (1964)
Group 3			0 (s) (pH 7.2)	Bavister (1971)
Group 3			4.8 (s) (pH 7.5)	Bavister (1971)
Group 3			41 (s) (pH 7.8)	Bavister (1971)
Pig				
Group 1	0	41.0 (s and g)		Hancock (1959)
Group 2	2	10 (s), 20 (g)		Thibault (1959)

Abbreviations—s: polyspermy; g: polygyny.
* With application of local heat.
† With induction of hyperthermia.

Table 8.3 also shows the striking increase in polygyny (fertilization following suppression of the second polar body) that has been observed following delayed mating or insemination in the hamster and pig. This too is a membrane effect, secondarily if not primarily, since polar body emission necessarily involves membrane changes.

Further responses on the part of the egg to sperm penetration include resumption of the second meiotic division and changes in metabolic activity. These will not be considered here since they have but an indirect relation to membrane function and they have been reviewed recently (Austin, 1974).

REFERENCES

ALLISON, A. C. and HARTREE, E. F. (1970). *J. Reprod. Fert.*, **21**:501.

ANDERSON, E. and BEAMS, H. W. (1960). *J. Ultrastruct. Res.*, **3**:432.

AUSTIN, C. R. (1948). *Nature, Lond.*, **162**:534.

AUSTIN, C. R. (1952). *Aust. J. scient. Res., Ser. B*, **5**:354.

AUSTIN, C. R. (1956a). *Expl Cell Res.*, **10**:533.

AUSTIN, C. R. (1956b). *J. exp. Biol.*, **33**:338.

AUSTIN, C. R. (1956c). *J. exp. Biol.*, **33**:348.

AUSTIN, C. R. (1961). *Proceedings 4th International Congress on Animal Reproduction, The Hague.*

AUSTIN, C. R. (1964). *Proceedings 5th International Congress on Animal Reproduction, Trento*, Vol. 7, p. 302.

AUSTIN, C. R. (1968). *Ultrastructure of Fertilization*, p. 47. New York; Holt, Rinehart & Winston.

AUSTIN, C. R. (1969). *Fertilization*, pp. 437–466. Ed. C. B. METZ and A. MONROY. New York; Academic Press.

AUSTIN, C. R. (1974). *Concepts of Development*, pp. 48–75. Ed. J. W. LASH and J. R. WHITTAKER. Stamford, Connecticut; Sinauer Associates.

AUSTIN, C. R. and BAVISTER, B. D. (1975). *The Functional Anatomy of the Spermatozoon*, pp. 83–87. Wenner–Gren Symposium, Aug. 1973. Ed. B. A. AFZELIUS. Oxford; Pergamon Press.

AUSTIN, C. R., BAVISTER, B. D. and EDWARDS, R. G. (1973). *Regulation of Mammalian Reproduction*, N.I.H. Conference, Sept. 1970, Washington. Ed. S. J. SEGAL, R. CROZIER, P. CORFMAN and P. CONDLIFFE. pp. 247–256. Springfield, Illinois; Charles C. Thomas.

AUSTIN, C. R. and BISHOP, M. W. H. (1958). *Proc. R. Soc., Ser. B*, **149**:241.

AUSTIN, C. R. and BRADEN, A. W. H. (1953). *Aust. J. biol. Sci.*, **6**:674.

AUSTIN, C. R. and BRADEN, A. W. H. (1954a). *Aust. J. biol. Sci.*, **7**:537.

AUSTIN, C. R. and BRADEN, A. W. H. (1954b). *Aust. J. biol. Sci.*, **7**:195.

AUSTIN, C. R. and BRADEN, A. W. H. (1956). *J. exp. Biol.*, **33**:358.

AUSTIN, C. R. and SHORT, R. V. (Eds) (1973). *Reproduction in Mammals*, Vol. 1, p. 47. Cambridge, England; Cambridge University Press.

AUSTIN, C. R. and WALTON, A. (1960). *Marshall's Physiology of Reproduction*, 3rd edn, Vol. 1, Part 2, pp. 310–416. Ed. A. S. PARKES. London; Longmans, Green.

BARROS, C. and AUSTIN, C. R. (1967). *J. exp. Zool.*, **166**:317.

BARROS, C., BERRIOS, M. and HERRERA, E. (1973). *J. Reprod. Fert.*, **34**:547.

BARROS, C. and GARAVAGNO, A. (1970). *J. Reprod. Fert.*, **22**:381.

BARROS, C. and YANAGIMACHI, R. (1971). *Nature, Lond.*, **233**:268.

BARROS, C. and YANAGIMACHI, R. (1972). *J. exp. Zool.*, **180**:251.

BARROS, C., BEDFORD, J. M., FRANKLIN, M. C. and AUSTIN, C. R. (1967). *J. Cell Biol.*, **34**:C1.

BAVISTER, B. D. (1971). *Doctoral Thesis*, University of Cambridge.

BAVISTER, B. D. (1974). *J. Reprod. Fert.*, **38**:431.

BAVISTER, B. D., EDWARDS, R. G. and STEPTOE, P. C. (1969). *J. Reprod. Fert.*, **20**:159.

BEDFORD, J. M. (1967a). *Am. J. Anat.*, **121**:425.

BEDFORD, J. M. (1967b). *J. Reprod. Fert.*, Suppl. **2**:35.

BEDFORD, J. M. (1970). *Biol. Reprod.*, Suppl. **2**:128.

BEDFORD, J. M. and CHANG, M. C. (1962). *Am. J. Physiol.*, **202**:178.

BISHOP, M. W. H. and AUSTIN, C. R. (1957). *Endeavour*, **16**:137.

BISHOP, M. W. H. and WALTON, A. (1960). *Marshall's Physiology of Reproduction*, 3rd edn Vol. 1, Part 2, pp. 1–129. Ed. A. S. PARKES. London; Longmans, Green.

BOVERI, T. (1891). *Ergebn. Anat. EntwGesch.*, **1**:386.

BRADEN, A. W. H. (1952). *Aust. J. scient. Res., Ser. B*, **5**:460.

BRADEN, A. W. H. (1958). *Fert. Steril.*, **9**:243.

BRADEN, A. W. H., AUSTIN, C. R. and DAVID, H. A. (1954). *Aust. J. biol. Sci.*, **7**:391.

CALVIN, H. I. and BEDFORD, J. M. (1971). *J. Reprod. Fert.*, Suppl. **13**:65.

CHANG, M. C. (1957). *Nature, Lond.*, **179**:258.

CHANG, M. C. and FERNANDEZ-CANO, L. (1958). *Anat. Rec.*, **132**:307.

CHANG, M. C. and HUNT, D. M. (1956). *Expl Cell Res.*, **11**:497.

COLWIN, A. L. and COLWIN, L. H. (1961). *J. biophys. biochem. Cytol.*, **10**:255.

COLWIN, L. H. and COLWIN, A. L. (1963). *J. Cell Biol.*, **19**:501.

COLWIN, L. H. and COLWIN, A. L. (1967). *Fertilization*, Vol. 1, pp. 295–367. Ed. C. B. METZ and A. MONROY. New York; Academic Press.

CONRAD, K., BUCKLEY, J. and STAMBAUGH, R. (1971). *J. Reprod. Fert.*, **27**:133.

DAN, J. C. (1967). *Fertilization*, Vol. 1, pp. 237–293. Ed. C. B. METZ and A. MONROY. New York; Academic Press.

EDELMAN, G. M. and MILLETTE, C. F. (1971). *Proc. natn. Acad. Sci. U.S.A.*, **68**:2436.

FAWCETT, D. W. (1965). *Z. Zellforsch. mikrosk. Anat.*, **67**:279.

FAWCETT, D. W. (1970). *Biol. Reprod.*, Suppl. **2**:90.

FAWCETT, D. W. (1975). *The Functional Anatomy of the Spermatozoon*, pp. 199–210. Wenner-Gren Symposium, Aug. 1973. Ed. B. A. AFZELIUS. Oxford; Pergamon Press.

FAWCETT, D. W. and ITO, S. (1965). *Am. J. Anat.*, **116**:567.

FLECHON, J. E. and THIBAULT, C. (1964). *J. Microsc.*, **3**:34.

FRANKLIN, L. E. (1968). *Anat. Rec.*, **161**:149.

FRANKLIN, L. E., BARROS, C. and FUSSELL, E. N. (1970). *Biol. Reprod.*, **3**:180.

FRANKLIN, L. E. and FUSSELL, E. N. (1972). *Biol. Reprod.*, **7**:194.

FRIEND, G. F. (1936). *Q. Jl microsc. Sci.*, **78**:419.

GORDON, M., DANDEKAR, P. V. and BARTOSZEWICZ, W. (1974). *J. Reprod. Fert.*, **36**:211.

GREEN, D. E. (1972). *Ann. N.Y. Acad. Sci.*, **195**:150.

GWATKIN, R. B. L. (1964). *J. Reprod. Fert.*, **7**:99.

GWATKIN, R. B. L. and HUTCHINSON, C. F. (1971). *Nature, Lond.*, **299**:343.

GWATKIN, R. B. L., WILLIAMS, D. T., HARTMANN, J. F. and KNIAZUK, M. (1973). *J. Reprod. Fert.*, **32**:259.

HANCOCK, J. L. (1959). *Anim. Prod.*, **1**:103.

HANCOCK, J. L. and TREVAN, D. J. (1957). *Jl R. microsc. Soc.*, **76**:77.

HARTMANN, J. F. and GWATKIN, R. B. L. (1971). *Nature, Lond.*, **234**:479.

HARTMANN, J. F., GWATKIN, R. B. L. and HUTCHINSON, C. F. (1972). *Proc. natn. Acad. Sci. U.S.A.*, **69**:2767.

HARTMANN, J. F. and HUTCHINSON, C. F. (1974). *J. Reprod. Fert.*, **36**:49.

HUMPHREY, J. H. and DOURMASHKIN, R. R. (1969). *Adv. Immun.*, **2**:75.

HUNTER, A. G. and NORNES, H. O. (1969). *J. Reprod. Fert.*, **20**:419.

JOHNSON, M. H. and HOWE, C. W. S. (1974). *Biol. J. Linn. Soc.*, 7, Suppl. 1:205.

JOHNSON, W. L. and HUNTER, A. H. (1972). *Biol. Reprod.*, **7**:332.

KAUFMAN, M. H. (1973). *J. Reprod. Fert.*, **35**:67.

KIRTON, K. T. and HAFS, H. D. (1965). *Science, N.Y.*, **150**:618.

KOEHLER, J. K. (1969). *J. Cell Biol.*, **43** (Abstr.):70a.

KOLTZOFF, N. K. and SCHRÖDER, V. N. (1933). *Nature, Lond.*, **131**:329.

KRZANOWSKA, H. (1960). *Folia biol., Kraków*, **8**:268.

LONGO, F. J. (1973). *Biol. Reprod.*, **9**:149.

NEVO, A. C., MICHAELI, I. and SCHINDLER, H. (1961). *Expl Cell Res.*, **23**:69.

NICOLSON, G. L., LACORBIERE, M. and YANAGIMACHI, R. (1972). *Proc. Soc. exp. Biol. Med.*, **141**:661.

NICOLSON, G. L. and YANAGIMACHI, R. (1972). *Science, Lond.*, **177**:276.

ODOR, D. L. and BLANDAU, R. J. (1956). *Fert. Steril.*, **7**:456.

OIKAWA, T., YANAGIMACHI, R. and NICOLSON, G. L. (1973). *Nature, Lond.*, **241**:256.

ONUMA, H. and NISHIKAWA, Y. (1963). *Bull. natn. Inst. Anim. Ind., Tokyo*, **1**:125.

OWNBY, C. L. and SHIVERS, C. A. (1972). *Biol. Reprod.*, **6**:310.

PETHICA, B. A. (1961). *Expl Cell Res.*, 22, Suppl. 8:123.

PIKO, L. (1958). *C. r. Séanc. Soc. Biol.*, **152**:1356.

PIKO, L. and TYLER, A. (1964). *Proceedings 5th International Congress on Animal Reproduction, Trento*, Vol. 2, p. 372.

SACCO, A. G. and SHIVERS, C. A. (1973). *Biol. Reprod.*, **8**:481.

SCHJEIDE, O. A., GALEY, F., GRELLERT, E. A., I-SAN LIN, R., DEVELLIS, J. and MEAD, J. F. (1970). *Biol. Reprod.*, Suppl. **2**:14.

SINGER, S. J. (1971). *Structure and Function of Biological Membranes*, pp. 145–222. Ed. L. J. ROTHFIELD. London; Academic Press.

SMITHBERG, M. (1953). *Anat. Rec.*, **117**:554.

SRIVASTAVA, P. N. (1973). *Biol. Reprod.*, **9** (Abstr.):84.

SRIVASTAVA, P. N., ADAMS, C. E. and HARTREE, E. F. (1965). *J. Reprod. Fert.*, **10**:61.

STAMBAUGH, R. and BUCKLEY, J. (1970). *Biol. Reprod.*, **3**:275.

SZOLLOSI, D. (1962). *J. Reprod. Fert.*, **4**:223.

SZOLLOSI, D. (1965a). *J. Cell Biol.*, **25**:545.

SZOLLOSI, D. (1965b). *J. exp. Zool.*, **159**:367.

SZOLLOSI, D. (1967). *Anat. Rec.*, **159**:431.

THIBAULT, C. (1959). *Annls Zootech.*, Suppl. **8**:165.

TOYODA, Y. and CHANG, M. C. (1974). *J. Reprod. Fert.*, **36**:9.

VACQUIER, V. D., EPEL, D. and DOUGLAS, L. A. (1972). *Nature, Lond.*, **237**:34.

VANDERKOOI, G. (1972). *Ann. N.Y. Acad. Sci.*, **195**:6.

WILLIAMS, W. L., ROBERTSON, R. T. and DUKELOW, W. R. (1970). *Schering Symposium on Mechanisms Involved in Conception*, pp. 61–72. Ed. G. RASPÉ. Oxford; Pergamon Press, and Braunschweig; Vieweg.

WOÒLLEY, D. M. and FAWCETT, D. W. (1973). *Anat. Rec.*, **177**:289.

WOTTON, R. M. and VILLAGE, P. A. (1951). *Anat. Rec.*, **110**:121.

YANAGIMACHI, R. (1969). *J. Reprod. Fert.*, **18**:275.

YANAGIMACHI, R. and CHANG, M. C. (1961). *J. exp. Zool.*, **148**:185.

YANAGIMACHI, R. and CHANG, M. C. (1964). *J. exp. Zool.*, **156**:361.

YANAGIMACHI, R. and NODA, Y. D. (1970a). *Am. J. Anat.*, **128**:429.

YANAGIMACHI, R. and NODA, Y. D. (1970b). *J. Ultrastruct. Res.*, **31**:486.

YANAGIMACHI, R. and NODA, Y. D. (1972). *Experientia*, **28**:69.

YANAGIMACHI, R. and USUI, N. (1974). *Expl Cell Res.*, **89**:161.

YANAGIMACHI, R., NODA, Y. D., FUJIMOTO, M. and NICOLSON, G. L. (1972). *Am. J. Anat.*, **135**:497.

ZAMBONI, L. (1970). *Biol. Reprod.*, Suppl. **2**:44.

ZANEVELD, L. J. D., POLAKOWSKI, K. L. and WILLIAMS, W. L. (1972). *Biol. Reprod.*, **6**:30.

9

Epithelial membranes and vitamin A

Luigi M. De Luca
National Cancer Institute, Bethesda, Maryland

The specialization and functional organization of epithelial tissues is made possible by epithelial membranes. Subcellular elements, such as Golgi apparatus, plasma membranes and the endoplasmic reticulum, as well as

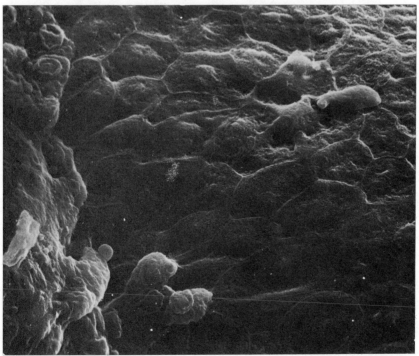

Figure 9.1   Scanning electron microscopy of 14-day-old chick embryo epidermis, kept 3 days in culture in a vitamin-A-free medium. × 990. (Courtesy of Dr Gary Peck)

other membrane structures such as villi or cilia, are necessary for the function and morphology of epithelial tissues.

Morphological studies have revealed the complexity of epithelial structures, and physiological studies have clarified their overall function, especially in the case of the absorption of nutrients. Usually epithelia contain three different cell types: (a) the *squamous cell*, which is present in the epidermis in normal states and in other epithelia after injury or transformation and is a flat hexagonal cell which has the capacity to synthesize specific products, namely keratins (*Figure 9.1*); (b) the *cuboidal cell*, which lines the surfaces of glands to form secreting units and ducts (*Figure 9.2*); and (c) the *columnar*

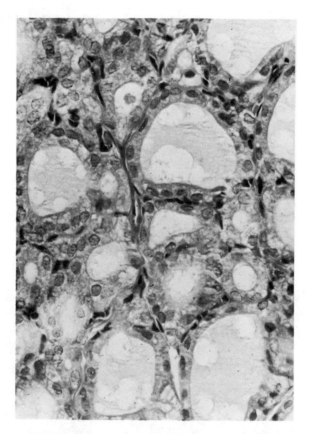

Figure 9.2   Human thyroid acini with epithelial cuboidal cells.
× 590

cell, an elongated cuboidal cell having secretory, protective and absorbing functions, which covers many epithelia of the body (*Figure 9.3*).

In membranous structures these epithelial cells are held together firmly by several specialized junctions and are supported by a basal lamina, which separates the epithelium from the underlying connective tissue (*Figure 9.4*). The epithelial lining is made up of orderly and oriented stacks of cells with very little intercellular material, whereas the connective tissue usually has vast intercellular spaces filled with ground substance, which is rich in acidic mucopolysaccharides.

The basal laminae of most epithelia stain with periodic acid–Schiff reagent (PAS) (*Figure 9.4*), indicating the polysaccharidic nature of some of their components. It is not clear at this point whether mucoprotein molecules, constituting the mesenchymal structures, have any selective function in maintaining the epithelial structures and their differentiation, although there are reports (Toole, 1972; Toole, Jackson and Gross, 1972) that they

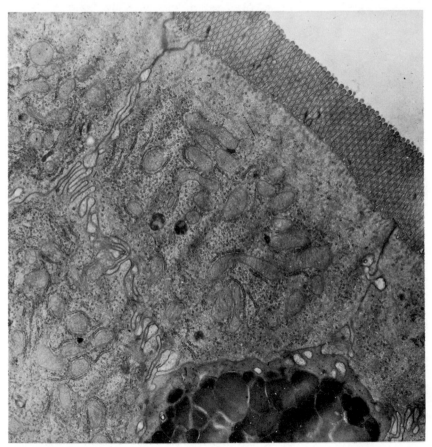

Figure 9.3   Electron microscopy of rat duodenal epithelium. The goblet cell with secretory granules and absorbing cell with microvilli are shown. × 7550

may function either as hormones which induce epithelial differentiation, or as filters to keep the correct balance of nutrients. It is of interest, in this context, that epithelia do not have blood vessels and thus derive their nutrients by diffusion from the mesenchyme.

Since a list of morphological descriptions is not within the scope of this chapter, we shall discuss epithelial tissues from the point of view of their phenotype, giving prominence to the interesting phenomenon of change in

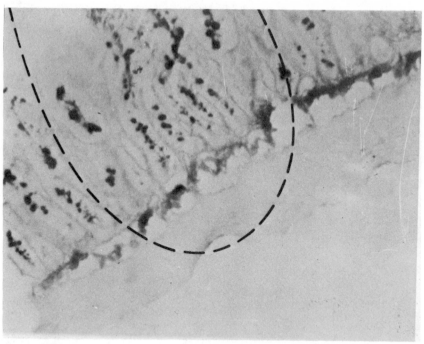

Figure 9.4   Rat small intestine. Periodic acid–Schiff reagent staining. × 330

Table 9.1   ACTION OF VITAMIN A ON EPITHELIAL TISSUES

Tissue	Phenotype	Vit. A deficiency	Vit. A excess
Rat small intestine	Columnar, mucus-secreting	Decrease in goblet cell	Not studied
Xenopus laevis small intestine	Columnar, mucus-secreting	Not studied	Increase in level of goblet cells
Rat testis	Columnar, germinal epithelium	Degeneration	Not studied
Rat pancreatic ducts	Columnar, mucus-secreting	Squamous metaplasia keratinization	Not studied
Urinary tract	Columnar	Squamous metaplasia keratinization	Not studied
Rat, hamster tracheal epithelium	Columnar, mucus-secreting	Squamous metaplasia keratinization	Increase in mucous cells to above normal level, increase in ciliated cells
Vagina	Columnar, mucus-secreting	Squamous metaplasia keratinization	Not studied
Chick epidermis	Keratinized	Not studied	Mucous metaplasia (*Figure 9.12*)
Mammalian epidermis	Keratinized with mucous sebaceous glands	Squamous metaplasia of glands	Ciliated, mucous metaplasia
Keratoachantoma	Keratinized	Not studied	Mucous metaplasia
Taste bud	Columnar, mucus-secreting	Squamous metaplasia keratinization	Not studied
Glandular tissues	Mucus-secreting	Squamous metaplasia keratinization	Not studied

differentiation in the adult from mucous type to keratinizing type or vice versa owing to deficiency or excess of the essential nutrient vitamin A (*Table 9.1*). We shall discuss the epidermis, a physiological example of keratinizing tissue, which under the influence of vitamin A can display the mucous phenotype and thus resembles the intestinal lining; the respiratory tract as a mucus-secreting tissue that can become keratinized both in vitamin A deficiency and under the influence of chemical carcinogens such as benzopyrene; and the intestinal mucosa, which responds to vitamin A deficiency by a reduction of mucus-secreting goblet cells, but without keratinization. We shall also discuss phosphoryl and glycosylphosphoryl derivatives of vitamin A and the role of polyisoprenols in sugar-transfer reactions in membranes as the biochemical basis for some of these morphological changes.

9.1   INTESTINAL MUCOSA

The intestinal mucosa belongs to the class of simple columnar epithelia. It contains two main cell types: the *absorbing cell* and the *goblet cell*. The absorbing cell's distinctive characteristic is the *microvillus* (*Figure 9.5*), a membranous structure which increases the absorbing area of the surface

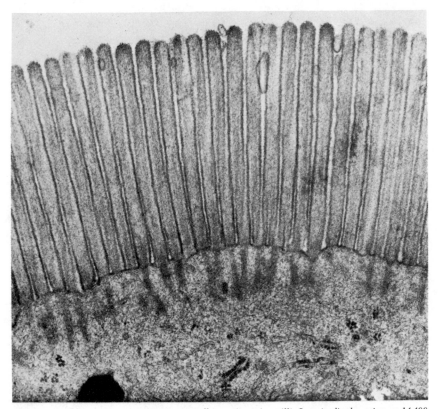

Figure 9.5   Electron micrograph of rat small intestine microvilli. Longitudinal section. × 14 400

by two orders of magnitude. A cross section (*Figure 9.6*) of the microvillar area reveals the mosaic structure of the microvilli. The absorbing cell has prominent rough and smooth endoplasmic reticular structures and a Golgi apparatus. Desmosomes ensure a stable stack of cells (*Figure 9.7*).

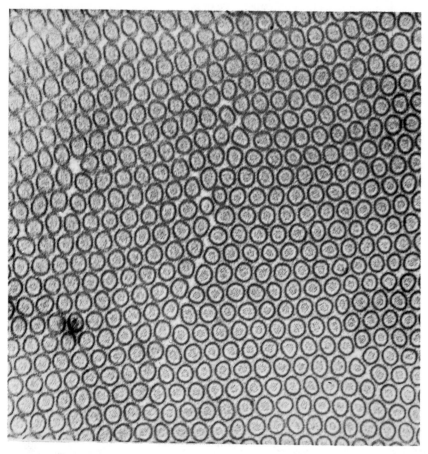

Figure 9.6   Cross section of rat small intestinal microvilli. × 15 000

The second most prominent cell type is the goblet cell (*Figure 9.8*). This type of cell is very rich in mucus and is thought of as a unicellular gland which discharges its contents into the lumen of the intestine. The goblet cell is essential to the maintenance of the moist mucous environment of the surface of several epithelia, including the respiratory and intestinal tracts. Moreover, the mucus has a protective function against alien particles that may injure the surface of the epithelium. In fact a faulty production of mucus may result in penetration by foreign substances which might otherwise be enveloped in the mucus and eliminated.

Since the goblet cell stains with PAS, it has been assumed to contain polysaccharidic compounds. In our studies on the mode of action of vitamin A in maintaining epithelial structures, we have found that vitamin A

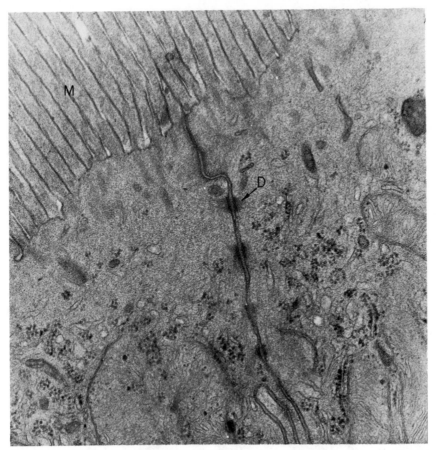

Figure 9.7 Section of absorbing cell of rat small intestine emphasizing microvilli (M) and desmosomes (D). × 12600

deficiency causes a great decrease in the number of mucus-secreting goblet cells in the rat intestine, whereas the total number of cells, as measured by DNA assay, remains relatively constant (*Table 9.2*). PAS staining of the goblet cells in a section of normal rat duodenum is compared with that of a pair-fed vitamin-A-deficient rat (*Plate 9.1*). Pair-feeding is necessary to remove any variation due to food intake, which is greatly decreased by

Table 9.2   DNA CONTENT OF NORMAL AND VITAMIN-A-DEFICIENT MUCOSAL SCRAPINGS OF RAT SMALL INTESTINE

	DNA, mg per gram of tissue	
	Normal	*Deficient*
Experiment I	109.2	120.2
Experiment II	50	45
Experiment III	54.4	45.9

Each measurement was performed on pooled samples from three rat small intestines.

vitamin A deficiency. Clearly, by the PAS technique there appear to be fewer goblet cells in vitamin A deficiency. This is capable of two explanations: first, that defective goblet cells are present in the epithelium without their usual endowment of mucus, owing to its faulty synthesis, and thus they cannot be revealed by PAS; secondly, that precursor cells do not mature to goblet cells. If this second possibility is correct, the vitamin would regulate the differentiation of goblet cells from precursor cells. Conversely, if the

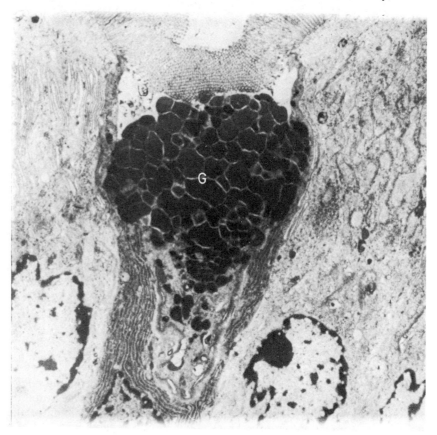

Figure 9.8   Electron micrograph of rat duodenal goblet cell with typical mucous granules, Golgi apparatus and rough endoplasmic reticulum. × 7800

change in phenotype is due to faulty synthesis of PAS-staining material, the vitamin must be involved somehow in the biosynthesis of glycoproteins. This second involvement would also affect the structure of membranes. Most mammalian membranes contain glycoproteins (Spiro, 1970; Bretscher, 1973; DePierre and Dallner, 1975), and a faulty synthetic mechanism for membrane constituents could affect cellular stacking (Burger and Noonan, 1970) and phenotype as well as the synthesis of secretory glycoproteins.

It is established that the mucous goblet cell needs vitamin A for its expression and survival (Wolbach and Howe, 1925; De Luca, Little and Wolf, 1969), hence we investigated the possibility that some specific molecular species of RNA is affected by vitamin A deficiency. Our studies revealed that such deficiency decreased the uptake of labeled orotic acid and uridine into 28-, 18- and 4-S RNA by 30 percent (De Luca *et al.*, 1971). This was not due to a change in pool size of precursors (*Table 9.3*). Interestingly, an analysis of the isoaccepting species of tRNAs for leucine, threonine and aspartic acid did not reveal any specific differences between normal and deficient

Table 9.3 INCORPORATION OF [6-^{14}C]OROTIC ACID AND [^{3}H-G]URIDINE INTO RAT INTESTINAL MUCOSAL RNA FROM NORMAL AND VITAMIN-A-DEFICIENT RATS

	Normal	*Deficient*	*% Deficient/normal*
Experiment 1			
	counts/min per 200 µg RNA		
Total RNA*	171	122	71.3
28-S + 18-S	216	149	69.2
tRNA	205	156	76.2
	mg per 3 rats		
Total DNA	109.2	120.2	
Total protein	650	520	
Experiment 2			
	counts/min per mg RNA		
Total RNA†	332	203	61.2
	mg per 3 rats		
RNA	48.2	54.9	
DNA	50.0	45.0	
Protein	455.0	442.0	
Experiment 3			
	counts/min per 500 µg RNA		
Total RNA*	219	142	64.8
28-S + 18-S	244	175	71.7
tRNA	592	446	75.3
	mg per 3 rats		
RNA	32.0	30.8	
DNA	54.4	45.9	
Protein	488.0	509.5	
	counts/min per 50 µg RNA		
RNA†	421	231	54.2

* Determined by the method of Hotchkiss (1957).
† Determined by the method of Fleck and Munro (1962).

Experiment 1.   40 µCi of [6-^{14}C]orotic acid (sp. act., 250 µCi/0.0541 mg) per 200 g of body weight was injected intraperitoneally into three normal and three vitamin-A-deficient rats 1 h before killing. Counting efficiency was 60 percent.

Experiment 2.   70 µCi of [^{3}H-G]uridine (sp. act., 1000 µCi/0.0428 mg) per 200 g body weight was injected intraperitoneally. Three normal and three vitamin-A-deficient rats were used for this experiment. Only incorporation into total RNA was measured. Methods were identical to those described for Experiment 1 except that RNA was quantitated after hydrolysis.

Experiment 3.   RNA was labeled with [^{3}H-G]uridine (sp. act., 1000 µCi/0.00812 mg) per 150 g of body weight, 100 µCi was used. Methods were identical to those described for Experiment 1. RNA was quantitated from its absorbance at 260 nm.

intestinal cells, but did show overall differences (*Table 9.3*). Studies on RNA synthesis in vitamin A deficiency were also conducted using liver, with similar results (Johnson, Kennedy and Chiba, 1969). The labeling of a high-molecular-weight RNA fraction is also reduced in the lining of the respiratory epithelium of the vitamin-A-deficient hamster (Kaufman *et al.*, 1972). Labeling of this fraction was restored to normal two weeks after administration of the vitamin; hence this was a late event, which occurred even after body weight had been restored to normal. Overall, the study of RNA synthesis in deficient animals did not appear to give any specific, readily interpretable insight into the molecular mode of action of vitamin A.

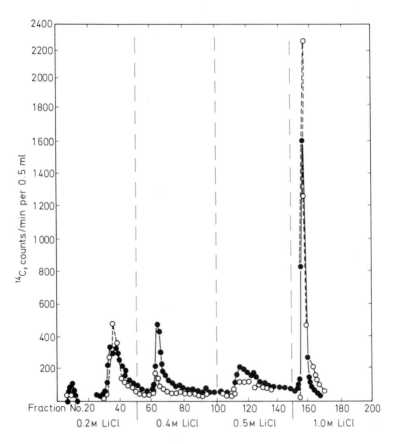

Figure 9.9   DEAE-Sephadex A-50 chromatography of the glycopeptides prepared from the intestinal mucosa of normal (●) and vitamin-A-deficient (○) rats. The rats were injected with [¹⁴C]glucosamine 2 h prior to killing. This experiment used eight normal and eight vitamin-A-deficient rats. In all, 122 mg of the normal glyco-peptide mixture and 120 mg of the deficient were obtained. The glycopeptide mixtures were dissolved separately in 27 ml of water, and 22 ml was applied to two identical columns (100 × 2 cm) of DEAE-Sephadex; 12.5-ml fractions were collected with an automatic fraction collector, and the radioactivity was assayed using 0.5-ml aliquots

9.2   BIOCHEMISTRY OF EPITHELIAL CELLS

9.2.1   Intestinal glycoproteins

The population of goblet cells of the intestinal mucosa in the rat constitutes about 10 percent of the total adult epithelial cell population. No method is available to separate this cell type from other cell types for biochemical studies. Any isolation procedure is made difficult by the fragility of the cell, which is full of mucus. To study whether a faulty mechanism exists in the biosynthesis of the goblet cell glycoprotein, we chose to start with a precursor, [1-^{14}C]D-glucosamine, which is incorporated into most glycoproteins; the different glycopeptide molecules were separated and those which were affected by vitamin A deficiency were selected. Their origin as constituents of the goblet cell was shown by immunofluorescence methods.

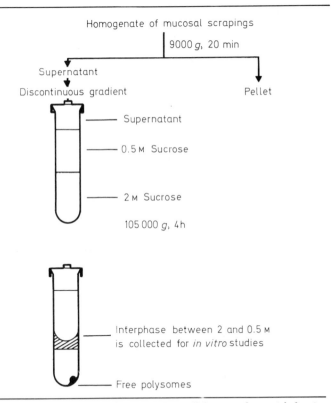

Figure 9.10   Fractionation scheme for preparing a cell-free, membrane-rich fraction from rat intestinal mucosa

Labeling of intestinal glycopeptides from normal and deficient rats *in vivo* was studied at different times after the injection of the label. Fractionation of the intestinal glycopeptides shows that the incorporation of [^{14}C]-glucosamine in the 0.4 M LiCl peak eluted from DEAE-Sephadex is decreased by 50–60 percent in vitamin A deficiency (*Figure 9.9*). Administration of

Table 9.4 EFFECT OF INJECTION OF VITAMIN A ON THE NUMBER OF GOBLET CELLS IN INTESTINAL MUCOSA OF VITAMIN-A-DEFICIENT RATS

Time after injection, h	Number of goblet cells per crypt
0	13.88±2.04*
4	14.35±2.08
10	15.11±2.19
18	19.58±2.41†
23	19.81±2.34

* Mean ± standard deviation of the mean.
† This represents a normal level of goblet cells.

Counts of goblet cells were performed as described previously (De Luca, Schumacher and Wolf, 1970). Ten sections were made for each intestine, representing one time point, and were then stained; 220 counts of goblet cells per crypt of Lieberkühn were made on each section. Statistical significance between zero time and 4 h, between 4 and 10 h, and between 10 and 18 h, is at the level of $p < 0.01$; between 18 and 23 h, $p < 0.2$.

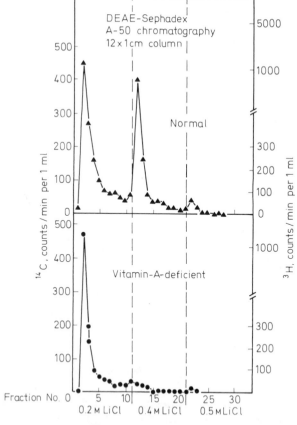

Figure 9.11 Synthesis in vitro of glycopeptides by rough endoplasmic reticulum and pH-5 enzyme fractions from normal and vitamin-A-deficient rats. DEAE-Sephadex A-50 chromatography of [^{14}C]glucosamine-labeled glycopeptides

vitamin A to deficient rats brings the incorporation back to normal levels within 15–20 hours at the same time as the number of goblet cells goes back to normal (*Table 9.4*). To avoid complications due to changes in the size of the pools of precursors, in vitamin A deficiency, a cell-free study was conducted in the presence of the membrane-rich fraction (obtained as described in *Figure 9.10*) using UDP–N-acetyl-[^{14}C]glucosamine as a labeled precursor (*Figure 9.11*). As is evident, very little labeling of the fraction eluted with 0.4 M LiCl occurred in vitamin A deficiency and the 1 M LiCl fraction from the study *in vivo* is not detected in the study *in vitro*. This latter fraction was labeled with radiosulfate, and was probably a mucopoly-saccharide synthesized by the mesenchyme. Gel filtration and chemical analysis of the 0.4 M LiCl fraction, affected by vitamin A deficiency, showed that this fraction contained three main glycopeptides. The only glycopeptide affected by vitamin A was eluted at the void volume. This glycopeptide contained fucose, as its distinctive component, and was termed 'fucose-glycopeptide'. It contains fucose, glucosamine, galactosamine, galactose and sialic acid in the respective molar ratios of 1:1.75:0.87:1.97:0.49 (De Luca and Wolf, 1972).

9.2.2   Immunofluorescence studies

The cellular location of the vitamin-A-dependent fucose-glycopeptide was determined by the indirect immunofluorescence method (Coons and Kaplan, 1950) using frozen sections of tissue, freeze-substituted with acetone (Chang and Hori, 1961) and chicken antiserum immunoglobulins. The fucose-glycopeptide was found to be in the goblet cell of the rat small intestine (De Luca, Schumacher and Nelson, 1972) (*Plate 9.2*). By using this technique, the capping of the goblet out of the surface of the epithelium is quite evident. Kleinman and Wolf (1974) have isolated and characterized the native goblet glycoprotein.

Studies in the hamster (Bonanni *et al.*, 1973) and in the rat (Bonanni and De Luca, 1974) have shown that vitamin A deficiency causes a decrease in the synthesis of specific glycopeptides of the tracheal epithelium, separated from the underlying mesenchyme. Similar results were obtained for the conjuctiva (Kim and Wolf, 1974). These tissues, when analyzed by the indirect immunofluorescence technique, cross reacted with the antiserum to the intestinal fucose-glycopeptide (De Luca *et al.*, 1972). The amount of cross-reacting material was appreciably reduced in the respiratory tract of rats on the vitamin-A-deficient diet.

9.2.3   Studies on excess vitamin A and the skin

Fell and Mellanby (1953) have shown that chick ectodermal explants from the metatarsal region, when cultured in the presence of vitamin A, give rise to PAS-staining globules in this, usually keratinizing, epithelium (*Figure 9.12*; cf. *Figure 9.1*). Mammalian skin, cultured in the presence of vitamin A, has been reported to develop cilia with perpendicular membrane processes, resembling a mucus-secreting epithelium (Barnett and Szabo, 1973).

Monolayer cultures of mouse epidermis incorporate labeled precursors into glycoproteins. These cultures, when treated with low doses of vitamin A, show proliferation of the endoplasmic reticulum and Golgi structures, but no increase in DNA synthesis (Yuspa and Harris, 1974). A study of the incorporation of labeled precursors into glycopeptides revealed an increase in the synthesis of specific glycopeptides and the appearance of a high-molecular-weight glycopeptide in the vitamin-A-treated cultures (De Luca and Yuspa, 1974).

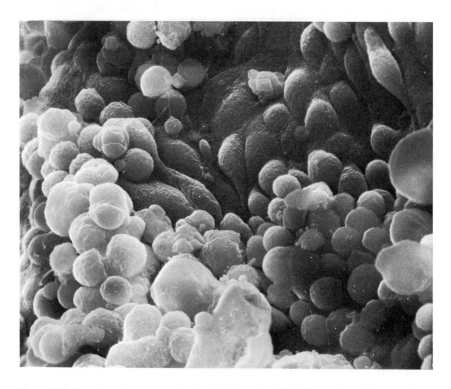

Figure 9.12 Scanning electron microscopy of 14-day-old chick epidermis, kept 3 days in culture in the presence of 7 μg ml⁻¹ of retinoic acid. × 1100. (Courtesy of Dr Gary Peck)

Keratoacanthomas, which are tumors obtained by painting 7,12-dimethyl-benzanthracene on rabbits' ears, can change their phenotype from keratin-izing to mucus-secreting upon topical application of retinoic acid, a compound that can replace retinol except in vision and reproduction (Prutkin, 1968). The incorporation of [^{14}C]glucosamine and [^{3}H]fucose into glycoprotein has been found to increase five- to tenfold in response to retinoic acid (Levinson and Wolf, 1972). The effect was not a result of changes in pool size of precursors, as argued from the fact that other glycopeptide fractions were not affected. Further purification gave several glycopeptide components, only one of which showed uptake of [^{3}H]fucose from retinoic-acid-treated tumors.

9.2.4   **Possible mode of action of vitamin A in membranes, the retinol and dolichol glycophospholipids**

Membranes of the endoplasmic reticulum and Golgi apparatus contain glycosyltransferases. The activated sugar is the donor of the sugar moiety to endogenous acceptor glycoproteins which are synthesized on the rough endoplasmic reticulum, and modern views tend to suggest that the protein moiety of a glycoprotein acquires its carbohydrate prosthetic group while traveling through the channels of the rough and smooth endoplasmic reticulum and the Golgi apparatus. Elegant autoradiographic studies by Leblond and associates (Neutra and Leblond, 1969; Merzel and Leblond, 1969) have clearly demonstrated this point.

The difficulty has always been to reconcile the extremely hydrophilic character of the sugar nucleotide with the hydrophobic character of the site of transfer on the membrane. Moreover, since the glycosylated proteins are synthesized on polysomes which are attached to membranes inside the channels of the endoplasmic reticulum, the sugar nucleotide would have to cross such membranes before becoming available for glycosylation.

In bacteria, polyisoprenols such as undecaprenol function in sugar transfer reactions and have been specifically studied in the synthesis of mannan in *Micrococcus lysodeikticus* (Scher, Lennarz and Sweeley, 1968; Lahav, Chiu and Lennarz, 1969), in the synthesis of O antigen in *Salmonella* (Wright *et al.*, 1967), and of cell wall polysaccharide (Strominger *et al.*, 1967). This work has been reviewed by Osborn (1969).

Caccam, Jackson and Eylar (1969) have reported that microsomal membranes from different mammalian tissues synthesize a mannolipid with the characteristics of the bacterial polyisoprenol–phosphate–sugar compound. The kinetics of uptake of radioactive mannose from GDP–mannose into the mannolipid and into the residual trichloroacetic-acid-precipitable pellet indicate that the mannolipid is a metabolic intermediate in mammalian tissues.

Since vitamin A deficiency causes selective decrease in the synthesis of specific glycoproteins, studies have been conducted in liver and in other tissues on the molecular involvement of vitamin A in the synthesis of lipid–phosphate–sugar compounds as intermediates in the synthesis of specific glycoproteins. Difficulties in working with vitamin A arise from its instability to oxygen and light, but are facilitated, with respect to other polyprenols, by the availability of vitamin-A-deficient animals which allow the demonstration of an effect on a specific enzymatic reaction. A compound containing the vitamin can be highly labeled because of the low endogenous pools in the deficiency state. Labeling of other polyisoprenols has caused problems because mevalonic acid is rapidly incorporated into steroids, but very little into polyisoprenols, at least in mammals (Gough and Hemming, 1970). We have now resolved this problem, as discussed later, by using regenerating liver, after partial hepatectomy, and stereospecifically labeled mevalonic acid as a precursor of dolichol.

9.2.5   Synthesis of mannolipids

A liver membrane fraction was found to incorporate [^{14}C]mannose and [^{3}H]retinol into a mannolipid fraction obtained by chromatography on DEAE-cellulose of the extract obtained with chloroform/methanol (2:1) (De Luca, Rosso and Wolf, 1970). Requirements for the synthesis of the mannolipids were ATP, retinol (in the case of vitamin-A-deficient animals), Mn^{2+} and EDTA. The retinol requirement disappeared when a normal animal was used and the ATP requirement was also reduced. Free polyribosomes did not catalyze the reaction between retinol and GDP–mannose; clearly the mannosyltransferase is on membrane-bound polysomes. The anhydroretinol-mannolipid was labile to mild acid and alkali, yielding retinol, mannose and mannose-1-P. When [β-^{32}P]GDP–[^{14}C]mannose was used to prepare the mannolipid, no ^{32}P was detected in the compound, showing that mannose alone was transferred to the lipid acceptor. Chromatography on silicic acid in a continuous gradient of chloroform and methanol yielded two ^{14}C-labeled mannolipids. The separation of the two compounds made it clear that retinol was associated with the second, more polar, compound, which was hydrolyzed more readily than the first. When the enzyme was obtained from vitamin-A-deficient livers, only the first compound was labeled and could be isolated by chromatography on silicic acid (De Luca et al., 1973).

Concurrently, reports began to appear from Hemming's laboratory (Richards and Hemming, 1972) on the involvement of dolichol in the biosynthesis of mammalian mannolipid. Hemming's work is particularly interesting in that it shows that the length of the polyprenol moiety is not crucial for recognition by the first transferase:

Polyprenyl-phosphate + GDP–mannose $\rightleftharpoons$
Polyprenyl-phosphate–mannose + GDP

This reaction can apparently use polyprenols of different chain length quite efficiently and dolichol itself is a family of compounds. Thus, it may be that specificity resides in the second reaction:

Polyprenyl-phosphate–mannose + Acceptor $\rightarrow$
Mannose–acceptor + Polyprenyl-phosphate

Since only dolichol and retinol mannolipids are detected under physiological conditions, it may well be that they are synthesized by the same transferase but serve different functions, with the retinol compound functioning in a more hydrophilic environment than the dolichol compound. Moreover the retinol derivatives are characterized by their lability to mild alkali and acids owing to the conjugation of five double bonds and the allylic arrangement in the terminal isoprenol while dolichol derivatives have a saturated terminal isoprenol and are stable to mild alkali.

For the remainder of this chapter we will compare the two systems and will discuss recent data which are beginning to show the distinctive role of the lipid intermediates.

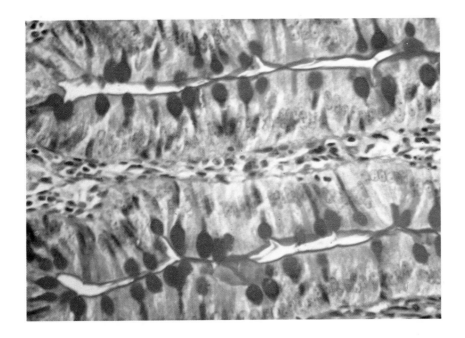

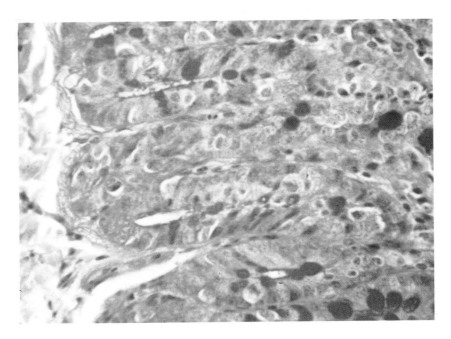

Plate 9.1   Comparison of normal (top) and vitamin-A-deficient (bottom) rat duodenal goblet cells. Periodic acid–Schiff reagent staining. × 1600

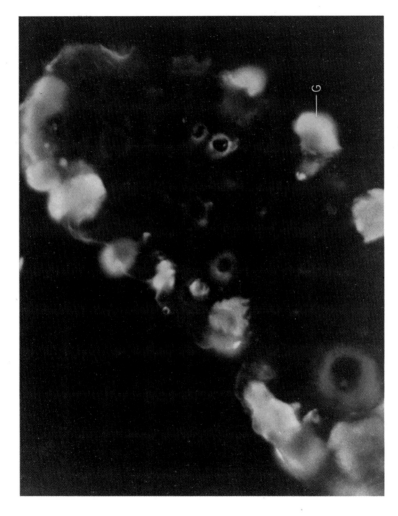

Plate 9.2   Rat intestinal mucosa villus. Goblet cells (G) are evidenced by the indirect immuno-fluorescence technique. Notice the capping of the goblet of mucus out of the surface of the villus

9.2.6   Synthesis of two mannolipids *in vivo* and *in vitro*

Progress in the characterization of the retinol and dolichol mannolipids has been accelerated recently by the use of chromatography on silica gel plates in chloroform/methanol/water (60:25:4). This solvent system separates the prenyl phosphate derivatives according to the length of the polyprenol chain (Tkacz *et al.*, 1974): longer-chain compounds, such as the dolichyl-phosphates, have R_f values of 0.4–0.5, while shorter compounds, such as retinyl-phosphate, have R_f values of 0.20–0.25. Thus the mannolipid, synthesized by purified liver membranes from GDP–[^{14}C]mannose and by the whole animal from [^{14}C]mannose, can be purified by DEAE-cellulose and silicic acid chromatography, and consistently gives two products on thin-layer chromatography in the chloroform/methanol/water system. However, 70 percent of mannosyl retinyl-phosphate partitions in the upper phase when the Folch extraction procedure is used (Silverman-Jones, Frot-Coutaz and De Luca, 1976).

Specific labeling of the retinol mannolipid can be achieved using [^{14}C-carbinol]retinol, *in vivo* and in the purified liver membrane system, but labeling of the dolichol moiety of the dolichol mannolipid presents problems. The brilliant research of Cornforth *et al.* (1966) on the stereochemistry of terpenoid compounds prompted us to use 3R,4S-[4-^{3}H]mevalonic acid, which labels *cis*-isoprenes (Hemming, 1970). The use of this stereospecifically labeled precursor of dolichol allowed the characterization of the dolichol mannolipid, made *in vivo*, and its distinction from the retinol compound, not labeled by 3R,4S-[4-^{3}H]mevalonate (Barr and De Luca, 1974). An important new facet emerged when the two products were studied for their resistance to mild alkali. This treatment is widely used to obtain dolichyl-phosphate–mannose, a molecule resistant to mild alkali by virtue of its non-allylic terminal isoprenol (Richards and Hemming, 1972). The retinol mannolipid is destroyed by 0.1 M KOH in propanol at 37 °C for 10 min. These results are in agreement with those of others (Baynes, Hsu and Heath, 1973) who, after three such treatments, could not find any retinol mannolipid.

These findings have been confirmed in several laboratories (Helting and Peterson, 1972; Martin and Thorne, 1974). Martin and Thorne found that [^{14}C]retinol and [^{3}H]mevalonic acid are both incorporated into rat liver and intestinal mannolipid, but were not successful in separating the retinol and dolichol mannolipids. Helting and Peterson characterized a retinyl-phosphate–galatose as a possible intermediate in the biosynthesis of mast cell glycoproteins.

9.2.7   Synthetic retinyl phosphate

To demonstrate that retinol was structurally involved in the synthesis of the mannolipid, chemical synthesis of the possible first intermediate, retinyl-phosphate, was clearly useful.

This first step has been achieved by a modification of the procedure of Popják *et al.* (1962). The synthetic retinyl-phosphate stimulates the synthesis of the retinol mannolipid when tested *in vitro* in the purified membrane system (Rosso *et al.*, 1973, 1975). Labeled retinyl-phosphate was also prepared and shown to be incorporated *in vitro* into the retinol mannolipid.

All the results presented strongly suggest that retinol is structurally involved in the synthesis of a mannolipid. We have not, as yet, achieved the chemical synthesis of retinyl-phosphate–mannose, nor have we established the function of the retinol mannolipid, although we have shown that the bulk mannolipid is a better donor of [^{14}C]mannose to endogenous acceptors than GDP–[^{14}C]mannose (Maestri and De Luca, 1973). However, the nature of the endogenous acceptor remains to be established.

9.3 CONCLUSION

The metabolic function of vitamin A, apart from in vision, has eluded many investigators for some sixty years. Studies involving excess vitamin A have been subject to criticism because the resulting morphological and biochemical changes may be the result of pharmacological events conceivably having little to do with normal physiology. Excess vitamin A induces breakage of membranes with all the imaginable side effects. Conversely, studies on vitamin A deficiency have been subject to criticism because the deficient animal is usually sick and infected, with many enzyme activities above or below physiological levels. Nevertheless, studies of vitamin A excess or deficiency may help in determining enzymatic events that depend on it, but the unequivocal demonstration of the molecular involvement of the vitamin is essential to demonstrate the postulated or suggested function.

Recent work has demonstrated the presence of retinyl-phosphate in intestinal epithelial membranes (Frot-Coutaz, Silverman-Jones and De Luca, 1975) and a dramatic decrease in [^{14}C]mannose incorporation into liver-membrane mannoconjugates caused by vitamin A deficiency (De Luca, Silverman-Jones and Barr, 1975). Moreover 1 mM retinyl-phosphate stimulates the synthesis of retinyl-phosphate–mannose by liver and intestinal microsomal membranes a hundredfold, without effect on the synthesis of dolichyl-phosphate–mannose.

Although these findings are suggestive of a role for retinyl-phosphate–mannose as an intermediate in mannosyl transfer reactions, a specific precursor–product relationship awaits elucidation.

Having stated the above, it is quite clear that much work still needs to be done in order to understand the mode of action of vitamin A. The lipid-soluble vitamins D, E and K have been equally elusive. This may well be due to the fact that the biochemical study of membranes is still a very difficult enterprise. When better techniques for biochemical studies of membranes become available, the fat-soluble vitamins will probably be less of a mystery.

REFERENCES

BARNETT, M. L. and SZABO, G. (1973). *Expl Cell Res.*, **76**:118.
BARR, R. M. and DE LUCA, L. (1974). *Biochem. biophys. Res. Commun.*, **60**:355.
BAYNES, J. W., HSU, A. and HEATH, E. C. (1973). *J. biol. Chem.*, **248**:5693.
BONANNI, F. and DE LUCA, L. (1974). *Biochim. biophys. Acta*, **343**:632.
BONANNI, F., LEVINSON, S. S., WOLF, G. and DE LUCA, L. (1973). *Biochim. biophys. Acta*, **297**:441.
BRETSCHER, M. S. (1973). *Science, N.Y.*, **181**:622.
BURGER, M. M. and NOONAN, K. D. (1970). *Nature, Lond.*, **228**:512.

CACCAM, J. F., JACKSON, J. J. and EYLAR, E. J. (1969). *Biochem. biophys. Res. Commun.*, **35**:505.

CHANG, J. P. and HORI, S. H. (1961). *J. Histochem. Cytochem.*, **9**:292.

COONS, A. H. and KAPLAN, M. H. (1950). *J. exp. Med.*, **91**:1.

CORNFORTH, J. W., CORNFORTH, R. H., DONNINGER, C. and POPJÁK, G. (1966). *Proc. R. Soc., Ser. B*, **163**:492.

DE LUCA, L., LITTLE, E. P. and WOLF, G. (1969). *J. biol. Chem.*, **244**:701.

DE LUCA, L., ROSSO, G. and WOLF, G. (1970). *Biochem. biophys. Res. Commun.*, **41**:615.

DE LUCA, L., SCHUMACHER, M. and NELSON, D. (1972). *J. biol. Chem.*, **248**:5762.

DE LUCA, L., SCHUMACHER, M. and WOLF, G. (1970). *J. biol. Chem.*, **245**:4551.

DE LUCA, L. and WOLF, G. (1972). *Agr. Food Chem.*, **20**:474.

DE LUCA, L. and YUSPA, S. H. (1974). *Expl Cell Res.*, **86**:106.

DE LUCA, L., KLEINMAN, H. K., LITTLE, E. P. and WOLF, G. (1971). *Archs Biochem. Biophys.*, **145**:332.

DE LUCA, L., MAESTRI, N., BONANNI, F. and NELSON, D. (1972). *Cancer, N.Y.*, **30**:1326.

DE LUCA, L., MAESTRI, N., ROSSO, G. and WOLF, G. (1973). *J. biol. Chem.*, **248**:641.

DE LUCA, L. M., SILVERMAN-JONES, C. S. and BARR, R. M. (1975). *Biochim. biophys. Acta*, **409**:342.

DEPIERRE, J. W. and DALLNER, G. (1975). *Biochim. biophys. Acta*, **415**:411.

FELL, H. B. and MELLANBY, E. (1953). *J. Physiol., Lond.*, **119**:470.

FLECK, A. and MUNRO, H. N. (1962). *Biochim. biophys. Acta*, **55**:571.

FROT-COUTAZ, J. P., SILVERMAN-JONES, C. S. and DE LUCA, L. M. (1976). *J. Lipid Res.*, **17**:220.

GOUGH, D. P. and HEMMING, F. W. (1970). *Biochem. J.*, **118**:163.

HELTING, T. and PETERSON, P. A. (1972). *Biochem. biophys. Res. Commun.*, **46**:429.

HEMMING, F. W. (1970). *Biochem. Soc. Symp.*, **29**:105.

HOTCHKISS, R. D. (1957). *Meth. Enzym.*, **3**:710.

JOHNSON, B. C., KENNEDY, M. and CHIBA, N. (1969). *Am. J. clin. Nutr.*, **22**:1048.

KAUFMAN, D. G., BAKER, M. S., SMITH, J. M., HENDERSON, W. R., HARRIS, C. C., SPORN, M. B. and SAFFIOTTI, U. (1972). *Science, N.Y.*, **127**:1105.

KIM, Y. L. and WOLF, G. (1974). *J. Nutr.*, **104**:718.

KLEINMAN, H. and WOLF, G. (1974). *Biochim. biophys. Acta*, **359**:90.

LAHAV, M., CHIU, T. H. and LENNARZ, W. J. (1969). *J. biol. Chem.*, **244**:5890.

LEVINSON, S. S. and WOLF, G. (1972). *Cancer Res.*, **32**:2248.

MAESTRI, N. and DE LUCA, L. (1973). *Biochem. biophys. Res. Commun.*, **53**:1344.

MARTIN, H. G. and THORNE, K. J. I. (1974). *Biochem. J.*, **138**:281.

MERZEL, J. and LEBLOND, C. P. (1969). *Am. J. Anat.*, **124**:281.

NEUTRA, M. and LEBLOND, C. P. (1969). *Scient. Am.*, **220**:100.

OSBORN, M. J. (1969). *A. Rev. Biochem.*, **38**:501.

POPJÁK, G., CORNFORTH, J. W., CORNFORTH, R. H., RYHAGE, R. and GOODMAN, D. S. (1962). *J. biol. Chem.*, **237**:56.

PRUTKIN, L. (1968). *Cancer Res.*, **28**:1021.

RICHARDS, J. B. and HEMMING, F. W. (1972). *Biochem. J.*, **129**:77.

ROSSO, G. C., DE LUCA, L., WARREN, C. D. and WOLF, G. (1973). *Fedn Proc. Fedn Am. Socs exp. Biol.*, **32**:Abstr. 947.

ROSSO, G. C., DE LUCA, L. M., WARREN, C. D. and WOLF, G. (1975). *J. Lipid Res.*, **16**:235.

SCHER, M., LENNARZ, W. J. and SWEELY, C. C. (1968). *Proc. natn. Acad. Sci. U.S.A.*, **59**:1313.

SILVERMAN-JONES, C. S., FROT-COUTAZ, J. P. and DE LUCA, L. M. (1976). *Analyt. Biochem.*, in press.

SPIRO, R. G. (1970). *A. Rev. Biochem.*, **39**:599.

STROMINGER, J. L., TSAKI, K., MATSUHASHI, M. and TIPPER, D. J. (1967). *Fedn Proc. Fedn Am. Socs exp. Biol.*, **26**:9.

TKACZ, J. S., HERSCOVICS, A., WARREN, C. D. and JEANLOZ, R. W. (1974). *J. biol. Chem.*, **249**:6372.

TOOLE, B. P. (1972). *Devl Biol.*, **29**:321.

TOOLE, B. P., JACKSON, G. and GROSS, J. (1972). *Proc. natn. Acad. Sci. U.S.A.*, **69**:1389.

WOLBACH, S. B. and HOWE, P. R. (1925). *J. exp. Med.*, **47**:753.

WRIGHT, A., DANKERT, M., FENNESSEY, P. and ROBBINS, P. W. (1967). *Proc. natn. Acad. Sci. U.S.A.*, **57**:1798.

YUSPA, S. H. and HARRIS, C. C. (1974). *Expl Cell Res.*, **86**:95.

10

Membrane specialization in the course of differentiation

Robert J. McLean

Department of Biological Sciences, State University College, Brockport, New York

10.1   INTRODUCTION

The plasma membrane acts as a discriminating barrier between the external environment and the cell contents including the genetic material. Any communication received or transmitted by the cell requires participation of the plasma membrane. The receptors or transmitters are molecules which are on, or embedded in, the membrane surface. The plasma membrane is programmed to receive specific incoming messages. For example, hormones will affect certain targeted cells while they have no influence on other cells. It would appear that only target cells are capable of receiving the hormone message.

In the course of differentiation and development, cell surface programming is under the control of the genetic material. Certain genes are turned on or off and they ultimately control the fate of the cell. Part of the differentiation process requires that the cell surface should express or reflect the cell's mission within the organism. This may be expressed as changes in adhesiveness and recognition abilities in cells that move to different positions, as during gastrulation or formation of primary mesenchyme. Another example is the production of hormone receptor sites on differentiating target cells.

This chapter is concerned primarily with changes in membrane programming that occur in the course of differentiation and development. These topics include origin and turnover of the plasma membrane, and discussions of membrane changes in specific differentiating cell systems and embryonic development. Literature since 1968 will be included; reviews by Steinberg (1964), Curtis (1967) and Weiss (1967) deal with work done prior to 1968. A more recent review on the cell surface in development has been edited by Moscona (1974).

10.2   ORIGIN OF PLASMA MEMBRANE AND SURFACE COATS

In defining plasma membrane specialization and changes during differentiation and development, it becomes necessary to determine the origin or source of the plasma membrane and its components. Membrane biogenesis is particularly difficult to study because of the continuity of the various membrane systems and the problems involved in obtaining pure preparations of a particular type of membrane such as smooth endoplasmic reticulum or plasma membrane. Present evidence indicates that the protein component of the plasma membrane is synthesized on the rough endoplasmic reticulum (RER) or free polysomes and moved to the Golgi complex where phospholipid is synthesized. These two components are then assembled *in situ* in the Golgi. Golgi vesicles budding off the cisternae migrate to fuse with the plasma membrane.

A cytochemical test for acyltransferase activity was used by Levine, Higgins and Barnett (1972) in studies on the salt glands of ducks. They determined that phospholipid synthesis occurs exclusively in the Golgi apparatus where it is incorporated into the membrane, packaged in vesicles and moved to fuse with the existing plasma membrane.

Synthesis of the protein moiety of glycoproteins occurs exclusively on the RER or free ribosomes (Jamieson and Palade, 1967a, b). The synthesis of the carbohydrate moiety may occur partly in the RER (Hirano *et al.*, 1972) but primarily takes place in the Golgi apparatus (Bennett, 1970; Neutra and Leblond, 1966). Labeled leucine has been traced by autoradiography from the RER to the Golgi to secretory vesicles that fuse with the cell's surface in pancreatic exocrine cells actively producing zymogen (Jamieson and Palade, 1967a, b). Fucose, which is a specific glycoprotein precursor, has been used in autoradiographic studies of a variety of secretory tissues such as colonic goblet cells (Neutra and Leblond, 1966) and columnar cells of the intestine (Bennett, 1970; Bennett and Leblond, 1970; Rambourg *et al.*, 1971) to demonstrate the Golgi origin of surface glycoproteins and the mechanism of their transport to the surface. Bennett and Leblond (1970) proposed that Golgi secretion is the general mechanism by which all cells may make their cell coats. A recent comprehensive study by Bennett, Leblond and Haddad (1974) of more than 50 different cell types of young rats revealed labeled fucose in all Golgi complexes in the early stages after pulsing. Subsequent stages showed labeling in lysosomes, secretory material and plasma membrane. Labeling appeared at the apical surfaces of some cell types or generally distributed over the surface, and amounts seemed indicative of turnover rates of the different tissues.

Glycosyltransferases located in the Golgi apparatus are responsible for synthesis of the carbohydrate moiety of glycoproteins (Morré, Merlin and Keenan, 1969; Schachter *et al.*, 1970; Wagner and Cynkin, 1971) and glyco-lipids (Keenan, Morré and Basu, 1974). These enzymes add a specific monosaccharide unit to the nonreducing termini of nascent saccharide chains. They are part of the Golgi membrane and are not free in the cisternal matrix (Lawford and Schachter, 1966; O'Brien *et al.*, 1966; Simkin and Jamieson, 1967). Glycosyltransferases have also been found on the external surface of the plasma membrane (Roth, McGuire and Roseman, 1971; Bosmann,

1972; Roth and White, 1972; Bosmann, Case and Morgan, 1974). Since they also occur in the Golgi apparatus one might assume that they are incidentally secreted as part of the Golgi vesicles and hence their position outside the cell is unrelated to their original function (Patt and Grimes, 1974). However, endogenous acceptors for these enzymes are also present on the cell surface (Roth, McGuire and Roseman, 1971; Bosmann, 1972; Roth and White, 1972; Bosmann, Case and Morgan, 1974), indicating possible extracellular functions such as cell recognition and adhesion (Roseman, 1970). Indeed, McLean and Bosmann (1975) detected glycosyltransferases on the surface of *Chlamydomonas* gametes. When the two mating types were mixed and entered gametic adhesion, glycosyltransferase activity increased three- to fivefold. Cell surface glycosyltransferases have been reviewed by Shur and Roth (1975). Complementary cell surface RNA (Mayhew, 1974) and sulf-hydryl groups (George and Rao, 1975) have also been implicated in cell adhesiveness.

The role of Golgi complexes in differentiation and development has become more apparent. Brown *et al.* (1970) demonstrated the participation of the Golgi apparatus in development of the cell wall of the unicellular alga *Pleurochrysis*. In studies on corneal epithelium development, Trelstad (1970) observed that the Golgi bodies of the basal epithelial cells shifted their polarity at two different times during development that correlated with the appearance of an acellular collagenous matrix beneath the epithelium. The first shift occurred when the primary corneal stroma formed under the epithelium and the second when the Bowman's membrane was forming. Microtubules appear to have an influence on cell polarity (Tilney and Gibbins, 1969). Colchicine and hydrostatic pressure were observed to randomize the position of the Golgi during development of the primary mesenchyme in sea-urchin. Flickinger (1968) demonstrated that transcription was necessary in the maintenance of Golgi apparatus and other cytoplasmic membrane systems in actinomycin-D-treated amebae. This was confirmed when he subsequently showed that development of Golgi complexes requires the presence of the nucleus (Flickinger, 1969). Enucleated amebae became devoid of Golgi complexes after 5 days but began to recover as early as 6 hours after renucleation. The information suggests that Golgi complexes can arise *de novo* under certain conditions (Flickinger, 1969). This also ties the developmental role of the Golgi apparatus more closely to control by the genetic material. If the turnover rate of the Golgi apparatus is a matter of days, as suggested by the data, this allows for a fairly rapid change in the transcriptional and translational information which goes into the synthesis of new Golgi. The subsequent appearance of fusing Golgi vesicles at the plasma membrane would change the surface information and, therefore, the developmental role or fate of the cell (*see* Whaley, Dauwalder and Kephart, 1972).

The RER and Golgi apparatus provide a pathway or mechanism for ex-pressing certain aspects of developmental information stored in the genetic material. DNA transcribes RNA which in turn translates into enzymes required for membrane synthesis and synthesis of glycoproteins and glyco-lipids. The new membrane components become part of the RER and Golgi apparatus. Enzymes may also become part of these two membrane systems and be secreted along with the glycoproteins and glycolipids. The enzymes may then have a specialized role on the external plasma membrane surface which is related to cellular communication and subsequently the fate of the cell and possibly other cells.

An additional mechanism for modifying the cell surface is shown by the work of Petit, Strominger and Soll (1968) and Roberts, Petit and Strominger (1968), in which interpeptide bridges of bacterial cell walls were formed. The cross-linking is accomplished by insertion of amino acids via an amino-acyl-tRNA intermediate. This represents a nonribosomal mechanism of protein modification.

Bosmann, Lockwood and Morgan (1974) and Bosmann (1974) found proteases and glycosidases on the cell surfaces of transformed cells. They speculated that these enzymes may function in modification of the cell surface by maintaining the transformed state. This presents another possibility for cell surface modification that is not determined in the Golgi apparatus.

10.3   TURNOVER OF THE CELL SURFACE

Cells and tissues go through several different stages of development before they reach G_0 or the final differentiated stage. If the cell surface expresses the different developmental stages allowing for communications and changes in associations between cells, alteration of the cell surface must be assumed. For instance, presumptive primary mesenchyme cells must alter their adhesiveness in order to leave the ectoderm of the gastrula.

Warren and Glick (1968) established that there is a continual turnover of surface membrane material in mammalian cells (L cells). Growing and non-growing cells seemed to synthesize surface material at approximately the same rate, but only in growing cells was there a net increase of material. Synthetic and degradative aspects appeared to be linked since use of metabolic inhibitors or omission of amino acids caused a decrease in both synthesis and degradation. Macromolecules that are characteristic of the cell surface were found in the medium and Warren and Glick speculated that a factor involved in turnover may be the ejection of surface membrane material into the medium.

Additional evidence for the release or shedding of surface membrane material has accumulated. Early on Lillie (1912) found macromolecular material in the medium around sea-urchin eggs which would agglutinate sperm. Presumably, the egg surface material, which normally causes the sperm to adhere, is continually released and renewed. Chick embryo fibroblasts were found to reconstitute their electrokinetic shear plane within 90 minutes following trypsin treatment (Kapeller *et al.*, 1973). Repair of wheat-germ agglutinin (WGA) sites was accomplished in 8 hours. Analysis of the trypsinate and medium from normally growing cells showed many similarities, indicating that cells naturally shed components of their cell coats. McLean and Brown (1974) demonstrated that *Chlamydomonas* gametes can regain their mating ability within 45 minutes of trypsinization. Additionally, these gametes normally release excess membrane material into the medium (McLean, Laurendi and Brown, 1974). Membrane material can be pelleted from the medium of one mating type and agglutinate gametes of the opposite mating type.

Continuous turnover and renewal of membrane material may be a necessity related to function. Confluent normal chick embryo fibroblasts become highly agglutinable by concanavalin A (Con A) 6 hours after treatment with inhibitors of protein synthesis (Baker and Humphreys, 1972). Growing (nonconfluent) cells treated similarly do not become agglutinable. Frequent contact between confluent cells must 'wear down' surface sites and the inhibitors prevent their renewal. A product of cell contact must be the Con A receptor, which would be replaced or otherwise processed by the normal cell. Bosmann (1974) and Bosmann, Lockwood and Morgan (1974) have detected proteolytic enzymes and glycosidases on the cell surfaces of a variety of cell lines. They suggested that such enzymes may form a mechanism for maintaining or modifying the cell coat for specific designated functions.

10.4 MEMBRANE SPECIALIZATION DURING DIFFERENTIATION

10.4.1 Cell cycle

The cell cycle is an obvious aspect of growth and development. It also might be considered a microsystem of differentiation where each stage in the cycle has its own characteristics and activities leading to the next stage. The surface morphology of synchronized mammalian cells shows definite characteristics related to the cell cycle stages. Scanning electron microscopy has contributed significantly to this aspect of membrane knowledge (Dalen and Scheie, 1969; Porter, Prescott and Frye, 1973; Rubin and Everhart, 1973). Synchronized liver cells at interphase were observed to flatten with numerous microextensions projecting from most of the surface. During the M phase, the cells rounded up and had fewer but longer extensions (Dalen and Scheie, 1969). Porter, Prescott and Frye (1973) studied the surfaces of transformed Chinese hamster ovary cells at confluency. They observed that many microvilli, blebs and ruffles were present at G_1 while the microvilli and blebs diminished during S until the cells were smooth. At G_2, the microvilli increased in number, the cells thickened and eventually rounded for the M phase. Normal cells in confluent cultures did not have microvilli, blebs or ruffles. Rubin and Everhart (1973) used the same transformed cell line grown in low-density cultures and reported that the cells under such conditions maintained the G_1 morphology through G_1, S and G_2. They concluded that cell contact was needed for the complete variation of expression of surface morphologies described by Porter, Prescott and Frye (1973).

Freeze-cleavage studies by Scott, Carter and Kidwell (1971) on L cells showed a high density of intramembranous particles from G_1 to metaphase. The particles decreased in number by a factor of two or three from metaphase to G_1. The observed decrease in particles may have been caused by an increase in insertion of nonparticle material or a temporary decrease in synthesis of components that maintain particle integrity. Presumably the particles were glycoprotein (Tillack, Scott and Marchesi, 1970) or glycolipid in character. Bosmann and Winston (1970) monitored glycoprotein and glycolipid synthesis during the cell cycle of mouse lymphoma cells. Glycoprotein synthesis was higher during S and secretion of glycoprotein was higher during late S and G_2. The nature of secreted glycoproteins was some-what different from that of the total glycoprotein fraction. Glycolipid release occurred at G_2 and M. It is difficult to correlate these studies since they were done with different cell systems. However, the above authors demonstrated that membrane changes during the cell cycle can be detected both morpho-logically and biochemically and this suggests the need for a coordination of these two approaches.

Antisera and lectins have been useful as probes for studying the cell surface. Neoplastic mast cells which display the H-2 antigen have a decreased sensitivity to H-2 antiserum during G_1–S, but sensitivity is restored in G_2 (Pasternak, Warmsley and Thomas, 1971). The investigators speculated that a decrease in H-2 synthesis plus an increase in cell size could lead to a lower H-2 density per surface area. H-2 and Moloney leukemia virus antigens were expressed more during G_1 in mouse bone-marrow cells (Cikes and Friberg, 1971). Fluorescein–WGA was observed by Fox, Sheppard and Burger (1971) to bind to normal 3T3 cells more readily at mitosis, while it bound to trans-formed cells during the entire cell cycle. Similar results were obtained by Noonan, Levine and Burger (1973) with Con A. If metaphase was blocked with colchicine, the cell surface still entered G_1, which indicated that surface events were independent of nuclear events to some extent.

Bosmann (1971) determined that the peak activities of five different glycosyltransferases on the surface of mouse lymphoma cells (L5178Y) were found during the S phase. A short half-life of these enzymes indicated a relatively rapid decay rate, which may be related to the completed function of the enzyme during that part of the cell cycle.

The above studies present evidence for morphological and biochemical differences in the plasma membrane during the cell cycle. The results vary in many instances with the cell system and no definite conclusions can be drawn with respect to the relationship between the morphological and bio-chemical data or between different sets of biochemical data. It does seem apparent, however, that certain membrane changes reflect the requirements or activities of the cell during a particular stage of the cell cycle.

10.4.2   Cellular differentiation and embryonic development

Embryonic tissues undergo changes in adhesiveness and aggregation abilities during development. Morphogenetic events and histogenesis require that cells be able to move (decreased adhesiveness) at the proper time and to aggregate with appropriate cells. These are surface-mediated phenomena and one should be able to detect cell surface changes from one developmental stage to the other.

Morphogenetic movement has been correlated with changes in the ad-hesiveness of embryonic cells (Kiremidjian and Kopac, 1972; Schaeffer, Schaeffer and Brick, 1973); presumably such changes are due to modifications of the cell surface, the modifications being dictated by the developmental stage of the organism. Cell surface determinants have been shown (Bennett, Boyse and Old, 1971) to be involved in development even in early stages. Adler (1970) studied reaggregation of dissociated neural tube cells of chick embryos at different stages of development. Cells from 3-day-old embryos showed little tendency to aggregate, while cells from 4-day-old embryos could aggregate. On days 5 and 6, cells from the same tissue would not form rosettes. Garber and Moscona (1972a) observed a similar age dependency in the aggregation of chick brain cells. Cells from the same embryonic age would aggregate more readily than cells from different ages. Saturation density (maximum number of cells grown on a culture dish surface) depends on various culture conditions, but if conditions are held steady one can observe cells from different stages of development. Saijo (1972) reported that embryonic mouse kidney cells had a higher saturation density than adult cells and the growth rate was also higher. Additionally, adult cells were more sensitive to contact inhibition than embryonic cells. Since contact inhibition is a surface-mediated phenomenon, it is obvious that modifications of the cell surface with age were responsible for the differences in behavior *in vitro*.

Hsie, Jones and Puck (1971) studied Chinese hamster ovary cells as they differentiated into fibroblasts in the presence of cAMP and hormones. These epithelial-like cells became less agglutinable by lectins and specific antibodies as differentiation progressed. Additionally they gained the characteristic of strict contact inhibition. The cell surface properties thus showed obvious changes during the course of differentiation.

Induction of glutamine synthetase in chick embryonic neural retina cells by corticosteroids has been studied by Moscona and co-workers (*see* Moscona, 1971; Sarkar and Moscona, 1973; Sarkar *et al.*, 1972) as a system in which cellular differentiation can be investigated. The response *in vitro* of cells to hormones depends on the level of histotypic organization (Morris and Moscona, 1971). Neural retina cells from different embryonic ages were induced with hormone and the induction of glutamine synthetase compared in intact retinal tissue, in aggregates of retinal cells and in monolayers. The level of induction was highest in intact tissue and lowest in monolayers, with aggregates intermediate regardless of embryonic age. Inducibility of glutamine synthetase increased with embryonic age in intact tissues but decreased with embryonic age in cell aggregates, where it depended on multicellular organization; that is, on the ability of dispersed cells to become reassociated and reorganized in aggregates receptive to inducer.

Cells from the same embryonic tissue tend to aggregate selectively and sort out in cultures of mixed cells (Armstrong, 1970; Roth, 1968). However, embryonic cells of a given tissue from different species do not possess the same cell-sorting properties since mouse and chick embryonic myocardial cells sort out *in vitro* (Burdick and Steinberg, 1969). Different tissues with related biological function can interact because of their similar cell surface properties (Barbera, Marchase and Roth, 1973). Neurones from the dorsal neural retina of chick embryos adhere preferentially to ventral optic tectum and vice versa. This selective adhesion has been observed *in vitro* but is believed to be related to biological function *in vivo* (Barbera, Marchase and Roth, 1973). Membranes from chick embryonic optic tectum and neural retina cells inhibit aggregation of homotypic cells *in vitro*. The membranes bind well to cells from the same embryonic age but weakly to cells from different ages (Gottleib, Merrell and Glaser, 1974). Tectal cells react weakly with retinal cell membranes but there is no reaction between retinal cells and tectal membranes. The strongest reaction is observed if the age of the tectal cells and the retinal cell membranes is the same. Gottleib, Merrell and

Glaser demonstrated that cell surface compositions change with age of development and that the change may be synchronous in the case of certain tissues.

A factor isolated from the culture medium of sponges (Moscona, 1968), embryonic chick neural retina cells (Lilien, 1968) and brain cells (Garber and Moscona, 1972b) has been observed to enhance the aggregation of these cell types *in vitro*. The factors were believed to be a product of the cell surface; indeed, of the adhesive sites themselves. McLean, Laurendi and Brown (1974) isolated a similar agglutinating factor from the medium of *Chlamydomonas* gametes. Electron microscope observations revealed the factor to be membrane vesicles which incidentally budded off the flagellar surface where they would normally be present. Aggregating factors have been obtained directly from plasma membranes but their mechanism of action has not been determined (McClay and Moscona, 1974; Merrell, Gottleib and Glaser, 1975).

Sheffield and Moscona (1969) studied the cellular aggregation phenomenon in detail. They found that initial adhesion was nonspecific and was followed by cell orientation, then aggregation or formation of concentric structures. Inhibitors of protein synthesis did not affect initial adhesion but did prevent aggregation.

Acetylcholine (ACh) acts as the synaptic transmitter between motoneurones and skeletal muscle fibers. Muscle fibers have ACh receptors which are clustered in the area of the neurone synapse. During development of the myotube, myoblasts fuse followed by synthesis of ACh receptors (Fambrough and Rash, 1971; Patrick *et al.*, 1972; Paterson and Prives, 1973) and actin and myosin filaments (Fambrough and Rash, 1971). Lack of adequate Ca^{2+} levels in the medium prevents myoblasts from fusing but does not inhibit production and appearance of the ACh receptor (Paterson and Prives, 1973). ACh sensitivity develops before innervation occurs (Fambrough and Rash, 1971; Sytkowski, Vogel and Nirenberg, 1973). Sytkowski, Vogel and Nirenberg (1973) showed that ACh receptors cluster on myotubes in the absence of neurones. Receptor density in the clusters was 9000 per μm^2 while it was 900 per μm^2 on other areas of the cell. Hartzell and Fambrough (1973) determined that the turnover rate of ACh receptors was slow. Cycloheximide and inhibitors of ATP synthesis prevented incorporation of new ACh receptors while actinomycin D had no effect. This may show the relative state of differentiation of the cell and that the cell surface now reflects the cell's mission. For a review on this differentiating cell system *see* Fambrough *et al.* (1974).

The cellular slime mold *Dictyostelium discoideum* has been used as a model system to study morphogenesis and differentiation. The unicellular stage (myxamebae) aggregates after starvation and upon a chemotactic stimulus (cAMP) differentiates into a multicellular fruiting body. This presents a system in which surface changes that lead to aggregation and differentiation can be studied. Surface components were shown to be involved in aggregation by Beug *et al.* (1970) when they prevented aggregation by the use of univalent antibodies. Surface charge density was not an important factor in these phenomena (Lee, 1972). Beug *et al.* (1973) demonstrated that specific surface sites are involved in aggregation and subsequently (Beug, Katz and Gerisch, 1973) showed a polarity to the adhesion sites. Sites responsible for end to end adhesion and side to side adhesion could be selectively blocked with specific antibodies.

Comparative studies of the cohesiveness of various stages of development in *Dictyostelium* showed that there was an abrupt increase at the aggregation stage. Changes in electrophoretic mobility paralleled the above data (Garrod, 1972). Katz and Bourguignon (1974) found that all cells in starved synchronized cultures were midway in G_2 when aggregation occurred. Weeks (1973) observed that Con A would bind more readily to exponential cells than at other stages of development. Freeze-etch studies revealed an increase in intramembranous particle size with each stage up to maturity (Aldrich and Gregg, 1973). Cyclic AMP, which causes aggregation, and Ca^{2+} changed the particle composition on the intramembranous surface of myxamebae within 2 hours (Gregg and Nesom, 1973). These investigators suggested that cAMP mobilizes intracellular Ca^{2+}, which may in turn be effective in changing plasma membrane structure. Rosen *et al.* (1973) found a carbohydrate-binding protein on the surface of cohesive but not vegetative cells that appeared 3–9 hours after starvation. The protein bound *N*-acetylgalactosamine, D-galactose and L-fucose. They could not directly implicate this protein in aggregation.

Dictyostelium is an organism whose study could contribute significantly to an understanding of surface specialization during differentiation. It is a relatively simple microbial system that goes through recognizable stages of differentiation. In addition, a considerable amount is known about it, as outlined above.

Studies of the chemical compositions of cell surfaces in developing systems show some differences but the functions of the different chemical components are difficult to determine. Qualitative and quantitative differences were reported in a protein fraction from developing sea-urchin eggs and embryos (Kondo, 1972) but the fraction was obtained from homogenized cells. Zalik and Scott (1972) observed that negatively charged groups are lost from the cell surface as iris de-differentiates into lens. The morphology of mouse embryo submandibular gland cells depended on the presence of a glycoprotein on the cell surface. Removal of the glycoprotein caused the cells to become rounded (Bernfield, Banerjee and Cohn, 1972). Cell surface glycoproteins on mouse embryos at the cleavage and blastocyst stage were observed to be different by Pinsker and Mintz (1973). The blastocyst stage was enriched in higher-molecular-weight components. The investigators compared the blastocyst with its developing trophoblast (an invasive body forming the placenta) to transformation and speculated that attachment and invasiveness capacity were related to the observed biochemical changes (Pinsker and Mintz, 1973). Brown (1971) studied the surface glycopeptide composition of neuroblastoma cells as they differentiated into neurone-like cells. A major glycoprotein was found on the differentiated cells and not on those that were undifferentiated. The inhibitor 5-bromodeoxyuridine promoted differentiation and production of the glycoprotein while the effect was reversed in the presence of thymidine. Although the appearance of certain cell surface glycoproteins in the above studies may manifest a particular stage of differentiation or development, in no case has the function of the glycoprotein been established.

Antibodies are useful as cell surface probes for determining the particular antigenic complement of cells or tissues at certain developmental stages. Surface antigens may be considered phenotypic expressions of tissues and cells (Goldschneider and Moscona, 1972). A tissue may have a broad range of antigens which are specific for the tissue (Merrell and Glaser, 1973; Bhargava and Bhargava, 1968) or the species (Bhargava and Bhargava, 1968; Clagett et al., 1973). Some antigens are interspecific or shared by related organisms, for example rat and mouse liver (Sheffield and Emmelot, 1972) and rat and mouse brain (Clagett et al., 1973). Even during embryonic development, differentiating tissues show strong qualitative differences in antigen composition (Goldschneider and Moscona, 1972). Some tissues of common embryological origin (mesenchyme) share some antigens. Goldschneider and Moscona (1972) speculated on the possible function of tissue-specific antigens in histospecific recognition during histogenesis. Such specific recognition factors would be important in the organizational maintenance of tissue integrity during morphogenetic interactions.

Lectins are glycoproteins, usually obtained from plants, that can react with other glycoproteins containing a specific sugar in a terminal position. Thus Con A reacts specifically with glucopyranosyl or mannopyranosyl residues (Goldstein, Hollerman and Merrick, 1965). Lectins can be used as probes to study the cell surface much as antibodies have been used, although lectins are not as specific. Nicolson and Singer (1974) demonstrated that lectin-binding sites for Con A and Ricinus communis agglutinin (RCA) are located on the plasma membrane exterior surface and are distributed randomly. Embryonic and fetal chick neural retina cells were studied by Kleinschuster and Moscona (1972) with different lectins to determine membrane changes during embryonic development. They observed that RCA reacts nearly equally with cells from all stages of development. WGA agglutinated only trypsinized cells regardless of the developmental stage. Retina cells from early embryos would agglutinate in the presence of Con A while cells from the fetal stage would agglutinate only after trypsinization. Lallier (1972) reported that Con A interfered with fertilization and differentiation in sea-urchin eggs and inhibited reaggregation of dissociated cells. Since Con A attaches to the cell surface, Lallier has demonstrated that the surface is involved in these phenomena. The disadvantage of using lectins is their specificity for a sugar rather than a whole molecule. Con A may react with glucose or mannose but this does not provide information on the remainder of the molecule. Although Kleinschuster and Moscona (1972) demonstrated that RCA agglutinates retina cells at all stages of development, this does not indicate the nature of the binding sites. Indeed, the binding sites on cells from older stages may differ from those on earlier stages except for the fact that the terminal sugar is the same. However, the presence of a terminal monosaccharide may be critical since Oppenheimer (1975) demonstrated that β-galactosidase could remove activity of a teratoma cell adhesion factor.

The net surface charge of cells may be measured by determining their electrophoretic mobility. Morphogenetic activity has been correlated with surface charge in Rana pipiens (Kiremidjian and Kopac, 1972; Schaeffer, Schaeffer and Brick, 1973). Pronephric cells change adhesiveness during pronephros construction. An experimentally induced decrease in surface charge leads to greater cell adhesiveness (Kiremidjian and Kopac, 1972).

Schaeffer, Schaeffer and Brick (1973) made an electrophoretic map of four regions of *Rana pipiens* embryos at four different times during development. They found that increased surface charge correlated with greater morphogenetic activity and decreased adhesiveness. Both studies suggest that increased surface charge leads to decreased adhesiveness and vice versa.

Hormones exert regulatory effects on growth, differentiation and metabolic activity. In the course of development, selected cells (target cells) acquire the ability to be influenced by hormones. Peptide hormones bind to the plasma membrane of the target cell to initiate their activities. Steroid hormones bind to a receptor in the cytosol or nucleus (Catt, 1970). In both cases the plasma membrane must be prepared to participate either by the presence of binding sites or by permeability to the hormone.

Insulin is known to bind to the plasma membrane of fat cells in the initiation of its action (Cuatrecasas, 1969, 1971), while thyroxine binds to the nuclear membrane (Samuels and Tsai, 1973). The estrogen receptor is a protein located in the nucleus (McGuire, DeLaGarza and Chamness, 1973; Mester *et al.*, 1973). Estrogen induces the differentiation *in vivo* of albumin-secreting glands of immature chick oviducts (Kohler, Grimley and O'Malley, 1969). Even in immature chicks, cells were found to be susceptible (permeable?) to the hormone although it was not normally present in the animal at an early age. Lissitzky *et al.* (1971) observed that thyroid cells associate *in vitro* into follicular structures in the presence of thyrotropin and cAMP. Verrier, Fayet and Lissitzky (1974) found that thyrotropin binds both to purified membranes and isolated thyroid cells. Binding of hormone stimulates adenylate cyclase activity in the membranes. This enzyme controls production of cAMP which in turn acts as a secondary messenger to carry out the hormone's action. Cyclic AMP appears to be involved in differentiation in this system, as has also been demonstrated in plant (Basile, Wood and Braun, 1973) and animal cell systems (Hsie, Jones and Puck, 1971).

Lymphocyte differentiation is particularly advantageous to study because it is a single-celled system that can be followed both *in vivo* and *in vitro*. Two cell types are involved: B cells, which arise from bone marrow, and T cells, which are thymus-derived. B cells have receptor immunoglobulins on their outer membrane to which antigen attaches. This event stimulates B cells to proliferate and differentiate into immunoglobulin-producing and -secreting cells (Miller and Mitchell, 1969). T cells show specificity for antigen but have no immunoglobulin receptors. After stimulation by antigens, T cells proliferate (Greaves and Moller, 1970).

Bach *et al.* (1971) observed that thymosin, a thymic hormone, causes differentiation of bone marrow cells into cells with T-cell characteristics. They believed that thymosin causes B precursors of T cells to differentiate into T cells in the thymus *in vivo*. The influence of thymosin on the development of immunological competence in stem cell populations *in vitro* and *in vivo* was studied by Goldstein *et al.* (1971). Immunological competence was accelerated by the hormone. Thymopoietin (thymin), a hormone derived from the thymus, was found to induce differentiation of thymus cells *in vitro* (Basch and Goldstein, 1974). The TL and Thy-1 (θ) antigens, which appeared after hormone treatment, are characteristic of early stages of thymocyte maturation. These two antigens disappeared as differentiation progressed while they were still in the thymus (Boyse, Old and Stockert, 1965). It was concluded that the thymus has an endocrine role in the development and maturation of lymphoid cells (Basch and Goldstein, 1974; Goldstein *et al.*, 1971). Membrane fluidity may be related to the ability of lymphoid cells to undergo differentiation caused by an external stimulus (Inbar *et al.*, 1973). Two myeloid leukemic cell lines were investigated. One (D^+) could be stimulated to differentiate into mature macrophages and granulocytes by Con A but the other (D^-) could not. Both cell lines bound the same amount of Con A per cell but D^+ formed caps. The capping phenomenon appeared to be directly related to membrane fluidity (Edelman, Yahara and Wang, 1973).

Precursors to lymphocytes go through several stages during differentiation. The stages can be detected by changes in the surface antigen and immunoglobulin composition. The thymus is involved in some aspects of lymphocyte differentiation.

10.4.3   Transformation

The process of transformation from a normal to a tumor cell may be considered a form of differentiation or de-differentiation depending on the viewpoint. In any case, definite changes are known to take place on the cell surface and are then reflected in behavioral changes during contact inhibition. Presumably the surface receptors or mediators of contact inhibition are either modified or absent on the transformed cell.

Martinez-Palomo, Braislovsky and Bernhard (1969) observed by staining with ruthenium red that the surface coats of transformed cells have more acid mucopolysaccharide than those of normal cells. Porter, Todaro and Font (1973), with the use of scanning electron microscopy, showed definite morphological differences between normal and transformed cells. These included changes from an epithelial to a fibroblastic form and changes in frequency of microvilli, blebs and ruffles.

Comparative biochemical studies of normal and transformed cells have shown differences in protein composition (Buck, Glick and Warren, 1971; Hynes, 1973; Warren, Fuhrer and Buck, 1972), levels of glycosyltransferases (Bosmann, Case and Morgan, 1974) and levels of proteases and glycosidases (Bosmann, Lockwood and Morgan, 1974). Antigen levels (Tarro and Sabin, 1970) and composition (Ting *et al.*, 1972; Tarro, 1973) were also different. Ting *et al.* (1972) provided evidence to show that tumor cells have some antigens common to the fetal stage of the organism.

Selective agglutinability of transformed cells by lectins (Burger, 1969) was considered of particular importance since Con A agglutinability was directly related to tumorigenicity of a cell line (Benjamin and Burger, 1970; Inbar, Ben-Bassat and Sachs, 1972). Con A binds to both normal and transformed cells (Mallucci, 1971) but the arrangement of the binding sites is the main difference determining agglutinability. Normal cells maintain their sites randomly on the surface while the sites become clustered on transformed cells (Nicolson, 1971; Rosenblith *et al.*, 1973). This was believed to be due to an increase in the fluidity of the membranes of transformed cells (Edelman, Yahara and Wang, 1973; Nicolson, 1973; Rosenblith *et al.*, 1973).

10.5 SUMMARY

Biochemical changes in cell surface composition are responsible for the changing behavior of cells during differentiation and development. The cell surface mediates cellular events during morphogenesis and histogenesis. The mature, differentiated cell also displays a cell surface composition that is consistent with its function in the organism.

Cell surface composition is dictated initially by the genetic material. Synthesis of surface components occurs in the RER and Golgi apparatus. The information is then sent to the plasma membrane via Golgi vesicles. During differentiation and development, cell surface components must be replaced or modified according to the changing role of the cell. Possible mechanisms for removing old material may include shedding the surface coat or release of membrane vesicles. Additional changes in surface components could take place since proteases, glycosidases and glycosyltransferases have been detected on the plasma membrane.

Stages of the cell cycle have definite morphological and biochemical characteristics that are apparent on the cell surface. Although each cell line has its own stage characteristics, the characteristics are consistent and probably reflect the requirements and activities of the cell during a particular stage of the cycle.

Morphogenesis and histogenesis require that there be continual and coordinated changes in the adhesiveness and recognition ability of cells and tissues. That the cell surface is changing during development can be shown by immunological and biochemical studies. An increase in the surface charge of the cell leads to decreased adhesiveness and vice versa. Surface components are responsible for recognition and aggregation but no specific component or mechanism of action have been directly implicated in these events. Glycosyltransferase–glycoprotein complexes are currently suspected as possibly being part of the mechanism of intercellular communication and interaction.

REFERENCES

ADLER, R. (1970). *Devl Biol.*, **21**:403.
ALDRICH, H. C. and GREGG, J. H. (1973). *Expl Cell Res.*, **81**:407.
ARMSTRONG, P. B. (1970). *J. Cell Biol.*, **47**:197.
BACH, J. F., DARDENNE, M., GOLDSTEIN, A. L., GUHA, A. and WHITE, A. (1971). *Proc. natn. Acad. Sci. U.S.A.*, **68**:2734.
BAKER, J. B. and HUMPHREYS, T. (1972). *Science, N.Y.*, **175**:905.
BARBERA, A. J., MARCHASE, R. B. and ROTH, S. (1973). *Proc. natn. Acad. Sci. U.S.A.*, **70**:2482.
BASCH, R. S. and GOLDSTEIN, G. (1974). *Proc. natn. Acad. Sci. U.S.A.*, **71**:1474.
BASILE, D. V., WOOD, H. N. and BRAUN, A. C. (1973). *Proc. natn. Acad. Sci. U.S.A.*, **70**:3055.
BENJAMIN, T. L. and BURGER, M. M. (1970). *Proc. natn. Acad. Sci. U.S.A.*, **67**:929.
BENNETT, D., BOYSE, E. A. and OLD, L. J. (1971). *International Lepetit Colloquium*, pp. 247–263. Ed. L. G. SILVESTRI. New York; American Elsevier.
BENNETT, G. (1970). *J. Cell Biol.*, **45**:668.
BENNETT, G. and LEBLOND, C. P. (1970). *J. Cell Biol.*, **46**:409.
BENNETT, G., LEBLOND, C. P. and HADDAD, A. (1974). *J. Cell Biol.*, **60**:258.
BERNFIELD, M. R., BANERJEE, S. D. and COHN, R. H. (1972). *J. Cell Biol.*, **52**:674.
BEUG, H., GERISCH, G., KEMPFF, S., RIEDEL, V. and CREMER, G. (1970). *Expl Cell Res.*, **63**:147.
BEUG, H., KATZ, F. E. and GERISCH, G. (1973). *J. Cell Biol.*, **56**:647.
BEUG, H., KATZ, F. E., STEIN, A. and GERISCH, G. (1973). *Proc. natn. Acad. Sci. U.S.A.*, **70**:3150.

BHARGAVA, K. and BHARGAVA, P. M. (1968). *Expl Cell Res.*, **50**:515.

BOSMANN, H. B. (1971). *Archs Biochem. Biophys.*, **145**:310.

BOSMANN, H. B. (1972). *Biochem. biophys. Res. Commun.*, **48**:523.

BOSMANN, H. B. (1974). *Nature, Lond.*, **249**:144.

BOSMANN, H. B., CASE, K. R. and MORGAN, H. R. (1974). *Expl Cell Res.*, **83**:15.

BOSMANN, H. B., LOCKWOOD, T. and MORGAN, H. R. (1974). *Expl Cell Res.*, **83**:25.

BOSMANN, H. B. and WINSTON, R. A. (1970). *J. Cell Biol.*, **45**:23.

BOYSE, E. A., OLD, L. J. and STOCKERT, E. (1965). *Immunopathology, 4th International Symposium*, pp. 23–40. Ed. P. GRABAR and P. A. MEISCHER. Basle; Schwab.

BROWN, J. C. (1971). *Expl Cell Res.*, **69**:440.

BROWN, R. M., FRANKE, W. W., KLEINIG, H., FALK, H. and SITTE, P. (1970). *J. Cell Biol.*, **45**:246.

BUCK, C. A., GLICK, M. C. and WARREN, L. (1971). *Science*, **172**:169.

BURDICK, M. L. and STEINBERG, M. S. (1969). *Proc. natn. Acad. Sci. U.S.A.*, **63**:1169.

BURGER, M. (1969). *Proc. natn. Acad. Sci. U.S.A.*, **62**:994.

CATT, K. J. (1970). *Lancet*, **ii**, 763.

CIKES, M. and FRIBERG, S. (1971). *Proc. natn. Acad. Sci. U.S.A.*, **68**:566.

CLAGETT, J., PETER, H. H., FELDMAN, J. D. and WEIGLE, W. O. (1973). *J. Immun.*, **110**:1085.

CUATRECASAS, P. (1969). *Proc. natn. Acad. Sci.*, **63**:450.

CUATRECASAS, P. (1971). *J. biol. Chem.*, **246**:6522.

CURTIS, A. S. G. (1967). *The Cell Surface: Its Molecular Role in Morphogenesis.* New York; Academic Press.

DALEN, H. and SCHEIE, P. (1969). *Expl Cell Res.*, **57**:351.

EDELMAN, G. M., YAHARA, I. and WANG, J. L. (1973). *Proc. natn. Acad. Sci. U.S.A.*, **70**:1442.

FAMBROUGH, D. M. and RASH, J. E. (1971). *Devl Biol.*, **26**:55.

FAMBROUGH, D. M., HARTZELL, H. C., POWELL, J. A., RASH, J. E. and JOSEPH, N. (1974). *Synaptic Transmission and Neuronal Interaction*, pp. 285–313. New York; Raven Press.

FLICKINGER, C. J. (1968). *Expl Cell Res.*, **53**:241.

FLICKINGER, C. J. (1969). *J. Cell Biol.*, **43**:250.

FOX. T. O., SHEPPARD, J. R. and BURGER, M. M. (1971). *Proc. natn. Acad. Sci. U.S.A.*, **68**:244.

GARBER, B. B. and MOSCONA, A. A. (1972a). *Devl Biol.*, **27**:217.

GARBER, B. B. and MOSCONA, A. A. (1972b). *Devl Biol.*, **27**:235.

GARROD, D. R. (1972). *Expl Cell Res.*, **72**:588.

GEORGE, J. V. and RAO, K. V. (1975). *J. cell. Physiol.*, **85**:547.

GOLDSCHNEIDER, I. and MOSCONA, A. A. (1972). *J. Cell Biol.*, **53**:435.

GOLDSTEIN, A. L., GUHA, A., HOWE, M. L. and WHITE, A. (1971). *J. Immun.*, **106**:773.

GOLDSTEIN, I. J., HOLLERMAN, C. E. and MERRICK, J. M. (1965). *Biochim. biophys. Acta*, **97**:68.

GOTTLEIB, D. I., MERRELL, R. and GLASER, L. (1974). *Proc. natn. Acad. Sci. U.S.A.*, **71**:1800.

GREAVES, M. F. and MOLLER, E. (1970). *Cell Immun.*, **1**:372.

GREGG, J. H. and NESOM, M. G. (1973). *Proc. natn. Acad. Sci. U.S.A.*, **70**:1630.

HARTZELL, H. C. and FAMBROUGH, D. M. (1973). *Devl Biol.*, **30**:153.

HIRANO, H., PARKHOUSE, B., NICOLSON, G., LENNOX, E. S. and SINGER, S. J. (1972). *Proc. natn. Acad. Sci. U.S.A.*, **69**:2945.

HSIE, A. W., JONES, C. and PUCK, T. T. (1971). *Proc. natn. Acad. Sci. U.S.A.*, **68**:1648.

HYNES, R. O. (1973). *Proc. natn. Acad. Sci. U.S.A.*, **70**:3170.

INBAR, M., BEN-BASSAT, H. and SACHS, L. (1972). *Nature, New Biol.*, **236**:3.

INBAR, M., BEN-BASSAT, H., FIBACH, E. and SACHS, L. (1973). *Proc. natn. Acad. Sci. U.S.A.*, **70**:2577.

JAMIESON, J. D. and PALADE, G. E. (1967a). *J. Cell Biol.*, **34**:577.

JAMIESON, J. D. and PALADE, G. E. (1967b). *J. Cell Biol.*, **34**:597.

KAPELLER, M., GAL-OZ, R., GROVER, N. B. and DOLJANSKI, F. (1973). *Expl Cell Res.*, **79**:152.

KATZ, E. R. and BOURGUIGNON, L. Y. W. (1974). *Devl Biol.*, **36**:82.

KEENAN, T. W., MORRÉ, D. J. and BASU, S. (1974). *J. biol. Chem.*, **249**:310.

KIREMIDJIAN, L. and KOPAC, M. J. (1972). *Devl Biol.*, **27**:116.

KLEINSCHUSTER, S. J. and MOSCONA, A. A. (1972). *Expl Cell Res.*, **70**:397.

KOHLER, P. O., GRIMLEY, P. M. and O'MALLEY, B. W. (1969). *J. Cell Biol.*, **40**:8.

KONDO, H. (1972). *Expl Cell Res.*, **72**:519.

LALLIER, R. (1972). *Expl Cell Res.*, **72**:157.

LAWFORD, G. R. and SCHACHTER, H. (1966). *J. biol. Chem.*, **241**:5408.

LEE, K. C. (1972). *J. Cell Sci.*, **10**:249.

LEVINE, A. M., HIGGINS, J. A. and BARNETT, R. J. (1972). *J. Cell Sci.*, **11**:855.

LILIEN, J. E. (1968). *Devl Biol.*, **17**:657.

LILLIE, F. R. (1912). *Science, N.Y.*, **36**:527.
LISSITZKY, S., FAYET, G., GIRAUD, A., VERRIER, B. and TORRESANI, J. (1971). *Eur. J. Biochem.*, **24**:88.
MALLUCCI, L. (1971). *Nature, New Biol.*, **233**:241.
MARTINEZ-PALOMO, A., BRAISLOVSKY, C. and BERNHARD, W. (1969). *Cancer Res.*, **29**:925.
MAYHEW, E. (1974). *J. theor. Biol.*, **47**:483.
MCCLAY, D. R. and MOSCONA, A. A. (1974). *Expl Cell Res.*, **87**:438.
MCGUIRE, W. L., DELAGARZA, M. and CHAMNESS, G. C. (1973). *Endocrinology*, **93**:810.
MCLEAN, R. J. and BOSMANN, H. B. (1975). *Proc. natn. Acad. Sci. U.S.A.*, **72**:310.
MCLEAN, R. J. and BROWN, R. M. (1974). *Devl Biol.*, **36**:279.
MCLEAN, R. J., LAURENDI, C. J. and BROWN, R. M. (1974). *Proc. natn. Acad. Sci. U.S.A.*, **71**:2610.
MERRELL, R. and GLASER, L. (1973). *Proc. natn. Acad. Sci. U.S.A.*, **70**:2794.
MERRELL, R., GOTTLIEB, D. I. and GLASER, L. (1975). *J. biol. Chem.*, **250**:5655.
MESTER, J., BRUNELLE, R., JUNG, I. and SONNENSCHEIN, C. (1973). *Expl Cell Res.*, **81**:447.
MILLER, J. F. A. P. and MITCHELL, G. F. (1969). *Transplantn Rev.*, **1**:3.
MORRÉ, D. J., MERLIN, L. M. and KEENAN, T. W. (1969). *Biochem. biophys. Res. Commun.*, **37**:813.
MORRIS, J. E. and MOSCONA, A. A. (1971). *Devl Biol.*, **25**:420.
MOSCONA, A. A. (1968). *Devl Biol.*, **18**:250.
MOSCONA, A. A. (1971). *Biochemistry of Cell Differentiation*, Vol. 24, pp. 1–23. Ed. A. MONROY and R. TSANEO. London; Academic Press.
MOSCONA, A. A. (Ed.) (1974). *The Cell Surface in Development*. Somerset, New Jersey; John Wiley.
NEUTRA, M. and LEBLOND, C. P. (1966). *J. Cell Biol.*, **30**:119.
NICOLSON, G. L. (1971). *Nature, New Biol.*, **233**:244.
NICOLSON, G. L. (1973). *Nature, New Biol.*, **243**:218.
NICOLSON, G. L. and SINGER, S. J. (1974). *J. Cell Biol.*, **60**:236.
NOONAN, K. D., LEVINE, A. J. and BURGER, M. M. (1973). *J. Cell Biol.*, **58**:491.
O'BRIEN, P. J., CANADY, M. R., HALL, C. W. and NEUFELD, E. F. (1966). *Biochim. biophys. Acta*, **117**:331.
OPPENHEIMER, S. B. (1975). *Expl Cell Res.*, **92**:122.
PASTERNAK, C. A., WARMSLEY, A. M. H. and THOMAS, D. B. (1971). *J. Cell Biol.*, **50**:562.
PATERSON, B. and PRIVES, J. (1973). *J. Cell Biol.*, **59**:241.
PATRICK, J., HEINEMANN, S. F., LINDSTROM, J., SCHUBERT, D. and STEINBACH, J. H. (1972). *Proc. natn. Acad. Sci. U.S.A.*, **69**:2762.
PATT, L. M. and GRIMES, W. J. (1974). *J. biol. Chem.*, **249**:4157.
PETIT, J. F., STROMINGER, J. L. and SOLL, D. (1968). *J. biol. Chem.*, **243**:757.
PINSKER, M. C. and MINTZ, B. (1973). *Proc. natn. Acad. Sci. U.S.A.*, **70**:1645.
PORTER, K., PRESCOTT, D. and FRYE, J. (1973). *J. Cell Biol.*, **57**:815.
PORTER, K., TODARO, G. J. and FONTE, V. (1973). *J. Cell Biol.*, **59**:633.
RAMBOURG, A., BENNETT, G., KOPRIWA, B. and LEBLOND, C. P. (1971). *J. Microsc.*, **11**:163.
ROBERTS, W. S. L., PETIT, J. F. and STROMINGER, J. L. (1968). *J. biol. Chem.*, **243**:768.
ROSEMAN, S. (1970). *Chem. Phys. Lipids*, **5**:270.
ROSEN, S. D., KAFKA, J. A., SIMPSON, D. L. and BARONDES, S. H. (1973). *Proc. natn. Acad. Sci. U.S.A.*, **70**:2554.
ROSENBLITH, J. Z., UKENA, T. E., YIN, H. H., BERLIN, R. D. and KARNOVSKY, M. J. (1973). *Proc. natn. Acad. Sci. U.S.A.*, **70**:1625.
ROTH, S. (1968). *Devl Biol.*, **18**:602.
ROTH, S., MCGUIRE, E. J. and ROSEMAN, S. (1971). *J. Cell Biol.*, **51**:536.
ROTH, S. and WHITE, D. (1972). *Proc. natn. Acad. Sci. U.S.A.*, **69**:485.
RUBIN, R. W. and EVERHART, L. P. (1973). *J. Cell Biol.*, **57**:837.
SAIJO, N. (1972). *Expl Cell Res.*, **72**:560.
SAMUELS, H. H. and TSAI, J. S. (1973). *Proc. natn. Acad. Sci. U.S.A.*, **70**:3488.
SARKAR, P. K. and MOSCONA, A. A. (1973). *Proc. natn. Acad. Sci. U.S.A.*, **70**:1667.
SARKAR, P. K., FISCHMAN, D., GOLDWASSER, A. and MOSCONA, A. A. (1972). *J. biol. Chem.*, **247**:7743.
SCOTT, R. E., CARTER, R. L. and KIDWELL, W. R. (1971). *Nature, New Biol.*, **233**:219.
SCHACHTER, H., JABBAL, I., HUDGIN, R. L., PINTERIC, L., MCGUIRE, E. J. and ROSEMAN, S. (1970). *J. biol. Chem.*, **245**:1090.
SCHAEFFER, B. E., SCHAEFFER, H. E. and BRICK, I. (1973). *Devl Biol.*, **34**:66.
SHEFFIELD, J. B. and EMMELOT, P. (1972). *Expl Cell Res.*, **71**:97.
SHEFFIELD, J. B. and MOSCONA, A. A. (1969). *Expl Cell Res.*, **57**:462.
SHUR, B. D. and ROTH, S. (1975). *Biochim. biophys. Acta*, **415**:473.

SIMKIN, J. L. and JAMIESON, J. C. (1967). *Biochem. J.*, **103**:153.

STEINBERG, M. s. (1964). *Cellular Membranes in Development*, pp. 321–366. Ed. M. LOCKE. New York; Academic Press.

SYTKOWSKI, A. J., VOGEL, Z. and NIRENBERG, M. W. (1973). *Proc. natn. Acad. Sci. U.S.A.*, **70**:270.

TARRO, G. (1973). *Proc. natn. Acad. Sci. U.S.A.*, **70**:325.

TARRO, G. and SABIN, A. B. (1970). *Proc. natn. Acad. Sci. U.S.A.*, **67**:731.

TILLACK, T. W., SCOTT, R. E. and MARCHESI, V. T. (1970). *J. Cell Biol.*, **47**:213a.

TILNEY, L. G. and GIBBINS, J. R. (1969). *J. Cell Biol.*, **41**:227.

TING, C. C., LAVRIN, D. H., SHIU, G. and HERBERMAN, R. B. (1972). *Proc. natn. Acad. Sci. U.S.A.*, **69**:1664.

TRELSTAD, R. L. (1970). *J. Cell Biol.*, **45**:34.

VERRIER, B., FAYET, G. and LISSITZKY, S. (1974). *Eur. J. Biochem.*, **42**:355.

WAGNER, R. R. and CYNKIN, M. A. (1971). *J. biol. Chem.*, **246**:143.

WARREN, L., FUHRER, J. P. and BUCK, C. A. (1972). *Proc. natn. Acad. Sci. U.S.A.*, **69**:1838.

WARREN, L. and GLICK, M. C. (1968). *J. Cell Biol.*, **37**:729.

WEEKS, G. (1973). *Expl Cell Res.*, **76**:467.

WEISS, L. (1967). *The Cell Periphery, Metastasis and Other Contact Phenomena*. New York; John Wiley.

WHALEY, W. G., DAUWALDER, M. and KEPHART, J. E. (1972). *Science, N.Y.*, **175**:596.

ZALIK, S. E. and SCOTT, V. (1972). *J. Cell Biol.*, **55**:134.

Index

Index